W9-CKU-134

FOOD WEBS AND CONTAINER HABITATS
The natural history and ecology of phytotelmata

The animal communities in plant-held water bodies, such as tree holes and pitcher plants, have become models for food-web studies. In this book, Professor Kitching introduces us to these fascinating miniature worlds and demonstrates how they can be used to tackle some of the major questions in community ecology. Based on thirty years' research in many parts of the world, this work presents much previously unpublished information, in addition to summarising over a hundred years of natural history observations by others. The book covers many aspects of the theory of food-web formation and maintenance presented with field-collected information on tree holes, bromeliads, pitcher plants, bamboo containers and the axils of fleshy plants. It is a unique introduction for the field naturalist, and a stimulating source treatment for graduate students and professionals working in the fields of tropical and other forest ecology, as well as entomology.

ROGER L. KITCHING holds the Chair of Ecology at Griffith University, Brisbane. A graduate of Imperial College, London, and the University of Oxford, he has spent the greater part of his working life in Australia, with wide-ranging interests in the ecology and natural history of insects, particularly in rainforests. He is a senior investigator within the Cooperative Research Centre for Tropical Rainforest Ecology and Management and was a Bullard Fellow at Harvard University in 1998. Professor Kitching has written over 120 books and papers, publishing from time to time in most of the world's ecological journals. He is author of *Systems Ecology*, co-author of *Insect Ecology* and has edited or co-edited four further books on topics ranging from the ecology of pests to the biology of butterflies. He has served as president of the Australian Entomological Society and the Australian Institute of Biology, and currently chairs the Biodiversity Advisory Council, the major federal government advisory body on biodiversity affairs in Australia.

FOOD WEBS AND CONTAINER HABITATS

The natural history and ecology of phytotelmata

R. L. KITCHING

Australian School of Environmental Studies, Griffith University

CAMBRIDGE
UNIVERSITY PRESS

PUBLISHED BY THE PRESS SYNDICATE OF THE UNIVERSITY OF CAMBRIDGE
The Pitt Building, Trumpington Street, Cambridge CB2 1RP, United Kingdom

CAMBRIDGE UNIVERSITY PRESS
The Edinburgh Building, Cambridge CB2 2RU, UK http://www.cup.cam.ac.uk
40 West 20th Street, New York, NY 10011–4211, USA http://www.cup.org
10 Stamford Road, Oakleigh, Melbourne 3166, Australia
Ruiz de Alarcón 13, 28014 Madrid, Spain

First published 2000

Printed in the United Kingdom at the University Press, Cambridge

Typeface Times 10/13pt. System QuarkXPress™ [wv]

A catalogue record for this book is available from the British Library

ISBN 0 521 77316 4 hardback

In memoriam:

Charles Elton,
Albrecht Thienemann.

Giants...

Contents

Preface

This book owes its inception to my fascination with the natural microcosms that are water-filled tree holes and, subsequently, the broader class of plant container habitats we call phytotelmata. That fascination was born, first, in a Somerset woodland, when my fellow undergraduate Alastair Sommerville pointed out to me a massive stump hole, commenting that such places were both entomologically special and of great potential as objects of ecological study. Accordingly, in 1966, when Charles Elton suggested to me that I make such habitats the subject of my DPhil at Oxford, his words fell on fertile ground. For me, the next three years were a time of learning and discovery under the supervision of Mick Southern and, unofficially, of Kitty Paviour-Smith. They taught me the fine art of combining natural history with ecology, scholarship with innovation.

This groundwork would have counted for nothing, and my primary preoccupation would have remained butterflies and bugs, but for two other persons. In 1979 Howard Frank invited me to participate in a symposium organised by Phil Lounibos and himself at the Kyoto Congress of Entomology. He suggested that I compare English and Australian tree-hole communities. Alas, at that point in time, I had not looked at Australian tree holes, but with an eighteen month lead time before the Congress, I set to and made a first study, with Cathy Callaghan, of tree-hole communities in the subtropical rainforests of south-east Queensland. The comparative approach that I developed for the Congress presentation established a framework that has guided much of the subsequent work that is described in this volume. Lastly the Kyoto Congress enabled me to re-establish contact with Stuart Pimm who showed me that such comparative studies had considerable bearing on some current debates in food-web theory: the start of a productive and exciting collaboration.

The observations and interpretations presented in this work owe much to the activities of my students, particularly Cathy Callaghan, Bert Jenkins and

Charles Clarke, who, in the finest tradition, educated their supervisor while receiving, I trust, some instruction from me. I take this opportunity to thank them sincerely for the stimulation, companionship and mutual trust we have enjoyed.

Rosemary Lott, Bert Jenkins, Stewart Jackson, Heather Mitchell and Peter Juniper have all acted as research assistants at one time or another during the course of my studies of phytotelmata. Without them the work would simply not have been possible as I became, progressively (a most inappropriate word), involved in more teaching, supervising, government advising, and administration over the years. I have benefited greatly, also, by those who have acted as hosts during a series of visits to their laboratories, spending their precious time to explain their work to me and discuss interpretations and theories concerning food webs, phytotelmata and ecological communties in general. A complete acknowledgement of all those in this category is impossible but some I must mention by name: Bill Bradshaw (University of Oregon) , Stuart Pimm (University of Tennessee), Howard Frank (University of Florida), Phil Lounibos (Florida Medical Entomology Laboratory), Phil Corbet, Norm Fashing (College of William and Mary), Joel Cohen (Rockefeller University) and Naomi Pierce (Harvard University). All provided intellectual, material and/or moral support at one time or another. Without this implicit and explicit support the book and the research on which it is based would not have been possible.

The research has been variously funded by the Australian Research Committee, the Australia/America Cooperative Science Scheme, the Ian Potter Foundation, the Wet Tropics Management Authority, the Australian Geographical Society, the University of New England and Griffith University. Without this material support it would simply have been impossible to carry out the sort of long-term comparative work on which the book is based.

Turning specifically to the five years or so during which the book has been in preparation I start by acknowledging Naomi Pierce and Ed. Wilson who acted as hosts during a sabbatical visit to the Museum of Comparative Zoology at Harvard University, during which the first half of this book was written. Bill and Mary Ann Bossert of Lowell House also made me welcome as a visiting fellow during this period. At Harvard, the librarians in the Museum of Comparative Zoology went far beyond the call of duty in helping me track down many of the more obscure references needed to make the first part of this work as complete as possible: I am grateful to them for these essential services. Stuart Pimm and Michael Fisher provided key advice on the actual publishing process without which the work might never have seen the light of day.

The preparation of Chapter 3 and the faunistic 'Annexe' required me to venture far into the specialist realms of others. Many of the sections on particular taxa have been read (and sometimes extensively revised) by specialist taxonomists in the relevant groups. These persons also provided me with key references on the phytotelm-inhabiting representatives of their chosen taxa. I am particularly grateful in this regard to the following: Peter Cranston (Diptera in general, Chironomidae in particular), Francis Gilbert (Syrphidae), Norm Fashing (Acari), John Cadle (Anura), Michael Lai (Anura) and Stephen Richards (Anura). I express my particular thanks to Stephen Richards who provided me with some key late references on the frogs of phytotelms ensuring that my treatment of them was less embarassingly deficient than it might otherwise have been.

This book has had a much longer than anticipated gestation having passed (for various reasons) through the hands of a number of editors. Early drafts benefited by comments from Bob May and Paul Harvey, and the final version owes much to the wise and encouraging comments of Michael Usher. Last but by no means least, Beverley Kitching has put up with my absences, distractedness and general neurosis about 'the food web book' for far longer than was reasonable and provided the appropriately encouraging responses to my frequently rhetorical questions during this period.

I thank the following copyright holders for permission to reproduce figures and other material from previously published work: Blackwell Scientific Publications (Figs. 10.1 to 10.7), the editors of *Oikos* (Fig. 9.3), Oxford University Press (Figs. 5.6, 11.6), the American Entomological Society (Figs. 4.11, 12.1), the Ecological Society of America (Fig. 12.2), the American Association for the Advancement of Science (Fig. 12.3), CSIRO and the editors of the *Australian Journal of Zoology* (Figs. 11.4, 11.5).

1

Introduction

Along the Labi Road in the tiny Sultanate of Brunei lies a series of sandy stream beds. Each is raised above the surrounding forest, presumably by eons of silt deposition during occasional periods of inundation and stream flow. But now, in May, the water flow is a mere trickle, winding around patches of tar seepage; here, not the product of human error and environmental insensitivity, but a natural phenomenon reflecting the oil-bearing strata that underlie this region of northern Borneo. But this is the perhumid tropics and wetness is the order of the day – any day. Treacherous quicksands lie centimetres below the scorched white sand and the peat-swamp forest on each side of the stream bed is permanently inundated with tea-coloured peaty water standing thigh-deep around the tangle of buttress roots, fallen trees, scrambling vines and creepers.

I first came to this nutritionally poor but biologically rich ecosystem in 1989. During the day the biological riches, at least of the more obvious kinds, are largely to be inferred rather than experienced directly. The clean smooth sand is criss-crossed with tracks of mammal and bird, reptile and insect: here the measured marks of a monitor lizard scavenging for carrion, eggs and nestlings; there, the dainty steps of forest rats. A honking flight of bushy-crested hornbills sweeping overhead and the distant exuberance of a gibbon troop call me back to nature immediate rather than nature previous, and direct my attention to the smaller scale concerns of the ecologist and the mundane necessities of field data collection.

For Odoardo Beccari, the egocentric Italian naturalist who visited this region from 1866 to 1868, this was the land of the 'maias' – the orang utans: he describes graphically the shooting, dissecting and killing of any number of the apes, from infants to magnificent old males (Beccari 1904). Alfred Russel Wallace, visiting Borneo in the 1850s in pursuit of beetles, butterflies and birds, noted the faunal contrasts between the Bornean animals and those

1

*of the islands to the east, in particular that of the Celebes (Wallace 1869).
The mass of natural historical information accrued by Wallace gelled in a
burst of inspiration that came to him, immured by rain, and light-headed with
malaria, in a Ternate bungalow. The end result of this concatenation of ideas
and observations was Darwin and Wallace's joint presentation to the Linnaean Society of London on July the 1st 1858:* On the Tendency of Varieties
to Depart Indefinitely from the Original Type. *The theory of evolution by natural selection had finally seen the light of day and biology would never be
the same again!*

*For these and many other less tangible reasons, Borneo, for me, was yet
another of those ecological destinations that had seemed impossibly distant
during my English incubation as a biologist. But by 1989 Borneo had become
in my mind the land of pitcher plants, and the opportunity to follow up this
impression with my graduate student Charles Clarke, an indomitable pitcher-
plant enthusiast and field worker, had been just too much to resist.*

*We walked gingerly through the quicksands of the Labi stream beds: cas-
cades of the delicate pitchers of* Nepenthes gracilis *hanging from their scrambling stems over exposed shrubs, or forming perfect reflections in the mir-
ror-like surfaces of the pitch pools. In the most exposed areas the larger and
more robust red and green pitchers of* N. mirabilis *marked out the parent
plants, free standing in the stream beds themselves. Scrambling among the
branches of the larger trees at the edge of the stream beds the fleshy green
leaves of* N. rafflesiana *terminated in dramatic trumpet-shaped upper pitchers hanging like red and yellow decorations on an equatorial Christmas tree.
In the peat swamp itself, we waded through the opaque water. Here the squat
pitchers of* N. ampullaria *hung in tight wreaths around the stems of their host
plants. Last, but by no means least, we stood engrossed by the extraordinary
giant pitchers of* Nepenthes bicalcarata, *which diverted even Beccari from his
pursuit of the inedible, by its extraordinary combination of features as both
an insectivorous pitcher plant, and a mutualistic ant plant.*

I shall return to the details of our pitcher-plant studies in due course: let them
now stand as an introduction to all of the so-called phytotelmata – plant-con-
tainer habitats. Pitchers, of course, act as traps for insects and other arthropods which drown in the fluid contained in the pitchers. They are digested
by plant exudates, and the nitrogen-rich nutrients so produced are absorbed
by the plant – enabling them, *inter alia*, to live in the nutrient-low substrates
of the kerangas forest. But these small perched water bodies offer themselves
as habitats for other species of animals which exploit the isolation and protection offered by the pitchers and cream off some of the nutrient base for

themselves. A complete food web exists within each of the pitchers and the species of organisms which make up the pitcher fauna are largely restricted to this very special environment, breeding nowhere else.

Phytotelmata – defined and described

Pitchers are by no means the only aquatic habitats that occur as plant-based containers and, in 1928, the German biologist Ludwig Varga coined the term *phytotelmata* for the whole class of such ecological situations. Varga carried out extensive work on the flora and fauna associated with the water-filled leaf axils of the European teasel *Dipsacus silvestris* and, in order to put his work into literary context, he drew attention to earlier work on comparable habitats. Varga's work confirmed and established a vigorous interest in the biology of phytotelmata by European, particularly German, ecologists which continued from the very early days of organised limnology and ecology through the 1950s to the present day.

Varga's interest in container habitats appears to have been stimulated by the work of Müller (1879 *et seq.*), Picado (1912, 1913) and van Oye (1921, 1923). These earlier writers focused on the 'tanks' formed by the overlapping leaf-bases of bromeliads which retain water and provide a habitat for a wide range of freshwater and terrestrial organisms. Varga, writing in 1928, was also aware of the existence of an aquatic fauna within the pitchers of the Old World genus *Nepenthes* and the American *Sarracenia* and he quotes the works of Sarasin & Sarasin (1905), Jensen (1910), Günther (1913) and van Oye (1921) as his sources. His designation 'phytotelma' also included water-filled tree holes in branch axils and stumps of the European beech tree *Fagus sylvatica* as well as the water bodies collected in a variety of leaf axils such as those he himself studied in the teasel. Russian work by Alpatoff (1922) on the water bodies collected within the inflated sheathing petioles around the inflorescences of *Angelica sylvestris* had also come to his notice. Varga was not the first author to attempt to draw general attention to this class of habitats but the earlier designations of Müller (1879) ('Miniatür-Gewässern'), Brehm (1925) ('Hängende Aquarien') and Alpatoff (1922) ('Mikrogewässern') were all superseded by Varga's 'phytotelma'.

Of course the mere bestowal of a name on an object does not indicate its first discovery and Frank and Lounibos (1983) have followed long tradition by tracing the earliest record of insects originating from water-filled plant containers to a classical Chinese source. Ch'en Ts'ang-ch'i, writing during the T'ang dynasty sometime between 618 and 905 AD, is quoted by them as follows:

Beyond the Great Wall there is a *wen mu t'sai* (mosquito-producing plant), in the
leaves there are living insects which change into mosquitoes.

(Pen T'sao Shih-yi)

Nevertheless Varga (1928) brought this class of habitats into the ken of sci-
entists and, in particular, to the attention of Albrecht Thienemann. Thienemann is deservedly regarded as one of the great organising forces in ecology
and limnology in Europe from the early part of the twentieth century until
the 1950s. As director of the Max-Planck Institüt in Plön he was influential
in directing a large part of German aquatic science throughout much of this
period. Thienemann organised a major German research effort in the East
Indies, the *Deutsche Limnologische Sunda-expedition*, the Proceedings of
which appeared as supplements to the *Archiv für Hydrobiologie*. It was here
that Thienemann drew attention to the fauna of *Nepenthes* pitchers in 1932.
This work was followed by an equally influential general treatment in 1934,
Die Tierwelt der tropischen Planzengewässer, which was encyclopaedic in
nature and was the basis for the extended treatment he gave the subject in
his massive work, *Chironomus*, published in 1954.

Phytotelmata occur in one form or another in a very wide range of ecosystems from subarctic bogs (*Sarracenia* pitchers) to anthropogenic road verges
(*Dipsacus* axils, *Angelica* axils, etc.) to deciduous woodlands (tree holes).
Container habitats occur on some of the most inhospitable and inaccessible
of locations on the face of the Earth. *Heliamphora* pitchers are commonly
found only on the remote Venezuelan outcrops known as *tepuyos* (see George
1989). The tiny Australian native pitcher plant *Cephalotus follicularis* shares
with other genera of pitchers its predilection for nutrient-poor swamplands.
But it is in tropical rainforests that phytotelmata display their full range and
ubiquity. To return, by way of example, to the forests and creek beds of the
Labi region of Brunei, I found, in addition to six species of *Nepenthes* pitcher
plants, water-filled tree holes, water-filled bamboo internodes, a wide range
of water bodies in the leaf axils of fleshy plants and the bract axils of inflorescences, together with pools of water in fallen palm leaves, horizontal logs
and animal-damaged woody fruits such as coconuts. This range, varying
slightly with location, can be found in tropical rainforests around the world.

Phytotelmata within modern ecology

There is little doubt that the early interest of naturalists in these habitats was
fuelled by the fascination with the unusual that drives most natural histori-
ans. The range of plants which host phytotelmata, the range of animals which

occur within them, and the sometimes curious and always interesting eco-logical interactions and other processes which occur within them form a legitimate base for their continued study.

But with the extension of community ecology from the descriptive to the predictive stage that has occurred in recent years, phytotelm studies have come to play a much more central role (Pimm 1982, Young 1997). The multitude of ideas on how natural communities have developed through evolutionary time, how they change through ecological time, and how and why they vary geographically call upon theoretical concepts relating to competition, predation and dispersal, as well as more holistic ideas which address structural limitations inherent (or otherwise) in the complex objects we know as food webs. It turns out that the communities which occur in phytotelmata are pre-eminent and predisposed to provide field tests of many of these ideas.

The reasons for this are fourfold:

- Phytotelm communities are aquatic but are located within terrestrial or semi-terrestrial ecosystems such as forests, woodlands, or swamps. Accordingly the aquatic organisms which occur within them encounter distinct edges which impose upon the community a discreteness not readily found in other more complex ecosystems.
- The communities of metazoan animals which occur in phytotelmata are relatively simple with the numbers of species ranging from one or two up to twenty or thirty. This relative simplicity allows the feeding links and hence the food web within them to be defined readily. This is in marked contrast to larger and more diffuse habitats such as lake beds or forest canopies in which species richness may be measured in hundreds of species and the number of potential feeding links is much larger.
- Most phytotelmata occur within larger ecosystems as a series of units scattered spatially within the forest, woodland, road verge or swamp. They are, to all intents and purposes, replicated across the ecological landscape. Of course, this is not a feature unique to phytotelmata and is comparable to the discrete habitat units provided by dung pats, macro-fungus fruiting bodies, fallen logs and so forth.
- The replicated, faunistically simple and (usually) accessible units provided by phytotelmata together with their structural simplicity allows them to be manipulated and/or imitated in the field so that experiments can be designed and carried out on manageable scales of space and time.

Drawing upon these features and the rich base of natural historical information available to us, a number of workers including myself have been using phytotelm communities to test ideas within community ecology.

This book, then, is devoted, first, to describing the natural history and basic biology of phytotelm communities and the excitement this engenders. Building on this I go on to describe recent and on-going work which uses these communities to test a range of ideas within community ecology. Most of the investigations have generated at least as many questions as they have answered. But this is not unusual.

The information base

Since the early 1960s my students and I have maintained an interest in phytotelm ecology. This received a major boost in the early eighties when Howard Frank, Phil Lounibos and Durland Fish called a Symposium which was held during the 16th International Congress of Entomology held in Kyoto in 1980 (Frank & Lounibos 1983). During this time I renewed contact with Stuart Pimm who added a critical theoretical dimension to my thinking and field work. This has resulted in a number of collaborative efforts and has underpinned the field research of two of my most recent students, Bertram Jenkins and Charles Clarke.

The observations that are used repeatedly to illustrate points within this book result from our field work in various parts of the world. In general the results have been published in the primary literature and details of study sites and circumstances are to be found in these works. These field investigations were carried out on different occasions during a period of nearly thirty years. Naturally my experience with and appreciation of phytotelmata and what I now see as the key questions to be addressed grew during this period. I summarise here the circumstances surrounding crucial phases in this development so that particular results can be placed in context.

Tree holes in Wytham Woods

During the period 1966 to 1969 I studied a set of tree holes located within beech trees in deciduous woodland at Wytham Woods near Oxford in the United Kingdom. This work formed the basis of a doctoral dissertation, an overview of the results of which is provided by Kitching (1971). Details of population fluctuations in two of the commonest species of Diptera from these sites are presented in Kitching (1972a,b).

Tree holes in Lamington National Park

In 1978 I began examination of the community found within water-filled tree holes in the subtropical rainforest of Lamington National Park in south-east

Queensland. This has remained a research site from that time but was studied intensively from 1978 to about 1984. Student collaborators at the time included Catherine Callaghan, Rosemary Lott, Bertram Jenkins and Stewart Jackson. Stuart Pimm also participated in elements of this work. The basic food web involved is described by Kitching & Callaghan (1982) and comparative analyses presented in Kitching (1983), Kitching & Pimm (1985) and Kitching & Beaver (1990). The process of community reassembly following complete disruption is described by Jenkins & Kitching (1990). Basic work on spatial and temporal variability in food-web structure was also based on Lamington data and is described in Kitching (1987a) and Kitching & Beaver (1990). Results of our earliest manipulative experiments, carried out here, are in Pimm & Kitching (1987).

Phytotelmata in Sulawesi

In January and February 1985 I participated in the Royal Entomological Society's centenary expedition ('Project Wallace') to northern Sulawesi, Indonesia. Field work on tree holes and bamboo containers was carried out in the tropical rainforest of the Dumoga-Bone National Park and work on the fauna of *Nepenthes maxima* at nearby Lake Mooat. Basic descriptions of the food webs encountered are in Kitching (1987b), a specific account of a tree-hole dragonfly that was encountered is Kitching (1986a), and Kitching & Schofield (1986) provide a semi-popular account of the pitcher-plant work.

Tree holes in New England

The northern Tablelands of New South Wales provide ready access to a variety of rainforest ecosystems and during the period 1986 to 1991 we studied water-filled tree holes in subtropical and, so-called, warm temperate rainforest in Dorrigo National Park, and cool temperate *Nothofagus*-dominated rainforest at New England and Werrikimbe National Parks. The basic results are incorporated in the comparative analyses presented later in this work. Experimental manipulations of energy supply patterns to analogues of these tree holes were made by Bertram Jenkins and are described in his thesis (Jenkins 1991) and in Jenkins *et al.* (1992).

Phytotelmata in the northern New Guinea

In January and February 1988 I received a research fellowship to work at the Christensen Institute just north of Madang in New Guinea. I was able to identify food webs from water-filled tree holes, bamboo containers and the bract axils of the zingiber, *Curcuma australasica*. These are described in Kitching (1990). Further results from this work appear for the first time in this volume.

Tree holes in northern Florida

In mid-1988, under the aegis of the US/Australia Cooperative Science Scheme, I spent an extended period working in northern Florida. Studies of the food web occurring in tree holes on the Tall Timbers Reserve of the University of Florida, Tallahassee, intended to build on the earlier work of Bradshaw and his colleagues, were only partly successful due to the extended drought being experienced at the time! Some of these results are presented here.

Phytotelmata in the Daintree

In January and February of both 1989 and 1990 the University of New England mounted expeditions to the Daintree region of far north Queensland based at Cape Tribulation. During this period Jenkins and I carried out the most extensive study of tree-hole communities that I am aware of, sampling and analysing the contents of over eighty sites. In addition we examined the water bodies in the leaf axils of an undescribed species of *Freycinetia* (Pandanaceae) and the bract axils of the inflorescences of the so-called 'backscratcher ginger', *Tapeinocheilus ananassae*. A few observations of animals occurring in fallen rat-opened coconuts were made, along with observations on the fauna of water-filled leaf bases of fallen palm fronds. Some of these results are presented in this volume for the first time.

Nepenthes *pitcher plants in northern Borneo*

As indicated in the opening paragraphs of this chapter, in May and June 1989, I assisted Charles Clarke in his studies of *Nepenthes* pitcher plants in northern Borneo. The results of these studies are presented in Clarke & Kitching (1993). A full account of the work comprises Clarke's doctoral dissertation (1992).

Tree holes in tropical Queensland

In the wet season of 1993, in preparation for writing this book, we undertook studies of the faunas of water-filled tree holes at a number of rainforest sites in Queensland selected to fill gaps in our knowledge, hitherto restricted to sites in the subtropical extreme south-east, and the lowland tropical North. Additional studies were carried out in higher altitude forests at Eungella National Park, on the Atherton Tablelands, and at Paluma; more southerly tropical lowland forests at Bellenden Ker National Park and in the subtropical forests of the Conondale Ranges north of Brisbane. This work was done with the assistance of Beverley Kitching, Charles Clarke, Peter Juniper and Heather Mitchell and the results appear for the first time in the present work.

Other studies

Of course the extensive work of other writers also forms a vital part of the information on which the present analyses are based. In particular the work of the following researchers form an indispensable part of the foundation for this work: Beaver on *Nepenthes* pitcher plants in West Malaysia; Bradshaw, Frank, Lounibos and many others on phytotelm mosquitoes in North, Central and South America; Seifert on the axil waters of *Heliconia* species in Central America; Kurihara, Mogi and other Japanese colleagues on phytotelm mosquitoes in Japan and south-east Asia; and, Machado-Alison and his students on Venezuelan phytotelms.

Structure and content

This work is designed to be encountered both as a whole and in parts. Accordingly, for the student of phytotelmata, I hope the traditional beginning-to-end read will be educational, profitable and thought-provoking. But, in addition, each chapter presents an essay within areas of natural history and ecology and those with more specialised interests will find value in reading them in at least partial isolation from the remainder of the work.

The basic structure and intended logic of the book are summarised in Figure 1.1. I have grouped the chapters into five parts after this Introduction. I have prefaced each of these parts with a word picture of some of the field experiences involved, deliberately to try to recreate some of the excitement of the naturalist, to offset in part the objectivity of the ecologist.

Part I contains three substantial 'background' chapters describing the plants, animals and physico-chemical environment of phytotelmata in general. These provide information on the underlying biology, natural history and environmental science which are the vital context of later discussions on the ecology of phytotelm communities. An extended annexe at the end of the book is a family-by-family treatment of the phytotelm fauna.

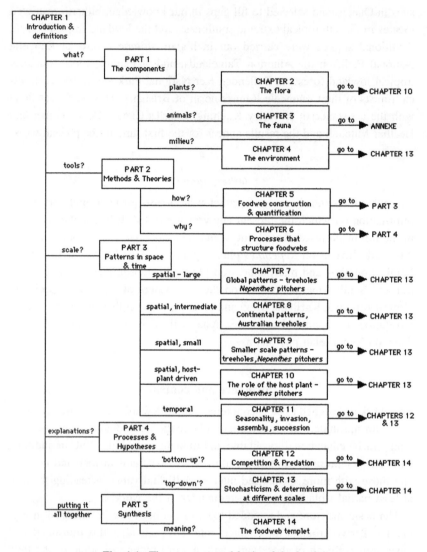

Fig. 1.1. The structure and logic of the book.

Part II of the work contains two important chapters. The first introduces the idea of the food web by presenting examples from each phytotelm type. The chapter defines key formalisms concerning the food-web statistics which will be used without further explanation in later chapters. The second chapter (Chapter 6) in this section reviews segments of the theory of food webs from the recent literature. In addition the chapter explores the biological mechanisms which may occur to structure food webs within phytotelmata at a variety of spatial and temporal scales, and in so doing poses a series of hypotheses which are revisited in Chapter 13.

Part III forms the great bulk of the remainder of the book presenting data from phytotelm studies to illustrate spatial and temporal patterns at a variety of scales. Its five chapters examine food-web variation at the global, continental, regional and local scales, the role of the host plant in structuring phytotelm communities, and the temporal processes observed within these habitats.

Part IV of the book brings together data and hypotheses and points the way ahead. Chapter 12 describes, more or less parenthetically, investigations on the processes of competition and predation in phytotelm systems or their analogues relevant to an understanding of community structuring. In Chapter 13, however, I revisit the hypotheses identified in Chapter 6 and evaluate them against the available data.

The final part of the work comprises a single synthesising chapter (Chapter 14) in which I combine and summarise the hypotheses into a templet structure. This provides a general theory concerning the determination of food-web structure in phytotelms and will act as a model within which future work may be designed and placed in a wider context.

The future

At the very outset of a work such as this it is entirely appropriate to look beyond the present state of knowledge. It must not be supposed that, even in the limited world of phytotelms, we have anything other than the beginnings of understanding how the ecological communities that they contain are maintained. Our understanding of food-web theory, food-web patterns and food-web processes are being added to continuously. Indeed it became necessary to draw a strict line in the sand, so to speak, in about mid-1997 in the preparation of the book. I apologise to those later authors whose important work has appeared since that time. Four key areas of future work are likely to occupy the minds of those who work on food webs – particularly using phytotelms as study objects – and I mention them briefly here.

The ontology of food webs

The way we draw food webs, currently, assumes that particular feeding links are established and do not change readily. So, for instance, within a food web from a tropical tree hole we may identify the presence of a particular species of predatory tadpole, or triclad or *Toxorhynchites* mosquito. We further identify those things which we know these predators feed upon in nature. On the basis of this information we draw links and produce a food web available for the extraction of statistics and further analysis. This overlooks the important fact that organisms develop.

When a *Toxorhynchites* female lays eggs within a tree hole, they emerge as tiny first instar larvae. Over the next several weeks these larvae, given sufficient food, will grow larger and may actually increase in size and volume by more than a hundredfold until the final massive (relatively) fourth instar finally pupates and ends its existence within the tree hole. During this period of time not only does its overall body size increase but its mandibles increase in strength and length, its demand for food increases and, most important in the present context, so does its capability of handling particular food items.

There has been relatively little definitive work on the way in which the natural diets of any insects change through their developmental period. One of the few exceptions is represented by the elegant studies of Marian Moore and her co-workers who examined the way in which the prey of larvae of *Chaoborus punctipennis* changed during the progression from first to fourth instar. This species has small predatory larvae which occur in freshwater systems such as ponds and lakes, not in phytotelms (although the related chaoborid *Corethrella appendiculata* does – see Annexe). Nevertheless Moore (1988) and Moore & Gilbert (1987) were able to show very clear changes in what prey items were consumed by larvae of the species through time. These must inevitably have consequences in considering the role of particular species within food webs.

The relatively few species that occur in most phytotelm food webs, their ease of handling and the readiness with which controlled feeding trials can be established, make phytotelm communities ideal for further studies on this third dimension of food webs.

Interaction strengths

Even when we represent a feeding link within a food web as a mere connecting arrow we are making many assumptions: not only assumptions which reflect our ignorance of ontological changes as discussed above but also just how important that particular link is within the food web.

So within a tree-hole food web we may identify predatory odonate larvae and connect this component of the community confidently to virtually all the insect larvae and other soft-bodied organisms within the food web. We do this as a result of feeding trials and our *a priori* knowledge of the generalist predatory nature of most odonate larvae. Food webs, as we draw them customarily, do not weight these links to reflect what proportion of the diet of a particular species of odonate is made up by each of the potential prey species. Nor do food webs indicate how much energy is actually flowing along each of the feeding links established.

The amount of energy flow along each of the feeding links is the *interaction strength* of the particular link. Quantifying these strengths would greatly increase our knowledge of the food-web dynamics in any system but represents a far from simple task. In our studies of the food webs in individual phytotelm units we have approximated the strength of interactions by taking into account the relative abundances of the species populations that sit at either end of a feeding link (see for example some of the examples in Chapter 9). Techniques using stable isotope labelling do offer the opportunity to quantify the strength of feeding links much more directly.

Again phytotelm food webs provide an ideal opportunity to quantify these interaction strengths using such techniques simply because they are relatively simple structures.

There is an unexploited opportunity to demonstrate interaction strength and the relative distribution of strong and weak links in food webs in phytotelms. Such studies could have important implications in studies of succession, community development and evolution. They await attention.

On experimentation

Active interventions and manipulations of community structure, environmental conditions and so forth is an exciting and productive field of study. Key results on the role of predators in phytotelm systems, on the effects of differing levels of energetic input to phytotelm units, and the construction of artificial microcosms to examine the role of competition and predation, all have followed elegantly designed manipulations.

There is little doubt that manipulative experimentation will continue to play an important role in resolving some of the outstanding questions relating to food-web dynamics in phytotelms. The occurrence of many units of particular phytotelm types within forest and other ecosystems provides the framework for performing such experiments in the field. The ability to reconstruct approximations of phytotelm food webs in the laboratory, also, will continue to play an important role.

The field ecologist, as no other, however, appreciates the limitations of such statistically beautiful experiments. This book aims to emphasise that the emergent properties of food webs reflect evolutionary and ecological actions and interactions on a variety of spatial and temporal scales. Only a few of these scales lend themselves to manipulative experiments: these are the local and short term. So, although I am firmly convinced that elegant experimentation will continue to shed light on how food webs work, I am even more convinced that it is in nature's laboratory that the really grand experiments are carried out. Our task is to read the results and fill in the 'Discussion' section.

Filling the faunistic gaps

Over the years we have accumulated a very large body of information which allows us to carry out analyses to examine the ways in which food-web pattern changes on various geographical scales. These data sets have been gathered at considerable cost in time and money over a great many years. Nevertheless they represent merely a minimal set for the analyses presented here. There is no doubt that continued study of these communities at more sites over longer periods and in more detail will be necessary if our appreciation of the ecology of these situations and the messages that they may have for ecology in general are to be exploited to the full.

We have for instance virtually no information on the communities of organisms in container habitats for the whole of the continent of Africa. There is no single food web that I can draw for any class of phytotelm for any African situation except that for *Nepenthes madagascariensis* for the great offshore island of Madagascar. The situation for South America is almost as bad. Although there have been many studies of tree-hole organisms in South America there is no data set publicly available which allows us to draw a single food web for this extremely widespread and important habitat type (but see Fincke 1999).

Phytotelm studies lend themselves, like no other set of natural systems, to the *comparative* study of communities and food webs. The gradual accumulation of appropriate faunistic data will open up new dimensions for analysis as patterns at different spatial scales emerge.

Darwin needed the Galapagos Islands in exactly the same way that a particle physicist needs a linear accelerator. As Darwin's intellectual descendants we ignore natural pattern – the result of immense natural experiments – at our peril. The world's rainforests are as much my laboratory as the ones with microscopes and benches. And the further prospect of making investigations in both sorts laboratories is both exciting and delightful!

Part I:

The container flora, fauna and environment

It is November in southern Ontario. An icy wind pierces clothing and trees and makes field work a chore to be hastened rather than the usual pleasure. We walk down to a boggy lakeside, ice crackling underfoot – making crossing a makeshift bridge more than usually difficult. On the lake itself the ice is spreading: not yet locking down the whole surface but forming a corona around the rim on which the black ducks stand disconsolately. The ground vegetation changes as we approach the lake: from leaf litter and the dried grass tussocks of winter, it becomes spongy sphagnum bog interspersed with a few dwarf pines. And it is deeply embedded in this sphagnum that Dolf Harmsen of Queen's University and I find what we are looking for. Just visible above the moss profile are the red and green leaves of Sarracenia *pitcher plants: still half-filled with liquid and, even in the winter cold, with one or two larvae of the pitcher-plant mosquito,* Wyeomyia smithii, *swimming within them. The rest of the bog is biologically silent, only the wind in the pines breaking the heavy winter pall which hangs over the northland – at least to the eye and ear of an antipodean naturalist!*

In contrast, I think of a village stand of bamboos just outside Madang in New Guinea. The air is hot and humid and filled with mosquitoes. The grove of bamboos is enormous and the stems themselves up to ten or even twenty centimetres across. This is a village resource held in common as a source of building material, scaffolding, piping, walking sticks, children's toys, and a hundred and one other uses. Each time the stand is cropped, the required stems are severed with the ubiquitous New Guinean 'bushknife' – as likely as not a wickedly sharp length of metal which began its life as the leaf of a car spring. Each cropping leaves a bamboo stump and, in consequence, a cup between the point of severence and the next nodal plate. In the perhumid climate these fill with water rapidly and are the home of a wide range of animals that feed either on the plant detritus which accumulates within the cups,

15

or on each other. But this particular stand comes to mind because here too the ground is covered with the inflorescences of what is sometimes called native turmeric, Curcuma australasica. *These are spikes of relatively inconspicuous white flowers each enclosed with a fleshy pink bract, and each of these bracts too is water filled. The bract water of each of these inflorescences is alive with the tiny white larvae of the mosquito* Uranotaenia diagonalis *and, occasionally, the voracious predatory larvae of the muscid fly* Graphomya. *The muscid larvae, unlike the mosquitoes on which they feed, are able to move freely from one bract axil to the next in order to keep up with the appetite that must be satisfied in order to complete the species' life cycle within the month or so that represents the life time of the inflorescence itself.*

Subarctic bogs and New Guinean rainforest may seem to have little in common but, in each, I have sought and found plants that provided water-filled container habitats for a variety of animal species. And I could have painted, instead, word pictures of English deciduous forest, Indonesian lakesides, Bornean mountain tops or cool temperate Gondwanic rainforest in Tasmania: each of these and most other ecosystems present examples and opportunities to students of phytotelmata.

In the next three chapters I review the widely scattered and sometimes obscure literature on the plants in which phytotelmata form (Chapter 2), the range of metazoan animals which occur within them (Chapter 3), and the physical and chemical properties of the environments themselves (Chapter 4). By including a more extended account of the animals, family by family, in an Annexe at the end of the work, I have attempted to make the review as complete as possible and, in this Annexe and the associated references, I have deliberately set out to save future workers a massive amount of delving in the older and non-English literature involved. The literature up to mid-1997 has been covered although, inevitably, some items will have been missed. In particular I know I have only scratched the surface of the immense literature on mosquito biology.

2

The container flora
The water-holding plants

Phytotelmata are formed whenever watertight hollows appear as part of the growth form of plants. They occur as five principal types with any number of additional minor categories. These five types – bromeliad 'tanks', pitcher plants, water-filled tree holes, bamboo internodes, and axil waters collected by leaves, bracts or petals – are all formed from parts of living plants. In a few instances water-collecting hollows are formed in the fallen parts of plants. For all these categories, in the vast majority of instances, the plants concerned are angiosperms. However, water-filled tree holes are recorded from a few 'lower' plants such as tree ferns and cycads, and Lounibos (1980) describes mosquito larvae from the rain-water pools collected in the concave tops of fungal basidiocarps.

In reviewing the plant groups from which an aquatic metazoan fauna has been recorded I draw heavily on the reviews of Thienemann (1934, 1954), Kitching (1971) and Fish (1983). Fish (1983) estimates that more than 1500 plant species may form phytotelmata. Given that water-filled tree holes may form, with greater or lesser frequency, in almost any species of tree, I suspect that this number is a significant underestimate.

I describe this special flora using the classification of phytotelmata alluded to above.

Bromeliad 'tanks'

According to Frank (1983) the Bromeliaceae contain about 2000 species of plants divided into three subfamilies and about 50 genera. All but one of these species occur in the warm temperate to tropical regions of the Americas, from Florida to central Argentina and Chile. The exception, *Pitcairnia feliciana*, occurs only in Guinea in West Africa. Bromeliads are herbaceous perennials which occur as terrestrial plants, although are better known as epiphytes in

17

the rainforest canopy. Many species may co-occur and plant densities can reach very high levels. Although essentially an American group, the bromeliads have long been popular in horticulture in warmer parts of the world and undoubtedly act as breeding sites for aquatic organisms well outside the plants' natural range. Thienemann (1934), for example, recorded organisms from bromeliads growing in the Bogor botanic garden in his survey of container habitats in South-east Asia.

All so-called 'tank' bromeliads impound water in their leaf axils which overlap tightly to form watertight cavities. The outer, more mature leaves form discrete water bodies; the younger, inner leaf axils combine to form a common pool (Beutelspacher 1971, Zahl 1975, Frank 1983) (Figure 2.1). Species of bromeliad from the two more advanced subfamilies, the Tillandsioideae and the Bromelioideae, have growth forms which may form such water-holding tanks (Pittendrigh 1948). Fish (1983) reports that about 40 of the 50 genera of bromeliads form water bodies of this kind.

The communities of animals which occur within bromeliads have been objects of fascination for many years, inspired originally perhaps by the extensive and detailed work of Picado (1913) who studied species of the *Aechmea, Billbergia, Guzmania, Tillandsia, Thecophyllum* and *Vriesia* in Costa Rica. Picado recorded all of the animals he found in and around bromeliads from Protozoa to fer-de-lance!

Table 2.1 reviews the literature on the aquatic communities which live in bromeliads. In this and other tables in this chapter I have omitted simple faunistic references to the occurrence of particular organisms in various phytotelmata. Some of these references are to be found in Chapter 3 dealing with phytotelm fauna: in other cases entry to the literature may be obtained through the more general ecological references reviewed here. In the case of the bromeliads, Frank (1983) provides an extensive set of such references. In general the literature on bromeliad communities comprises general, semi-popular accounts (e.g. Zahl 1975, Beutelspacher 1971) and accounts of studies of bromeliad-inhabiting mosquitoes, either as vectors of malaria (Downs & Pittendrigh 1946, Pittendrigh 1948 *et seq.*) or as nuisance pests (Frank & Curtis 1977 *et seq.*). Very few accounts deal with the community of organisms within the contained water bodies. Of these few that of Laessle (1961), dealing with the bromeliad genera *Guzmania, Tillandsia* and *Hohenbergia* in Jamaica, is outstanding and rivals Picado (1913) in completeness. The production of more complete accounts of the faunistics of bromeliad tanks allowing the construction of food webs for comparative analysis remains one of the major opportunities within phytotelm studies which was taken up, more recently, by Cotgreave *et al.* (1993).

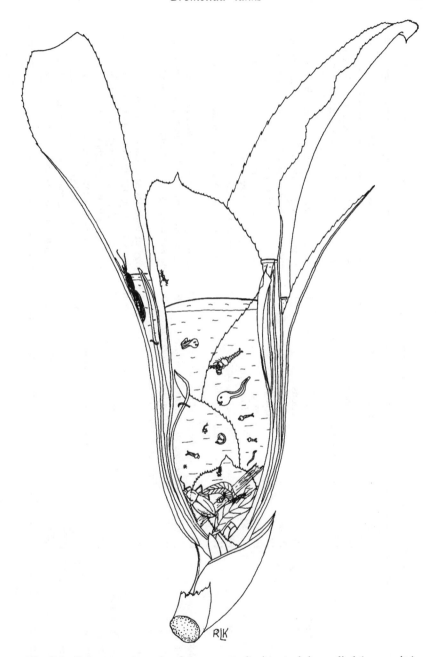

Fig. 2.1. Cut-away schematic of the community in a tank bromeliad (composite).

Table 2.1. *Key works on the aquatic communities associated with the Bromeliaceae*

Author(s)	Year(s)	Bromeliad genera or species	Subject matter
Beutelspacher	1971	*Aechmea bracteata*	The biota associated with the species in Mexico
Cotgreave *et al.*	1993	*Aechmea, Quesnelia, Neoregelia, Vriesia*	Body size and abundance of bromeliad fauna in Brazil
Downs & Pittendrigh	1946	Various	Bromeliads as sources of malaria vectors in Trinidad
Dunn	1937	Various	Herpetofauna of bromeliads in Costa Rica and Panama
Fish	1976	*Tillandsia* spp.	Aquatic invertebrates from bromeliads in Florida
Fish & Beaver	1978	General	Bibliography
Frank	1983	General	Review of bromeliad phytotelmata and their biota
Frank & Curtis	1977–81	*Tillandsia* spp.	Mosquito biology in Florida bromeliads
Frank & O'Meara	1985	*Tillandsia, Catopsis*	Habitat preferences of mosquitoes
Laessle	1961	*Guzmania, Tillandsia, Hohenbergia*	Macro- and micro-organisms in Jamaican bromeliads
Little & Hebert	1996	*Aechmea, Hohenbergia*	Population dynamics of eleven species of ostracod
Lounibos *et al.*	1987a	*Aechmea nudicaulis, A. aquilegia*	Faunal associates, particularly Odonata, in Venezuela
Picado	1913	*Guzmania, Tillandsia, Billbergia, Thecophyllum, Vriesia, Aechmea*	Fauna associated with bromeliads in Costa Rica
Pittendrigh	1948–50	Various	Ecology of malarial mosquitoes from bromeliads
Poinar	1996	None specified	A fossil damselfly supposedly associated with bromeliads
Polhemus & Polhemus	1991	Various	A review of the veliids (Heteroptera) of bromeliads
Pugialli Domingues *et al.*	1989	Various	Ecology of fauna from bromeliads in degraded forests
Reid & Janetsky	1996	*Aechmea, Hohenbergia*	Colonization dynamics especially by copepods
Tressler	1941, 1956	Various	Ostracods in bromeliads, Puerto Rico, Florida and Jamaica
Veloso *et al.*	1956	Various	Anopheline mosquitoes from bromeliads in Brazil
Zahl	1975	Various	Popular article 'Hidden Worlds in the Heart of a Plant'

Pitcher plants

Those insectivorous plants which, collectively, are referred to as pitcher plants, belong to one of five genera spread across three unrelated families in three different biogeographical regions. All species possess modified leaves or extensions of leaves which form watertight bodies which collect liquid. These act as pitfall traps for insects and other invertebrates. These organisms drown and are digested by extracellular enzymes secreted by the plants into the pitcher liquid together with the autolytic enzymes released by the drowned victims. In addition to the hapless captives, however, the pitchers and their organically rich contents provide a situation which has been exploited by a wide range of aquatic organisms. All species of pitcher that have been examined contain a more or less rich in-fauna living exclusively within them (Figure 2.2).

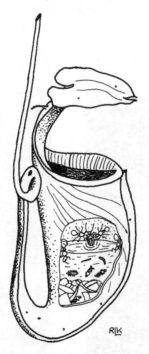

Fig. 2.2. Cut-away schematic of the community found in an aerial pitcher of *Nepenthes bicalcarata* (Clarke & Kitching 1993).

Nepenthaceae

Members of the great Old-World tropical genus *Nepenthes* are the pitcher plants *par excellence*. The seventy of so members of this monogeneric family range from Madagascar to New Caledonia, from China to the far north of Australia. The genus reaches peaks of species diversity in Borneo (31 species), Sumatra (21 species), West Malaysia (11 species) and New Guinea (10 species) (Clarke 1997, Phillips & Lamb 1996, Kurata 1976).

Species of *Nepenthes* occupy a wide range of habitats in the humid tropics occurring as low-growing scramblers on exposed mountain tops, as large epiphytic vines in lowland rainforests, as spreading thickets on floating sphagnum bogs, and as free-standing herbaceous shrubs in exposed situations (see the preamble to Chapter 1). All are species of nutrient-stressed situations and many are associated with disturbed-ground light gaps in forests. All begin life as ground-germinating rosettes. The pitchers themselves are modified tendrils at the tips of the plants' compound leaves. Virtually all species possess larger, more rounded lower pitchers and more elongate upper pitchers of lower internal capacity. All have enzyme-secreting glands on their inner surfaces. The upper pitchers in particular have evolved a variety of more or less fantastic shapes (Figures 2.3–2.8). Pitchers vary in internal capacity from a few cubic centimetres, to two litres or more in some of the giant ground pitchers of Bornean species.

Figs. 2.3–2.8. Pitchers of six co-occurring species of *Nepenthes* from the lowlands of Brunei on the island of Borneo: 2.3, *N. albomarginata*, 2.4, *N. ampullaria*, 2.5, *N. bicalcarata*, 2.6, *N. gracilis.*, 2.7, *N. mirabilis*, 2.8, *N. rafflesiana*. Photos: C. Clarke.

2.5

2.6

2.7

2.8

Pitchers present a widespread and readily available aquatic environment in some tropical ecosystems. The pitchers themselves, depending upon the species, may be relatively long lived (Clarke 1992) and numerous within particular locations. Accordingly the rich organic soup which develops within the pitchers of *Nepenthes* has been widely exploited as an environment by a variety of in-faunal species from mites and rotifers to midges, mosquitoes, dragonflies, and even frogs.

Key references on studies of the communities which occur within *Nepenthes* pitchers are given in Table 2.2. Albrecht Thienemann (1932) carried out the first critical investigations of *Nepenthes* faunas working on species in Sumatra and Java. The work of Beaver, both in describing the insect associates of pitchers from West Malaysia (1979a, 1980) and in synthesising other published information in order to carry out comparative analyses of the food webs contained within pitchers (1983, 1985), stimulated a new phase of work on *Nepenthes* communities. Kitching & Pimm (1985) and Kitching & Beaver (1990) presented further comparative analyses together with information on spatial and temporal variation in food-web structure. As indicated in Chapter 1, Charles Clarke spent an extended period of time in the field in northern Borneo studying the communities within pitchers of six co-occurring species of *Nepenthes* (Clarke 1992, 1997; Clarke & Kitching 1993, 1995). Other important intensive studies have been made by Kato *et al.* (1993) on a set of ten *Nepenthes* species in Sumatra, and by Mogi & Chan (1996, 1997) on the fly faunas of three pitcher species in Singapore. Lastly, of special note, the Madagascan species, *N. madagascariensis*, has been the subject of a detailed and insightful study by Ratasirarson & Silander (1996).

Sarraceniaceae

In the New World the three genera of Sarraceniaceae all possess leaves which have become modified into more or less conical, upwardly directed pitchers.

The ten species of *Sarracenia* occur in eastern North America from eastern Canada to Florida. They are plants of exposed bogs. The widespread *S. purpurea* maintains liquid-filled pitchers all year and as such presents a potential habitat for aquatic inquilines that is both persistent and predictable. Other, more southern species of *Sarracenia* with which *purpurea* overlaps, such as *alata*, *flava* and *leucophylla*, have pitchers which show a tendency to dry up in summer and senesce and die in winter (Bradshaw 1983). It appears that *S. purpurea* is the sole species to host an animal community of any complexity (McDaniel 1971).

Darlingtonia californica is the single western equivalent of the eastern *Sar-*

racenia. It occurs, again in acid bogs, from south-western Oregon to the Sierra Nevada of California. The pitchers are characteristically hooded, giving rise to the popular epithet, 'cobra-plant'. The Sarraceniaceae are completed by the six described species of Venezuelan *Heliamphora* from the sandstone outcrops known as 'tepuyos'. These little-known species have pitchers completed by tiny 'lids' reminiscent of those of the Old-World genus, *Nepenthes*.

As is summarised in Table 2.2, the in-fauna of *Sarracenia purpurea* has been much studied. This has focused on the mosquito *Wyeomyia smithii* and, to a lesser extent, the chironomid midge *Metriocnemus knabi*. Few authors have directed their attention at the complete fauna which also contains sarcophagid larvae (Aldrich 1916) and may well include mites and microcrustacea. *Darlingtonia* pitchers appear to have a depauperate fauna containing only the chironomid *Metriocnemus edwardsi* (Jones 1916, Szerlip 1975, Fish 1983). The pitchers of *Heliamphora* have been studied recently by a number of authors (see Jaffe *et al.*, 1992, for a general account). Barrera *et al.* (1989) recorded *Wyeomyia* mosquitoes from *H. heterodoxa* and *H. nutans*. Jaffe *et al.* studied *H. heterodoxa*. *H. minor*, *H. ionasii* and *H. tatei* in the Guyana Highlands of Venezuela and added species of the chironomid *Metrocnemus*, and the ceratopogonid *Dasyhelea* to the tally of in-faunal species from this genus of pitcher plant.

Cephalotaceae

The curious Western Australian pitcher plant, *Cephalotus follicularis*, is the only member of its family. The plant occurs from the east coast peat swamps of Western Australia inland to Mt Manypeaks, east of Albany in the extreme south of the state (Erickson 1968). This rosette plant possesses normal, bladed, leaves in addition to those that are modified into small ribbed pitchers, again with distinctive raised lids.

S. A. Clarke studied the fauna of *Cephalotus follicularis* pitchers. She observed a rich micro-fauna including copepods, mites and the larvae of two species of fly, a ceratopogonid and a micropezid (see Chapter 3) (Clarke 1985).

Tree holes

A tree hole is any cavity or depression existing in or on a tree. These may be divided, almost without exception, into two distinct categories: those which maintain an unbroken bark lining throughout their existence, and those which lack this lining and penetrate, ultimately, deep into the wood of the tree.

Table 2.2. *Key works on the aquatic communities associated with pitcher plants (excluding those concerning mosquitoes – see Table A.7)*

Author(s)	Year(s)	Species	Subject matter
Addicott	1974	*Sarracenia purpurea*	Effects of predation on protozoan assemblage
Beaver	1979a	*Nepenthes* spp.	Insect associates of *N. ampullaria, N. albomarginata* and *N. gracilis*
Beaver	1980	*Nepenthes* spp.	Insect associates of *N. ampullaria, N. albomarginata* and *N. gracilis*
Beaver	1983	*Nepenthes* spp.	Review of fauna and food webs in pitchers
Beaver	1985	*Nepenthes* spp.	Review of food-web dynamics
Bradshaw	1983	*Sarracenia purpurea*	Interaction between *Wyeomyia* and *Metriocnemus* in North America
Buffington	1970	*Sarracenia purpurea*	Ecology of *Wyeomyia* and *Metriocnemus* in North America
Cameron *et al.*	1977	*Sarracenia purpurea*	Effects of oxygen levels on *Metriocnemus knabi*
Clarke & Kitching	1993	*Nepenthes* spp.	Food webs from six Bornean species
Clarke & Kitching	1993	*Nepenthes bicalcarata*	Mutualism between pitcher plants and ants
Clarke, S.	1985	*Cephalotus follicularis*	General treatment of west Australian pitchers and fauna
Disney	1982	*Nepenthes distillatoria*	*Megaselia* (Phoridae) from pitchers
Erber	1979	*Nepenthes reinwardtiana*	Associated community in northern Sumatra
Fish & Beaver	1978	General	Bibliography
Fish & Hall	1978	*Sarracenia purpurea*	Succession and stratification of insects in pitchers in USA
Günther	1913	*Nepenthes distillatoria*	Insect associates of pitchers in Sri Lanka
Hamilton	1904	*Cephalotus follicularis*	Records of larvae and micro-organisms from Western Australia
Hegner	1926	*Sarracenia purpurea*	Protozoa from pitchers
Jones	1916	*Darlingtonia californica*	Records of midge larvae from California
Judd	1959	*Sarracenia purpurea*	Studies on 'inquilines and victims' in Ontario
Kato, Hotta, Tamin & Itino	1993	*Nepenthes* spp.	Community structure in 10 Sumatran *Nepenthes* spp.
Kingsolver	1979	*Sarracenia purpurea*	Environmental heterogeneity in pitcher plants
Kitching	1987b	*Nepenthes maxima*	Food web from northern Sulawesi
Kitching & Beaver	1990	*Nepenthes* spp.	Review of food-web dynamics
Kitching & Pimm	1985	*Nepenthes* spp.	Review of food-web dynamics
Mogi & Chan	1996	*Nepenthes* spp.	Predatory habits of Diptera in Singapore *Nepenthes*

Mogi & Chan	1997	*Nepenthes* spp.	Diptera assemblages in Singapore *Nepenthes*
Mogi & Yong	1992	*Nepenthes ampullaria*	Community analysis for pitchers in Malaysia, effect of pitcher age
Paterson	1971	*Sarracenia purpurea*	Overwintering in pitcher-plant faunas in New Brunswick
Ratasirarson & Silander	1996	*Nepenthes madagascariensis*	Community dynamics in Madagascar
Rymal & Folkerts	1982	*Sarracenia* spp.	General review of biology and conservation status
Shinonaga & Beaver	1979	*Nepenthes albomarginata*	Sarcophagid fly from pitchers in West Malaysia
Thienemann	1932	*Nepenthes* spp.	General account of habitat and fauna
Wirth & Beaver	1979	*Nepenthes* spp.	*Dasyhelea* midges from pitchers in South-east Asia
Wray & Brimley	1943	*Sarracenia purpurea*	Studies on 'inquilines and victims' in North Carolina
Yeates et al.	1989	*Nepenthes mirabilis*	Sarcophagid fly from pitchers in northern Australia

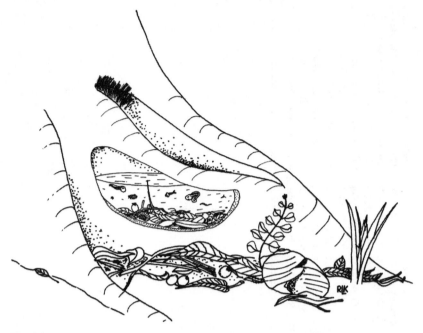

Fig. 2.9. Cut-away schematic of the community found in water-filled tree holes, based on that occurring in Lamington National Park, south-east Queensland (Kitching & Callahan 1982).

I have followed Röhnert (1950) in referring to the first of these as 'pans' and the second as 'rot holes' respectively. This distinction has come to be widely used. Figure 2.9 presents a schematic of a 'root pan' based on the community encountered in Lamington National Park, south-east Queensland.

Pans are formed in buttress roots and branch axils by the growing together of parts of the tree itself. This may be the result of some physical distortion of the growing tree or may reflect a growth form characteristic of the species of tree itself. Rot holes, in contrast, need some external agency to initiate them. Park *et al.* (1950) discuss the formation and development of such holes at length. All start with some damage to the bark either though storm, animal or human damage. Once the bark is penetrated then fungal rot will generally ensue until a larger cavity is formed. Park *et al.* (1950) describe two alternative courses of events. In the event that the cavity formed is not watertight then what they call the 'terrestrial microsere' ensues – in which case no phytotelma is formed. If the resulting cavity is watertight then an ecological succession they refer to as the 'aquatic microsere' follows. Where the size of the initial hole formed through the bark is relatively small, rot holes may heal over through callus formation by the plant. Rot holes may be formed accidentally but also form in cut stumps and following pollarding, coppicing or pruning. Laird (1988) draws particular attention to the artificially constructed tree holes in the bases of palm trees on certain Pacific islands. These are known as 'tungu' and are constructed to act as water-collecting devices. Like natural tree holes, they form habitats for a variety of organisms, including the vector mosquito *Aedes polynesiensis*. Watertight hollows may also be maintained, or form anew, when trees have fallen and some authors have referred to these as 'log holes' (A. Fanning pers. comm., Lee 1944, Kitching & Orr 1996).

In all types of tree holes layers of leaf litter accrue and rot down through the action of fungi and bacteria, following comminution by the in-fauna. Holes near the ground commonly contain mineral material formed either as a result of inundation or soil splash during heavy rainfall.

Tree holes, not suprisingly, come in almost all sizes. I have sampled the fauna of holes containing no more than three or four millilitres of liquid, and from holes containing more than twenty or thirty litres of material. In some instances the rotting following bark penetration may result in a hollowing of nearly the whole tree and, in exceptional cases, such immense hollows may fill with water. Tiny tree holes tend to be short lived as aquatic habitats, emptying and refilling to reflect the vicissitudes of rainfall and evaporation, but somewhat larger holes, say between 10 and 20 centimetres in diameter, are suprisingly persistent as aquatic habitats, especially when they form as pans. Some sites sampled in 1967 in beech woodland near Oxford, England, remain today and, although visited by me only at very irregular intervals, have contained water whenever visited. Certainly some sites contained within *Lophostemon conferta* trees in south-east Queensland have contained water continuously for fifteen years. Accordingly, to think of these as 'temporary' habitats may be misleading in the extreme.

Water-filled tree holes have been recorded from every continent except Antarctica and, if Northern Hemisphere knowledge is any basis, may occur in virtually any species of tree. Röhnert (1950) studied water-filled tree holes and their fauna in northern Germany. She recorded sites from *Quercus robur, Fraxinus excelsior, Ulmus laevis, U. scabra, Tilia cordata, T. platyphyllos, Alnus glutinosus, Fagus sylvatica, Carpinus betulus, Acer platanoides, A. campestre, Aesculus hippocastanum, Betula pendula, Abies alba* and *Picea abies* – in other words in a large proportion of the tree species which occur in the area! A similar analysis for Great Britain and North America confirms this generalisation (Kitching 1969).

This having been said, however, there can be no doubt that water-filled holes are much more frequent in some trees than others. In my early work in the United Kingdom, I studied holes in parkland beech trees (*Fagus sylvatica*). Bradshaw & Holzapfel (1986b) again found holes predominantly in beech trees in southern England. In parallel studies in France these authors found many holes in pollarded plane trees (*Platanus acerifolia*). In south-east Queensland we found many more water-filled holes in the massive exposed roots of *Lophostemon conferta* than elsewhere, whereas in tropical rainforest in north Queensland trees of *Xanthostemon javanicum* contained multiple holes, especially where they grew overhanging streams and creek beds. In the forest of the Dumoga-Bone National Park in northern Sulawesi rot holes in palm stumps were predominant, reflecting recent tree-harvesting practice. I have no doubt that in any particular location one or two species of tree will dominate in providing the majority of water-filled cavities as habitats for insects and other fauna. The species of tree will reflect topography, forest history, and management practices.

In general, authors have been unable to show that the species of tree in which tree holes are contained have any impact on the composition of the infauna. In contrast, factors relating to the physical location and form of water-filled tree holes certainly do contribute to the composition and relative abundance of the fauna they contain. Bradshaw & Holzapfel (1986b) identify exposure, size, orientation, pH, conductivity and tannin–lignin content as factors contributing to the occurrence of mosquitoes in European tree holes. Copeland & Craig (1990) have demonstrated clear patterns in the distribution of mosquitoes in the Great Lakes region of the United States reflecting differences in tree hole type (i.e. pans versus rot holes), depth and height of the hole from the ground (see Chapter 12). Sinsko & Grimstad (1977) show that differential selection of tree holes for oviposition on the basis of the height of the aperture results in spatial separation of individuals of the two North American sibling species of mosquito, *Aedes triseriatus* and *Ae. hendersoni*. Finally, Mattingly (1969), amongst other authors, identifies the tendency of sabethine mosquito species to seek out smaller tree holes, often with cryptic entrances.

Although there is a very large literature on the mosquito fauna of water-filled tree holes (see Chapter 3), there are many fewer cases where the whole metazoan fauna has been examined. These largely result from my work and those of my associates but, in a few other cases, food webs can be constructed from species lists presented by other authors, often as results incidental to their mosquito studies. Table 2.3 reviews this literature.

Table 2.3. *Key works on the aquatic fauna associated with water-filled tree holes (excluding those dealing exclusively with mosquitoes – see Tables A.5, A.6)*

Author(s)	Year(s)	Subject matter
Beattie & Howland	1929	Fauna and micro-organisms from English tree holes
Brandt	1934	Environment and organisms in beech tree holes, Germany
Fashing	1975	Tree-hole mites and associates in Kansas and elsewhere
Fashing	1976	Tree-hole mites and their associates in Virginia
Fincke	1999	Predator assemblages in Neotropical tree holes
Jenkins *et al.*	1992	Experimental manipulations of environment, Australia
Jenkins & Kitching	1990	Food-web reassembly after disturbance, Australia
Karstens & Pavlovskij	1927	Water analysis in ash tree holes in Russia
Keilin	1927	Fauna of a single hole in *Aesculus hippocastanum* in UK
Kitching	1971	Community in southern UK in beech trees
Kitching	1983	Comparison of English and Australian food webs
Kitching	1987a	Food-web structure from northern Sulawesi
Kitching	1987b	Spatial and temporal variation in food webs, Australia
Kitching	1990	Food-web structure from northern New Guinea
Kitching & Beaver	1990	Variation in food-web structure at different scales
Kitching & Callaghan	1982	Food web from holes in subtropical rainforest, Australia
Kitching & Orr	1996	Food-web from rainforest sites in Borneo
Kitching & Pimm	1985	Comparative food-web analysis
Lackey	1940	Microbiota of tree holes in Alabama
Mayer	1938	Fauna of tree holes in Czechoslovakia
Paradise & Dunson	1997	Scirtid larvae as facilitators of midge populations
Park *et al.*	1950	Tree-hole habitats with special reference to Pselaphidae
Pimm & Kitching	1987	Experimental manipulation of energy regime, Australia
Röhnert	1950	Environment and fauna in Schleswig-Holstein, Germany
Snow	1958	Arthropod stratification in a stump hole in Illinois

The container flora

Table 2.3. (*cont.*)

Author(s)	Year(s)	Subject matter
Sota	1996	Habitat size and resource base in tree holes on Tsushima Is.
Sota	1998	Habitat size and community structure on Iriomote Is.
Thienemann	1934	General account, introduction to the fauna
Woodward *et al.*	1988	General account, fauna of tree holes in oaks, California

Bamboo internodes

The bamboos are a tribe of grasses characterised by their formation of woody stems which are hollow and which have impervious nodal plates. They form about 80 genera and are cosmopolitan in occurrence although largely restricted to more tropical regions (where they may occur up to high altitudes). Bamboos are much used for construction and many thickets close to settlements are managed although seldom directly cultivated.

The series of watertight compartments that occur within the stems predispose bamboos to form phytotelmata once some means of water penetration is achieved. Such penetration is commonplace, either by natural or anthropogenic means (Figure 2.10).

Amongst the 'natural' agencies which lead to the penetration are cerambycid beetles which chew holes through the internode wall and lay their eggs within the stem hollow. After developing, the emerging beetles chew their way out of the stem low down on the internode, leaving a relatively large hole which allows water penetration and the establishment of a phytotelma. Damir Kovac has recently documented this process in Malaysia where he studied the lamiid cerambycid, *Abryna regispetri* (Kovac & Yong 1992, Kovac 1994, Kovac & Streit 1996). In Peru, Adele Conover (1994) has described a similar process where the bamboo stem is 'opened up' not by cerambycid damage but by the oviposition activities of a species of katydid. In bamboo thickets close to human habitation, internodes are opened when bamboo stems are collected for domestic purposes and the remaining stumps rapidly fill with water. Such situations may also arise following storm damage.

Bamboo internodes as phytotelma have water-holding capacities varying from a few cubic centimetres to over a litre depending on the species and age of the bamboo.

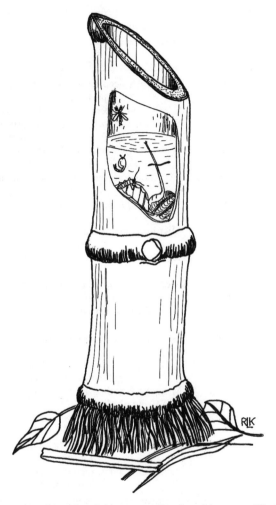

Fig. 2.10. Cut-away schematic of the community found in water-filled intenodes of bamboos, based on that occurring in northern New Guinea (Kitching 1990).

Table 2.4 reviews key works on the aquatic in-fauna of bamboos. Again most work has been done on species of mosquitoes which occur in these locations (Kovac recorded over 100 species from bamboo internodes in West Malaysia, 1994). I have examined the metazoan food web which occurs in these situations in both Sulawesi and New Guinea (Kitching 1987b, 1990) and Mogi & Suzuki (1983) published a complete study of the internode fauna from bamboos in Japan. Sota and Mogi (1996) followed this up with more detailed study on bamboo internodes in Sulawesi, focusing on the changes

Table 2.4. *Key works on the aquatic communities associated with bamboo internodes*

Author(s)	Year(s)	Subject matter
Conover	1994	Popular account of community in *Guadua weberbaueri*, Peru
Kato & Toriumi	1951	Mosquito assemblages in bamboos in Japan
Kitching	1987b	Food webs from internodes in northern Sulawesi
Kitching	1990	Food webs from internodes in northern New Guinea
Kitching & Beaver	1990	Local variation in Sulawesi food webs
Kovac	1994	Faunal study in West Malaysia
Kovac & Streit	1996	Definitive account of bamboo communities in Malaysia
Kovac, Pont & Skidmore	1997	*Graphomya* in Indonesian bamboos
Kurihara	1954	Micro-organisms and dipterous larvae in bamboo containers
Kurihara	1959	Relationship between pH and dipterous larvae in containers
Kurihara	1983	Succession of dipterous larvae in bamboos in Japan
Leicester	1903	First notice of the habitat
Louton, Gelhaus & Bouchard	1996	Aquatic macrofauna of sites in Peru
Macdonald & Traub	1960	*Armigeres* species, from internodes
Mogi & Suzuki	1983	Biota from internodes in Japan with special reference to mosquitoes
Nakata *et al.*	1953	Mosquito breeding in internodes in Kyoto, Japan
Sota & Mogi	1996	Effects of altitude in bamboos in Sulawesi
Sota *et al.*	1992	*Aedes riversi* and *A. albopictus* breeding in bamboo stumps in Japan
Sota *et al.*	1998	Habitat size and resource base in bamboos on Tsushima Is.
Thienemann	1934	General account, records large fauna from Java and Sumatra

that occurred in the community with changing altitude. Kovac & Streit (1996) presented the food web from these situations in peninsular Malaysia & Louton *et al.* (1996) from lowland tropical forest in Peru. In general the food webs which develop in bamboo internodes are closely similar to those found in neighbouring tree holes and many of the species of in-fauna are common to the two types of phytotelmata.

Axil waters

Water accumulates in the watertight axils of leaves and floral bracts as well as within some species of flower themselves. These axil waters are wide-spread within the plant kingdom and undoubtedly occur in many more situations than have been recorded. Fish (1983) has compiled a list of plants from which axil waters had been recorded to that date. I have re-arranged and updated this as Table 2.5, organising the entries into plant taxonomic order following Heywood (1978). Although there is no way in which such a list could be considered even near complete, nevertheless some interesting patterns do emerge from the set of records. The table lists 77 records. Of these, 66 (86%) are from monocotyledonous plants belonging to one of three superorders, the Commelinidae (28 records), Arecidae (23 records) and Lili-idae (15 records). The bromeliads (not included in this table) also belong to the Commelinidae. The plants appearing in the list are, for the most part, robust with sheathing leaf bases or floral bracts, and include many familiar ornamentals such as *Strelitzia, Cordyline, Dracaena, Dieffenbachia* and *Crinum*. The large majority of the records represent tropical or subtropical species and the temperate records which exist are, atypically perhaps, dominated by dicotyledonous plants. These include species such as the teasels and angelicas.

Axil waters are often small and have a relatively simple fauna. I recorded a mean liquid content of 21.3 cm^3 (±11.07) for inflorescences of *Curcuma australasica* from northern New Guinea, and each inflorescence comprised a series of ten to twenty separate bract axils. In contrast Haddow (1948), working on water-filled leaf axils in western Uganda, recorded average water contents, *per axil*, ranging from 1.5 cm^3 (*Sanseveria* spp.) through to 166.9 cm^3 (wild banana *Musa* (*Physocaulis*) sp.). He recorded a single axil of wild banana containing an astonishing 1754.8 cm^3! The question of the capacity of these and other phytotelmata is revisited in Chapter 4.

Table 2.6 reviews key papers on the faunas of axil waters from around the world. Of particular interest is the influential series of papers by the late Richard Seifert on the ecology of the bract axils of *Heliconia* inflorescences (Seifert 1975, 1980, 1982; Seifert & Seifert 1976a, b, 1979a, b) which first brought together knowledge of the natural history of any phytotelm communities with key theoretical questions within community ecology (see also Figure 2.11).

Table 2.5. *Plants from which axil waters have been recorded*

Taxon	Type
ROSIDAE	
Rafflesiales	
RAFFLESIACEAE	
Rafflesia tuanmudae	Flower
Euphorbiales	
EUPHORBIACEAE	
Euphorbia kamerunica	Leaf axil
Umbellales	
UMBELLIFERAE	
Eryngium floribundum	Leaf axils
Myrrhidendron sp.	Leaf axils
Angelica sylvestris	Leaf axils
ASTERIDAE	
Scrophulariales	
GESNERACEAE	
Cyrtandra glabra	Leaf axil
Campanulales	
CAMPANULACEAE	
Lobelia keniensis	Leaf axil
Dipsacales	
DIPSACACEAE	
Dipsacus fullonum	Leaf axil
D. laciniatus	Leaf axil
D. sylvestris	Leaf axil
Asterales	
COMPOSITAE	
Dubautia laxapseudoplantaginea	Leaf axil
COMMELINIDAE	
Commelinales	
COMMELINACEAE	
Commelina obliqua	Leaf axil
Commelina sp.	Leaf axil
Eriocaulales	
ERIOCAULACEAE	
Eriocaulon vaginatum	Leaf axil
Restionales	
FLAGELLARIACEAE	
Hanguana malayana	Leaf axil
Poales	
GRAMINEAE	
Coix lacrymajodi	Leaf axil
Saccharum officinarum	Leaf axil
Cyperales	
CYPERACEAE	
Cyperis grandis	Leaf axil
Scirpus sylvaticus	Leaf axil

Table 2.5. (*cont.*)

Taxon	Type
Typhales	
TYPHACEAE	
Typha sp.	Leaf axils
Zingiberales	
CANNACEAE	
Canna orientalis	Leaf axil
Canna sp.	Leaf axil
COSTACEAE	
Tapeinochilus ananassae	Flower bract
MARANTACEAE	
Calathea lutea	Floral bracts
Calathea sp.	Floral bracts
MUSACEAE	
Musa sp.	Leaf axil
Ensete sp.	Leaf axil
STRELITZIACEAE	
Phenakospermum guianense	
Strelitzia sp.	Flower bracts
Heliconia aurea	Flower bracts
H. bihai	Flower bracts
H. brasiliensis	Flower bracts
H. caribaea	Flower bracts
H. imbricata	Flower bracts
H. wagneriana	Flower bracts
ZINGIBERACEAE	
Curcuma australasica	Flower bracts
Curcuma sp.	Flower bracts
Zingiber sp.	Flower bracts
ARECIDAE	
Arecales	
PALMAE	
Manicaria saccifera	Leaf axil
Mauritia	Leaf axil
Raphia taedigera	Leaf axil
Veitchia sp.	Leaf axil
Pandanales	
PANDANACEAE	
Freycinetia arborea	Leaf axil
Freycinetia sp.	Leaf axil
Pandanus chiliocarpus	Leaf axil
P. tectorius	Leaf axil
P. thurstoni	Leaf axil
P. rabaiensis	Leaf axil
P. veitchii	Leaf axil
Pandanus sp.	Leaf axil

The container flora

Table 2.5. (*cont.*)

Taxon	Type
Arales	
ARACEAE	
Alocasia indica	Leaf axil
A. macrorhiza	Leaf axil
Colocasia antiquorum	Leaf axil
C. esculenta	Leaf axil
Colocasia sp.	Leaf axil
Cyrtosperma chamissonis	Leaf axil
Dieffenbachia sp	Leaf axil
Pleurospa arborescens	Leaf axil
Xanthosoma sagittifolium	Leaf axil
X. violaceum	Leaf axil
Zantedeschia aethiopica	Leaf axil
LILIIDAE	
Liliales	
AGAVACEAE	
Cordyline terminalis	Leaf axil
Dracaena brasiliensis	Leaf axil
D. fragrans	Leaf axil
D. imperialis	Leaf axil
D. reflexa	Leaf axil
D. steudneri	Leaf axil
D. ugandensis	Leaf axil
Sanseveria spp.	Leaf axil
AMARYLLIDACEAE	Leaf axil
Crinum asiaticum	Leaf axil
C. giganteum	Leaf axil
C. hybridum	Leaf axil
Crinum sp.	Leaf axil
Hymenocallis macleana	Leaf axil
LILIACEAE	
Astelia solandri	Leaf axil
Astelia sp.	Leaf axil

Modified from Fish (1983).

Minor categories

In addition to the five interrelated categories of phytotelmata already alluded to, a number of authors have examined aquatic communities in other plant-held water bodies, notably those in fallen fruits.

Thienemann (1934) describes the habit of the plantain squirrel, *Callosciurus notatus,* of biting holes in fallen coconuts and eating out the contents. In Java and Sumatra the resulting husks fill with water and form a breeding

Fig. 2.11. Cut-away schematic of the community found in the water-filled inflorescence bracts of *Heliconia caribaea*, based on that occurring in Venezuela (Machado-
Allison *et al.* 1993).

place for culicids, ceratopogonids, psychodids and other larvae. Wodzicki
(1968) reports a similar situation from the Tokelau Islands where rats gnaw
fallen coconuts. The white tailed rat, *Uromys caudimaculatus*, deals with
fallen coconuts in like manner in the lowland rainforests of far northern
Queensland, producing receptacles invariably containing mosquito larvae.

Surtees (1959), in Nigeria, noted fallen coconut husks and decaying gourd
shells as suitable habitats for mosquitoes. Indeed, in follow-up work in
Nigeria, Service (1965) used pierced gourds of *Lagenaria siceraria* as

Table 2.6. *Key works on the aquatic communities associated with plant axils*

Author(s)	Year(s)	Plant species	Plant part	Subject matter
Alpatoff	1922	*Angelica sylvestris*	Leaf bases	General natural history and review
Disney & Wirth	1982	*Dipsacus sylvestris*	Leaf bases	Ceratopogonid midge breeding
Fish	1983	Many	Various	Comprehensive list of animal records from axil waters
Haddow	1948	General	Various	Mosquitoes in plant axils in Uganda
Kitching	1990	*Curcuma australasica*	Flower bracts	Food webs from New Guinea
Lang & Ramos	1981	*Musa* sp.	Leaf bases	Ecology of mosquitoes in the Phillipines
Lee	1974	*Euphorbia kamerunica*	Leaf bases	Mosquito breeding
Lounibos	1979a	*Pandanus rabaiensis*	Leaf bases	Mosquito faunistics in Kenya
Machado-Allison *et al.*	1983	*Heliconia caribea*	Flower bracts	Associated insect communities in Venezuela
Maguire *et al.*	1968	*Heliconia bihai*	Flower bracts	Control of community structure, Puerto Rico
Means	1972	*Dipsacus laciniatus*	Leaf bases	Mosquito breeding
Mogi	1984	*Alocasia odora*	Leaf bases	Distribution and crowding in mosquito larave
Mogi & Sembel	1996	*Alocasia* sp.	Leaf bases	Predator–prey system within taro axils in Sulawesi
Mogi & Yamamura	1988	*Curcuma, Zingiber*	Flower bracts	General faunistics and regulation of mosquito populations
Mogi *et al.*	1985	*Alocasia* sp.	Leaf axils	Community dynamics especially of mosquitoes
Seifert	1975	*Heliconia* spp.	Flower bracts	Communities as ecological islands
Seifert	1980	*Heliconia aurea*	Flower bracts	Mosquito fauna
Seifert	1982	*Heliconia* spp.	Flower bracts	General review of Neotropical communities
Seifert & Barrera	1981	*Heliconia aurea*	Flower bracts	Studies on mosquito cohorts in Mexico
Seifert & Seifert	1976a	*Heliconia* spp.	Flower bracts	Natural history of associated insects
Seifert & Seifert	1976b	*Heliconia* spp.	Flower bracts	Community matrix analysis
Seifert & Seifert	1979a	*Heliconia* spp.	Flower bracts	*Xenarescus moncerus* (Coleoptera) in communities
Seifert & Seifert	1979b	*Heliconia* spp.	Flower bracts	A *Heliconia* community in a Venezuelan cloud forest
Strenzke	1950	*Scirpus sylvaticus*	Leaf bases	General natural history and review
Swezey	1936	*Freycinetia arborea*	Leaf bases	Insect fauna from axil waters in this plant in Hawaii
Teesdale	1957	*Musa* sp.	Leaf bases	Breeding of *Aedes* (*Stegomyia*) *simpsoni* in Kenya

Thienemann	1934	General	Various	Communities in axil waters (and other phytotelmata)
Vandermeer et al.	1972	*Heliconia* spp.	Flower bracts	*Paramecium* in axil waters
Varga	1928	*Dipsacus sylvester*	Leaf bases	General natural history and review
Williams	1944	General	Various	Diptera in axil waters (and other phytotelmata)
Wirth & Soria	1981	*Calathea* sp.	Flower	Ceratopogonid midge breeding

experimental substitutes for water-filled tree holes in his studies of mosquito assemblages in the northern Guinea savannahs.

Lounibos (1978) reported on water-filled fruit husks acting as breeding sites for species of *Eratmapodites* mosquitoes. Species of fruit involved were of *Strychnos innocua, S. spinosa* (Loganiaceae), *Saba florida* and *Landophia* sp. (Apocynaceae) (Lounibos 1978).

Lounibos (1980) also records mosquito larvae in water pools formed within the concave basidiocarps of the fungi *Microporus xanthopus* and *Lentinus sajor-caju*, and in water bodies found in fallen leaves. Such 'leaf pools' are also commonplace in Australian rainforests associated with the sheathing bases of fallen palm fronds.

Of peripheral interest from the work of Beaver (1972, 1973) and Lounibos (1980) are the water bodies contained within empty snail shells, which acted as breeding sites for several of the mosquito and other dipteran species also found within plant containers. I forbear from naming such habitats 'zootelms'!

3

The container fauna
The animals of phytotelmata

The aquatic fauna of phytotelmata ranges from unicellular Protozoa to Amphibia, from rotifers to dragonflies. This chapter focuses on the multicellular organisms most commonly encountered in container habitats. The chapter looks at the fauna on a guild-by-guild basis, leaving a much fuller, taxonomically organised account for the Annexe.

Guilds are sets of species within ecosystems which perform related ecological roles (Root 1967, Simberloff & Dayan 1991). Usually these have been defined in terms of trophic positions within food webs, sometimes modified by the size of the organisms involved ('small herbivores', middle-sized predators' etc.). Phytotelms contain allochthonous, detritus-based communities and hence the guilds can be clearly divided into the saprophages, predators and top predators. I use these categories here, variously qualified by the size of the species, the particle size of the detritus they exploit, and, for the predators and top predators, by aspects of their foraging strategy. Of course top predators are likely also to be predators but, to avoid repetition, I assign them to the former category. Additional minor categories encountered rarely in the field and the literature include herbivores and parasitoids.

Whenever an aquatic organism is encountered within a plant container a question is posed. Does this organism habitually seek out and inhabit the particular type of habitat unit, or is its occurrence merely accidental? The decision associated with the question determines, for instance, to what extent that organism is to be regarded as a member of the specialist community from the bromeliad, pitcher plant, tree hole or what-have-you. In this regard the Plön School in Germany, under Albrecht Thienemann, introduced a useful terminology.

Röhnert (1950), for instance, referred to the animals she encountered in tree holes as dendrolimnetobionts, dendrolimnetophils or dendrolimnetoxenes. I have used this classification subsequently (Kitching 1971), attaching the following meanings to the terms:

Dendrolimnetobionts are those organisms which are specialist inhabitants of tree holes which seldom if ever breed elsewhere. As the term implies, they are *biologically* committed to the habitat type and may be supposed to have evolved to become efficient exploiters of the milieu.

Dendrolimnetophils are those organisms which, although not restricted to tree holes, might be expected to occur in any similarly still, small water body. They can live and reproduce in tree holes but do occur elsewhere. The term suggests that they find tree-hole conditions equable, but in evolutionary terms have presumably adapted to a range of comparable habitat types.

Dendrolimnetoxenes are present in tree holes by mere accident. They do not thrive in the conditions presented by the habitat unit and, presumably, will not establish persistent populations there. As the term implies they are foreign to the ecosystem.

Thienemann (1932) had used a similar terminology with regard to the fauna of *Nepenthes* pitcher plants – nepenthebionts, nepenthephils and nepenthexenes (Beaver 1983) and Mogi & Yamamura (1988) have adapted the same compounds, somewhat less euphoniously, for the axil fauna.

These three somewhat flexible categories of association are helpful in considering animals from any sort of phytotelmata. There is no need to clutter the literature with more long Germanic compounds: suffice it to say that referring to the organism/habitat relationship as *biontic, philic* or *xenic* summarises the above arguments succinctly. The accounts in this chapter will deal, in general, with species having 'biontic' or, at least, 'philic' relationships with phytotelmata of one sort or another.

Saprophages

Filter feeders

A wide range of animals exploit suspended particle matter, alive or dead, within the water column of aquatic ecosystems. In container habitats, based, as they almost always are, on packs of decomposing organic matter, there is a rich fauna of filter-feeding organisms. The basic food resource for these animals is the suspended particles of detritus that are generated by the progressive finer and finer comminution of the detritus base. Undoubtedly the actual source of nutrient in many instances is generated by micro-organismic activity centred on these particles.

There is a subguild of the smaller filter feeders which have received relatively little attention in container habitats. Indeed I have no doubt they have frequently been overlooked. The rotifers are almost universal in aquatic habitats and have been recorded in all five principal types of phytotelms. In bromeliads Torales *et al.* (1972) confirmed the earlier records of Picado (1913). Rotifers have been noted frequently in European tree holes (Hauer 1923, Beattie & Howland 1929, Röhnert 1950) and eight genera are known from this habitat. No records of which I am aware exist for tree holes elsewhere although we have not generally searched for them in our studies in Australia and Asia. Bamboo internodes contain rotifers but these have not been studied (Mogi & Suzuki 1983). Some of the earliest work on phytotelm communities – on axil waters of *Angelica, Dipsacus* and *Scirpus* – recorded a range of species of rotifers (Alpatoff 1922, Varga 1928, Strenzke 1950). Again they have seldom been sought in axil waters elsewhere. Finally pitchers on plants of both *Sarracenia* (in North America, Addicott 1974) and *Nepenthes* (in South-east Asia, Thienemann 1932) contain a range of rotifer species. Beaver (1983) has suggested that rotifers in *Nepenthes* pitchers are non-specialist aquatic animals merely exploiting any aquatic ecosystem that they encounter: that is, they are philic species using the terminology introduced earlier. Röhnert (1950) certainly designated one of the species she encountered in German tree holes as a tree-hole specialist. Any resolution of this point will require far more information than we currently have.

Copepods are the commonest group of microcrustaceans encountered in phytotelmata. Although some are predatory (and, indeed, have been trialled as biocontrol agents for mosquitoes in parts of the Pacific – Marten 1984) most are thought of as filter-feeding decomposers (Dussart 1967). Both cyclopoid and harpacticoid copepods are frequent elements in tree-hole communities and the former group also occurs in bromeliads, bamboo internodes and axil waters (see Table A.2 for references).

Perhaps the most common and visible of filter feeders in container habitats are anopheline and aedine mosquitoes. Most anophelines are regarded as obligate filter feeders while aedines may combine this habit with browsing on smaller detritus particles (Mattingly 1969). The subgenus *Anopheles* (*Anopheles*) is recorded from tree holes and bamboos in the Oriental region. Another subgenus fills this ecological role in bromeliads and bamboos in the Neotropics. A wide range of subgenera of *Aedes* have been found in all classes of phytotelm around the world. Species of *Finlaya* are specialists in container habitats occurring in *Nepenthes* pitchers, bamboos, axil waters, bamboos, bromeliads and tree holes. This trophic class includes the extremely well known American species *Aedes triseriatus*, and the 'celebrated' immigrant species *A. albopictus*. This Asian species is well established in North America and has considerable vector potential.

Some eristaline syrphid larvae – familiar 'rat-tailed maggots' – are supposedly filter feeders and have been recorded from tree holes, bamboo internodes, bromeliads and leaf axils – although in this last instance it seems more likely that they feed by 'rasping' at the non-woody plant surface. Hartley (1963) suggested that this rasping habit – whether of plant surface or detritus particles – was a widespread habit in these species. The sometimes large larvae of these hoverflies are among the best known of container inhabitants. Their large size allows them to be equally accurately designated as 'macrosaprophages'.

Last among the frequently occurring filter feeders in container habitats are the histiostigmatid mites. Fashing (1994a) has examined the mouthparts of one North American tree-hole species, demonstrating their special role in straining out microparticles from the surrounding water. Other species of this family are restricted to pitcher plants in Asia (*Nepenthes* spp.) and North America (*Sarracenia* spp.). Nymphs of this species are phoretic upon the legs of emerging adult Diptera and, presumably, move among phytotelm habitat units in this fashion (Fashing 1975).

Fine detritus feeders

By far the most diverse group of organisms within phytotelms are those that ingest small particles of detritus which they scrape or excise from the detritus pack of the habitat unit. These include some of the most familiar, abundant and ubiquitous of phytotelm inhabitants. I review these briefly in this account and allude to a few other groups where the association with particular kinds of containers appears to be an obligate one.

Like the filter-feeding rotifers, nematodes are ubiquitous yet sufficiently small to have eluded the attention of most students of phytotelmata. Trophically nematodes are highly diverse, ranging from free-living predators to obligate endoparasites. In phytotelms, however, most records are of free-living, saprophagous species. Records of phytotelm nematodes are most complete for tree-hole species in Europe (Thienemann 1934), in *Sarracenia* pitchers (Goss *et al.* 1964) and in water-filled leaf axils in bananas (Thienemann 1934). This rather idiosyncratic list undoubtedly reflects the very modest amount of attention devoted to these animals in container habitats where, I am sure, they are ubiquitous.

Of the microcrustacea, the ostracods are commonly encountered in phytotelms – sometimes in huge numbers. They utilise a feeding method intermediate between filter feeding and substrate feeding. According to Henderson (1990) they stir up particles with their antennae before filtering them out through the cavity of their bivalved 'shells'. Two families of the Ostracoda have been recorded from water-filled tree holes in Europe and Australia and from bromeliads in the New World.

Five families of Diptera dominate the detritus packs in a wide range of phytotelms. Culicine mosquito larvae are the best known of these (Horsfall 1955). All of the Tribe Sabethini are container breeders occurring in tree holes, pitchers, bamboos, bromeliads and axil waters as well as fallen nuts, leaf bases and other less familiar plant containers. *Wyeomyia smithii* is probably almost as well known as *Aedes triseriatus*. A nuisance species with vector capabilities, it is common in *Sarracenia* pitchers. Other congeners are the best known mosquitoes of bromeliads in the Americas. The taxonomically complex genus *Tripteroides* also belongs here and its highly setose larvae are readily recognised inhabitants of tree holes and the full range of other plant containers in the Old World. The Tribe Culicini also contains many container breeders, although some members of this subgroup do breed in other water bodies. Various subgenera of *Culex* are among the larger free-swimming inhabitants of containers throughout the world. The delicate larvae of species of *Uranotaenia* belong here and are recorded from a range of plant axils, *Nepenthes* pitchers and tree holes as well as water accumulating in basidiocarp fruiting bodies. The cosmopolitan genus *Orthopodomyia*, also a culicine, is a further genus which is an obligate breeder in plant containers – from tree holes to bromeliads.

Larvae of non-biting midges, the Chironomidae, are principally saprophages feeding on small detritus particles. The subfamily Tanypodinae is the exception and these voracious predators are discussed below. Saprophagous chironomid larvae belong to the Podonominae, Chironominae and Orthocladiinae. Of these subfamilies a few genera within the Orthocladiinae appear to have adopted container habitats most frequently. The genus *Metriocnemus* is often an abundant inhabitant of tree holes in Europe and occurs commonly in *Sarracenia, Darlingtonia, Heliamphora* and *Nepenthes* pitchers, and a range of axil waters. The genus is absent from both North American tree holes (in which chironomids in general are rare), and from the Australian continent. Other more ubiquitous genera that nevertheless become established in containers include the chironomines *Polypedilum* (in a full range of phytotelms in both Old and New Worlds), *Tanytarsus* (in bromeliads in the West Indies, and bamboos in Indonesia) and *Chironomus* (in tree holes, bamboos and *Nepenthes* pitchers at many locations), and the orthocladiine *Compterosmittia* (in Malaysian bamboo internodes and the Tasmanian leaf axils of the giant epacrid, *Richea*.

The biting midges, the Ceratopogonidae, also breed readily in container habitats. Their larvae feed within the detritus pack but, in contrast to those of the Chironomidae, are capable of rapid undulating swimming within the water column. Almost all of the abundant records of larvae of this family in containers are of *Dasyhelea, Culicoides* or *Forcipomyia*. Species of *Dasyhelea* are commonest in tree holes and *Nepenthes* pitchers, with fewer records from bamboos and leaf axils. There is a single record from bromeliads (Winder & Silva 1972). The nuisance genus *Culicoides* is similarly known from a large range of phytotelms and is cosmopolitan. The last superabundant genus, *Forcipomyia*, is best known from bromeliad tanks, within which a radiation of this genus appears to have occurred. The very few non-bromeliad habitats recorded include *Pandanus* axils, *Saccharum* axils, and *Nepenthes* pitchers.

Moth flies or psychodids are familiar in the adult stage as associates of detritus of all kinds. The larvae have a characteristic 'hairy' appearance and a range of genera have been bred, variously, from tree holes, bromeliads, bamboos, and leaf axils but not from pitchers of any genus. The family is cosmopolitan.

The Phoridae, scuttle flies, are the largest of the Dipteran families (Disney 1994). Larvae from two subfamilies act as saprophages within container habitats. Most records are of larvae belonging to the metopinine mega-genus *Megaselia* which is often encountered in tree holes, bamboos, and the pitchers of *Nepenthes*. All of these records are from the Old World. The phorine genus *Dohrniphora* occurs in bromeliads in Brazil (Winder 1977) and the pitchers of *Sarracenia* in South Carolina (Rymal & Folkerts 1982).

Two other families of small Diptera deserve special mention here, not because of their range or abundance but because of the very 'tight' relationship which appears to exist between species within them and container habitats. The larvae of the micropezid fly *Badisis ambulans* are known only from pitchers of the West Australian pitcher plant, *Cephalotus follicularis* (Yeates 1992), where it occurs within most pitchers. The larvae are generalist saprophages. Similarly larvae of the Neurochaetidae – the 'upside-down flies' – are now known to be widespread in the Old-World tropics and subtropics but always occur within the water-filled axils of a range of fleshy plants. McAlpine (1993) suggests that the whole family are phytotelm specialists feeding on micro-organisms associated with the fine detritus which accumulates in axil waters.

Leaving the Diptera (but only for a while), I turn to the beetle family, Scirtidae, which have saprophagous larvae. The distinctive flattened larvae have characteristic long antennae and burrow into the leaf pack. Here they sweep up detritus particles using their highly specialised mouthparts. Larvae of the scirtids, or marsh beetles, are common, sometimes highly abundant in tree holes throughout the world, in bamboo internodes wherever they have been studied, and in bromeliads in the New World. The generic structuring of the container fauna is in need of substantial taxonomic attention.

Finally, among the commoner groups of fine-particle detritus feeders are two families of astigmatid mites: the Acaridae and the Algophagidae. The first of these include two fully aquatic species from North American tree holes (Fashing 1975), to which company Nesbitt (1985) added a related bromeliad-inhabiting species from the Neotropics. Acarids skeletonise leaves within the detritus pack and, it is suggested, move among habitat units as deutonymphs on the legs of syrphids. Algophagid mites are known from water-filled tree holes in North America and Australia. The American species *Algophagus pennsylvanicus*, grazes on the detritus within tree holes (Fashing & Campbell 1992).

Macrosaprophages

A few groups of phytotelm organisms join the larger larvae of eristaline syrphids under the guild designation 'macrosaprophage'. These I designate as larger animals involved in the process of detritus comminution from the moment that dead or dying material enter the containers.

Oligochaete worms of six families occur commonly in tree holes, bamboo internodes, bromeliads and axil waters. Like some other groups they have been little studied in container habitats. Thienemann (1934) noted their presence in many South-east Asian locations in tree holes, bamboos and the axils of *Colocasia*. Picado (1913) first noted them in bromeliads in the West Indies. Oligochaetes have been present in virtually all the tree holes we have ever investigated.

Larvae of tipulids – crane flies – are often the largest organisms encountered in tree holes. The flaccid larvae of *Tipulodina* species may be up to four centimetres long. They occur in crevices around the edges of tree holes, and in bamboos. Several saprophagous species occur in bromeliads (Picado 1913; Alexander 1915, 1920; Laessle 1961) in the West Indies. Some smaller species occur in leaf-axil habitats such as those of *Colocasia, Musa, Curcuma* and *Astelia*. Tipulid larvae in general are probably 'philic' in phytotelms and thrive in such habitats as they would in any other still, detritus-rich, small water bodies.

The Calliphoridae are quintessential saprophages associated with moist or semi-aquatic environments. Only two species appear to have become phytotelm specialists – both in Asia and both within the pitchers of *Nepenthes* species. The curious larvae of *Wilhelmina nepenthicola* from Borneo and those of *Nepenthomyia malayana* from peninsular Malaysia burrow into and feed on the closely packed prey masses in pitchers of *Nepenthes*.

Finally among arthropodan saprophages are larvae of so-called flesh flies – the Sarcophagidae – which have become specialised inhabitants of pitcher plants in both Old and New Worlds. The genera *Pierettia* and *Sarcosolomonia* are specialists in *Nepenthes* pitchers in Asia and the western Pacific (Shinonaga & Beaver 1979, Yeates *et al.* 1989, Souza-Lopez 1958). An unrelated suite of species belonging to the genera *Fletcherimyia, Sarraniomyia* and *Wohlfartiopsis* exploits the prey mass in pitchers of the New World genus *Sarracenia*.

Some frog tadpoles may also act as macrosaprophages within tree holes. Unfortunately the detailed feeding habits of few phytotelm anurans are known: they are known to have very variable feeding modes. This issue is discussed further in the Annexe.

Predators

Detritus-based predators

I use this designation to collect together those predators which forage in and on the detritus pack itself. Some do so actively; others adopt a sit-and-wait strategy, remaining motionless until a potential prey organism comes within striking or short pursuit range. Many of these predators are capable of free-swimming behaviour in the water column and, no doubt, the distinction between this and other categories of predators will be somewhat arbitrary.

Planarians, flatworms, prey upon a wide range of larvae and annelids by wrapping themselves around their prey, then enclosing them in part with an oral frill within an envelope of mucus. Digestion begins externally. This feeding habit, of course, also lends itself to the exploitation of moribund or newly dead prey. Accordingly, within phytotelm communities, planarians sit somewhere in the limbo between the obligate predators and saprophages. In my experience, in tree-hole systems they act frequently as active predators and for this reason I include them at this point of my narrative. A variety of flatworms in several families are recorded from phytotelms: principally bromeliads (e.g. Laessle 1961). In addition we have noted them frequently in water-filled tree holes in tropical Queensland, and Thienemann (1934) adds the axil waters of bananas in Indonesia to the tally of known habitats.

The tally of vermiform invertebrates in phytotelms is completed by namanereidine polychaetes of the genus *Lycastopsis* which are recorded both from tree holes and banana axils in New Guinea and the East Indies, respectively. I presume these active predators are generalists attacking soft-bodied insect larvae and, probably, the oligochaetes they encounter in these situations (Glasby *et al.* 1990).

The voracious larvae of chaoborid midges – close relatives of the mosquitoes – are deemed by Colless (1986) to be generalist predators on other insect larvae. Their small larvae are dorso-ventrally flattened, presumably insinuating themselves among the detritus pack in search of prey. They are recorded commonly from subtropical tree holes in Florida, from bromeliads in the Neotropics, and from a range of species of *Nepenthes* pitchers in northern Borneo (Bradshaw & Holzapfel 1984, Frank & Curtis 1981, Clarke & Kitching 1993).

The Tanypodinae are an exclusively predatory subfamily of the Chironomidae, although this general designation has been disputed by some (Oliver 1971). In Old-World phytotelms they *are* undoubtedly predators and the species *Anatopynnia pennipes* is an important, frequently, abundant, structuring force in Australian tree hole communities (see *inter alia*, Chapter 12). The related genera *Paramerina* and *Pentaneura* occur – much less commonly – in New Guinean and South-east Asian tree holes, bamboo stumps and *Colocasia* axils (Thienemann 1934, Kitching 1990). Other species are known from bromeliads and a range of axil waters in the Neotropics (see, for example, Laessle 1961, Winder & Silva 1972, Naeem 1990).

Predatory mites of the family Anisitsiellidae are abundant in the leaf axils of *Richea* in Tasmania and prey upon the chironomid larvae that also occur within these highly specialised near-alpine phytotelms.

Fixed substrate predators

A couple of species of larger predators move actively around the habitat units in which they occur, often anchoring themselves on the container plant itself before striking at prey.

Larvae belonging to the family Cecidomyiidae have become specialist predators within *Nepenthes* pitchers and bamboo axils in South-east Asia (Clarke & Kitching 1993, Kovac & Streit 1996). They feed on phorid larvae and other small dipterous prey. Kovac & Streit (1996) note a complex of species in Malaysian bamboo internodes where they also act as top predators (see below). There are also records of this family from bamboos in Peru and bromeliads in Brazil.

The larvae of the muscid fly *Graphomya* were the only predators found in the axil waters held by the inflorescences of *Curcuma australasica* in New

Guinea, moving from one bract axil to another in the water film which frequently enveloped the plant. They preyed upon larvae of mosquitoes and midges. *Graphomya* larvae have also been noted, but rarely, from rot holes in Western Europe (Skidmore 1985).

Surface predators

Three small groups of Hemiptera live in loose association with some phytotelms where they swim on the water surface, striking through the meniscus at free-swimming potential prey in the water column beneath. Kovac & Streit (1996) record species of hydrometrids and of gerrids living within water-filled bamboo stems in Malaysia, and we have noticed veliid bugs acting in similar fashion around tree holes in the lowland rainforest of tropical Australia.

Free-swimming predators

A number of predators encountered in phytotelms swim freely in the water column in search of similarly free-swimming prey. No doubt, though, they also pick less mobile prey off the surfaces within particular habitat units.

I have noted already that some predatory copepods are known from tree holes and similar situations.

The immature stages of damselflies occur widely in plant containers, prowling the surfaces and swimming freely in the water column. The larvae also penetrate deep into the detritus mass and are, like most odonates, generalists. Like other odonates they are commoner in larger phytotelms – particularly tree holes, bamboos and bromeliads – but are also recorded from *Nepenthes* pitchers and leaf axils. Records spanning four families of Zygoptera from four continents exist and are detailed in the Annexe. Corbet (1983) has reviewed the occurrence and biology of phytotelm odonates in general.

Larvae of mosquitoes belonging to the subgenus *Culex (Lutzia)* are predatory – probably on other mosquito and midge larvae – within the tree-hole community in Sulawesi. Other smaller mosquitoes belonging to the genera *Topomyia* and *Zeugnomyia* are predators within South-east Asia – in bamboo internodes and *Colocasia* axils in the first instance, and in the water-filled bases of fallen leaves in the second. Yet other mosquito larvae may be occasional predators, turning to moribund prey facultatively. Such species include the American *Anopheles barberi* and the African *Eratmopodites* (Petersen *et al.* 1969, Lounibos 1980). I deal with the quintessential phytotelm predator – larvae of *Toxorhynchites* – as a top predator, below.

Larvae of periscelid flies belonging to the genus *Stenomicra* are specialist predators within bromeliads and a few axil waters (Khoo 1984). Fish (1976) notes them feeding on larvae of the bromeliad mosquitoes, *Wyeomyia* spp.

Dytiscid larvae are frequent visitors to the more highly visible bromeliads (Laessle 1961) and tree holes in the tropics. There are no records of dysticid larvae in plant containers and I assume their relationship with the habitat is a 'philic' one. Undoubtedly, though, when a number of dytiscid adults are present in a phytotelm they may potentially have considerable impact upon the insect larvae and oligochaetes on which they likely prey.

The highly distinctive red mites of the family Arrhenuridae are a common element in tree holes within subtropical rainforest in Australia. Out of line with other members of the family, this species does feed on dipterous larvae (Kitching & Callaghan 1982, H. Proctor, pers. comm.). Laessle (1961) also notes the presence of arrhenurid mites within bromeliads in Jamaica.

Top predators

Free-swimming predators

Best known and most important of this category of top predators within phytotelmata are the ferocious larvae of the mosquito *Toxorhynchites*. Container specialists, these larvae are keystone predators within tree-hole food webs (see Chapter 12). They are cannibalistic and, except in the largest habitat units, large larvae will eliminate rivals. Usually set down as predators on other culicid larvae, they prey upon a range of insect larvae while remaining generally within the water column. They have been widely studied in North America and the Pacific for their potential as biocontrol agents for vector mosquitoes. *Toxorhynchites* spp. are also recorded from bamboos, bromeliads and pitcher plants. Steffan & Evenhuis (1981) present a wide-ranging review of the biology of the genus.

Last among the free-swimming predators of phytotelms are the tadpoles of carnivorous frogs. In Australian subtropical tree holes, tadpoles of the myobatrachid frog *Lechriodus fletcheri* are the top predators within the aquatic food web and may have dramatic impacts on the rest of the fauna (see, for example, Jenkins 1991, Pimm & Kitching 1987).

Sit-and-wait predators

Among the most dramatic of top predators within phytotelm webs are the species of larval odonate which have been encountered within these habitats.

These sit-and-wait predators use their extrusible labia to seize passing prey which may range from small insect larvae to large *Toxorhynchites* or even tadpoles. The anisopterans found within containers are more likely to act as top predators than the smaller and more gracile zygopterans. The large aeshnids occur within tree holes in Borneo and Central America but the smaller larvae of libellulids have been encountered more frequently. They are recorded from tree holes, bamboos and pitchers. Geographically these occurrences are principally in South-east Asia but a few records exist from Africa, Central America and Australia. There is even a single record from *Sarracenia* pitchers in the USA (Corbet 1983).

I have noted already the complex of cecidomyid larvae identified by Kovac & Streit (1996) in Malaysian bamboo internodes. These authors record the cecidomyids explicitly preying upon larave of *Toxorhynchites* and tanypodines within these systems.

Semi-terrestrial predators

A number of predators associated with phytotelms actually spend time preferentially in the drier parts of the habitat unit rather than within the water body. They may be significant in the present context, however, if they venture, partly or wholly, into the water body in search of prey. The carabid *Colpodes* undoubtedly does this in bromeliads, where Laessle records them feeding on scirtid larvae. I suspect that like many other carabids they will prey upon any item large enough to be grasped by their mandibles, including other aquatic predators such as tanypodines.

The ant *Camponotus schmitzi* also falls into this category and Clarke has recorded it preying upon larvae of the predatory *Toxorhynchites*, as well as other dipterous larvae, within pitchers of *Nepenthes bicalcarata*. Of course this species also feeds upon prey items which, in concert, the ants extract from within the pitchers (Clarke 1997, Clarke & Kitching 1995).

Minor categories

Herbivores

A small numbers of organisms have been noted as confirmed (or highly suspected) herbivores within phytotelm systems. Those that are phloem feeders or which merely scrape the surface of the container plant obviously are compatible with the persistence of the aquatic milieu. Others, which actually chew the substance of the container plants themselves, will, sooner or later, breech the container so that it drains of liquid.

So the cercopid *Aphrophora* sp. encountered commonly within pitchers of the *Sarracenia* in eastern Canada are phloem feeders but, according to Laird (1988) exhibit a preference for feeding beneath the surface of the liquid in the pitchers. A similar cercopid also occurs in *Heliconia* bracts in Costa Rica (Fish 1977).

The food webs we studied within the bract axils of *Curcuma australasica* in New Guinea commonly contained the larvae of a small eristaline syrphid which appeared to feed by rasping the epidermis of the plant. We speculate later (Chapter 11) that this may be a vital activity in generating the discarded plant fragments which allow a more complex food web to develop within these inflorescences.

Larvae of the richardiid flies *Beebiomyia* feed on floral parts within the water-filled bract axils of *Heliconia* flowers but may in fact be no more than accidental elements within the aquatic food web: perhaps more 'xenic' than 'philic' even.

Lastly among the herbivores I note the range of Lepidoptera recorded from phytotelmata. Most undoubtedly are only accidentally aquatic but a few – such as the larvae of the noctuid genus *Eublemma* in *Nepenthes* pitchers – appear to prefer a subaquatic existence. In general the presence of lepidopterous larvae leads to destruction of the water-holding capacity of plant containers, but those that restrict their activities to feeding on the epidermis of pitchers or axil waters may coexist with a flourishing aquatic food web for some time within the same container.

Trophic egg feeders

Some tree frogs, such as species of *Dendrobates*, which deposit their fertile eggs within tree holes, bromeliads, leaf axils or even larger *Nepenthes* pitchers, also lay non-fertile but protein-rich 'trophic' eggs within the same containers. These act as food for the emerging tadpoles which remain, accordingly, more or less independent from the food web within the container. Some adopt a mixed feeding strategy with trophic eggs forming but part of their diet and, of course, they are themselves potential prey for larger predators (such as odonates) within these systems.

Many other species of animals have been encountered from time to time in phytotelms and the Annexe deals with most of these records. Some groups have been little studied; these include the spiders which may from time to time pluck prey from within aquatic food webs (see, for example, Clarke 1997) or the hymnopterous larvae which may have a parasitic association with some of the aquatic species within phytotelms (see, for example, Beaver 1983). As in many other aspects of community ecology, phytotelms will provide convenient model systems for the study of this key area of the interaction between adjacent aquatic and terrestrial communities.

4

The phytotelm environment
The container milieu

The ecologist who delves deep within a large tree hole will find that there is a close-packed layer of sediment at the bottom of the deeper holes which, when disturbed, emits pungent bubbles of hydrogen sulphide. In other holes the surface may be glazed with an oily film derived from particular fruits or even animal cadavers which happen to have come to rest within them. In one upper pitcher of *Nepenthes rafflesiana* in Brunei, Charles Clarke and I measured the pH as 1.5 – the tiny water body was greasy to the touch but still contained mosquito larvae! Some bamboo cups or leaf axils contain litres of water: in other cases, only a few cubic centimetres of liquid. And at a different level I once spent a whole day searching in cool temperate rainforest in western Tasmania to find but two water-filled tree holes, whereas a month later in Borneo I encountered about a dozen in a couple of hours' walk. The environment at both macroscales and microscales presented to organisms which inhabit phytotelmata may be extreme or highly variable or both. This chapter examines aspects of the physical and chemical environment presented by phytotelmata.

Foodwebs made up of animals feeding directly or indirectly upon detritus within water bodies of phytotelmata of whatever kind live within a particular physical and chemical context. Those that are regular specialist inhabitants of the water bodies have presumably evolved to be able to cope with the ranges and variability displayed by the set of environmental parameters within each habitat type. In this chapter I examine a number of these parameters across the range of phytotelm types, insofar as these have been described in the literature. The availability of such data is patchy, scattered and altogether hard to find. In part this is because such data are often inserted parenthetically in accounts which focus on the fauna.

I shall be concerned here with five principal aspects of the phytotelm environment:

- The frequency and pattern of occurrence of habitat units within the wider ecosystem in which they occur.
- The dimensions and volumetric capacity of the habitat units within a particular region.
- The dynamics of detritus including the temporal pattern of leaf fall and turnover.
- The quantity and quality of the water-body within the habitat units.
- Temporal variation of the physical properties, particularly the temperature regimes and dissolved oxygen regimes, within the water bodies.

Information exists on each of these aspects for at least some phytotelmata although there are many gaps in our information base.

Patterns of occurrence of habitat units

Phytotelmata occur nested within larger ecosystems – in most cases, forests but sometimes swamps, grasslands or even alpine heathlands. In some instances, whenever the containing plant is present, so is the container: pitcher plants and bromeliads are obvious cases in point. In contrast, water-filled tree holes, open bamboo internodes, and even water-filled inflorescence bracts occur in only some of the plants that potentially may contain them. In all of these instances, however, the units of phytotelm habitat themselves have a spatial distribution within the wider ecosystem. Tree holes, for instance, may be abundant in one area, but extremely scarce in another, reflecting *inter alia* the age structure of the forest, its recent history in terms of cutting or storm damage, and the growth form of the tree species of which the particular area of forest is composed. Similarly, plants of *Nepenthes* may be abundant and well grown, each with many pitchers, along one creek bed in Borneo but may occur only as rare scattered plants in another.

In very few cases known to me has any attempt been made by phytotelm workers to describe the abundance and spatial pattern of habitat units in this fashion. There remain significant open questions about how the general abundance of phytotelm units within a forest do or do not affect the species richness and relative abundances of the in-fauna. I present here the three partial exceptions to this general statement.

Bromeliads in Trinidad

Shortly after the 1939–45 World War a series of papers appeared describing the ecology, epidemiology and vector biology of 'bromeliad' malaria in

Trinidad in the West Indies. The disease in this case is transmitted by *Anopheles* (*Kerteszia*) *bellator*, a species which breeds exclusively in bromeliads (Downs *et al.* 1943, Downs & Pittendrigh 1946). Colin Pittendrigh made very detailed, exemplary studies of the distribution of bromeliads in selected study sites in central Trinidad and the distribution and population dynamics of both *A. bellator* and the co-occurring *A. homunculus* (Pittendrigh 1948, 1950a, b).

Pittendrigh (1948) presents detailed information on the distribution of bromeliads in three strips of vegetation from Tamana. All three strips were almost adjacent, one to the other, but two were of forest and the third was a cacoa plantation. The two forest strips represented a non-riparian forest with abundant vine growth and accordingly poor light penetration and a riparian, vine-free forest with much better light penetration. Pittendrigh mapped every bromeliad within these three 200 × 25 feet strips by the simple expedient, in the case of the two forest strips, of felling every tree within the plot and mapping the epiphytes in detail! Figure 4.1 is taken from his paper.

The generic and species diversity of the bromeliads Pittendrigh observed peaked in the more open riparian forest, although in sheer numbers of plants the plantation had almost twice as many present as the richest of the two forests (Table 4.1). Pittendrigh classified the species of bromeliads on the basis of their places of occurrence. High in the canopy of both forest types and the cacao plantation, fully exposed to the sun, he encountered 'exposure' species such as *Catopsis sessiliflora* and *Vriesia procera*. All species in this group had water-containing tanks. The so-called 'sun' group of species contained the largest number of species (eight species of five genera) and occurred throughout the canopy. They dominated both the more open forest and the plantation assemblages of bromeliads. Again these were all tank species and included the largest species with greatest water-holding capacity. Lastly Pittendrigh recognised a small group of 'shade' species which comprised species of *Tillandsia, Guzmania* and *Vriesia*. These occurred in the lowest levels of the two forest plots and were absent from the plantation. This group represented over 60% of all bromeliads in the more densely shaded forest. Although all of 'tank' varieties, this group contained the species that Pittendrigh classified as having 'ephemeral' tanks of small capacity.

Having established this basic framework Pittendrigh went on to compare the bromeliad loads in cacao plantations in five different rainfall zones within Trinidad (Figures 4.2 and 4.3). Essentially the distributions of both total numbers of bromeliads and total number of species involved are bimodal (Figure 4.2) with high values for both variables in the very wet and relatively dry regions, with a decline in numbers at intermediate levels (although, curiously, the two nadirs do not coincide precisely). Although not apparent from the

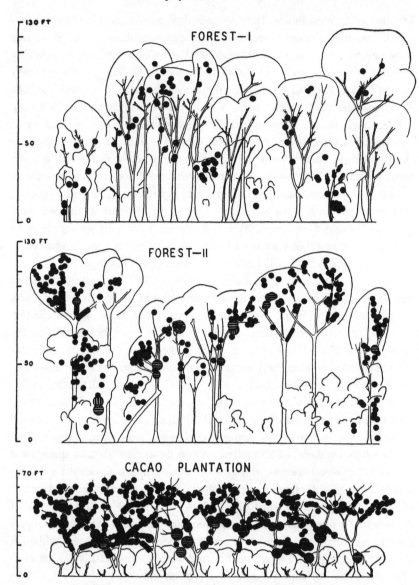

Fig. 4.1. The distribution of bromeliad plants in a non-riparian forest (top), a riparian forest and a cacao plantation in Trinidad (from Pittendrigh 1948).

figures, the particular species which reached peaks at the two ends of the rainfall gradient overlapped little. Pittendrigh further analysed these results using the 'exposure', 'sun' and 'shade' categories he had established in his first survey and was able to show that the bimodality is principally the result of vari-

Table 4.1. *The distribution of bromeliads in three vegetation types in Trinidad*

Statistic	Forest 1 [heavily shaded]	Forest 2 [open, riparian]	Cacao Plantation [very open, drier]
Number of bromeliad genera	4	7	6
Number of bromeliad species	10	17	9
Total number of bromeliad plants	88	382	708
Number of 'exposure' plants	11	45	201
Number of 'sun' plants	23	188	507
Number of 'shade' plants	54	149	0

After Pittendrigh (1948).

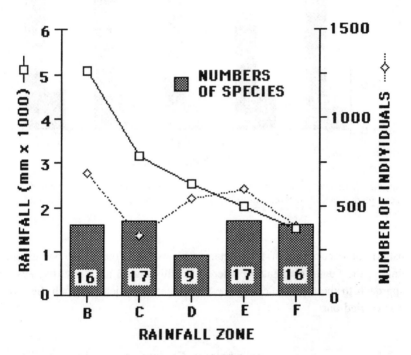

Fig. 4.2. Total numbers of species and individuals of bromeliads in different rainfall zones in Trinidad (redrawn from Pittendrigh 1948).

ations in numbers and abundances of the 'sun' species (Figure 4.3) – no doubt simply because this is by far the largest of his three categories.

Pittendrigh's (1948) analysis is remarkable among phytotelm studies because it does take such a comprehensive 'top-down' approach, first establishing the

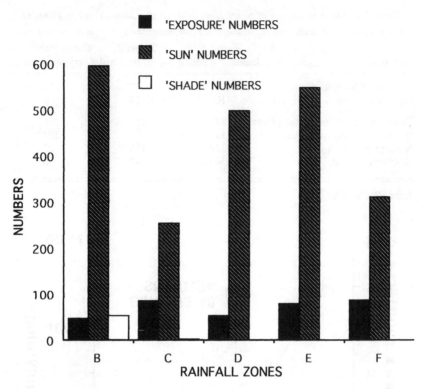

Fig. 4.3. Variation in the numbers of 'exposure', 'sun' and 'shade' bromeliads in cacao plantations in different rainfall zones in Trinidad (redrawn from Pittendrigh 1948).

pattern of occurrence of the container plants, then the water bodies and, finally, the faunal elements of interest. In this regard it clearly displays an approach to the topic reflecting the public health milieu in which the work was carried out.

Tree holes in England

Between September 1966 and June 1969 I carried out a survey of the fauna of water-filled tree holes in Wytham Woods near Oxford, England. As well as providing the information base for a doctoral dissertation, this study also formed part of the wider ecological survey of Wytham Woods directed by Charles Elton (Elton 1966).

As part of this study I mapped all the beech trees (*Fagus sylvaticus*) over 30 cm in diameter at 1 m in a region of the Wytham Estate and counted all

Table 4.2. *The number of tree holes in a patch of deciduous woodland, Wytham Woods Estate, Oxford, England*

Tree-hole type	Tree-hole size	Vertical position	Number in random subsample of trees	Estimated number in 3000 ha of woodland
Rot holes	–	Field layer	0	0
Rot holes	–	Canopy layer	6	48
Pans	Small	Field layer	50	397
Pans	Small	Canopy layer	33	262
Pans	Medium	Field layer	18	143
Pans	Medium	Canopy layer	19	151
Pans	Large	Field layer	6	48
Pans	Large	Canopy layer	9	71
Total	–	–	141	1119

'Rot holes' and 'pans' are defined in Chapter 2.
'Field' and 'canopy' layers are below and above about 1 m from ground level, respectively.
Size categories are: 'small' < 15 cm diam., 'medium' 15–30 cm diam., 'large' > 30 cm diam.

water-containing holes within a random sample of 30 of these 238 trees. The holes were classified as either rot holes or pans (see Chapter 2); as small (<15 cm maximum diameter), medium (15–30 cm diameter) or large (>30 cm diameter); and, as occurring in the buttress roots of the trees (the 'field' layer) or in the canopy above (the 'canopy' layer). This survey was unusual inasmuch as it allowed some estimate of the frequency of occurrence of water-filled tree holes in the woodland at large. The woodland was very simple and the trees relatively low growing. Nevertheless the survey that I did make in Wytham was both arduous and time-consuming. It would be manifestly impossible in most rainforest situations that I have encountered subsequently!

The results obtained are summarised in Table 4.2. The survey of the random sample of trees resulted in a total of 141 tree holes leading to an estimate of over 1000 sites in the patch of woodland overall. By far the greatest number of these are bark-lined 'pans' as opposed to rot holes. More occur in the buttress roots of the beech trees than in the canopy and most are less than 15 cm in diameter. Nevertheless this is a large number of habitat units which, I believe, would be seldom matched in other locations. In this case the growth form of the trees reflecting the intrinsic characteristics of the species, the woodland management practices and the age of the stands of trees all contribute to producing the high figure.

Banana axils in East Africa

Haddow (1948) and Teesdale (1957) both made extensive surveys of axil habitats, particularly in varieties of banana (*Musa* spp.) in East Africa. Haddow (1948) worked in the extreme west of Uganda as part of a project investigating breeding places of the yellow-fever vector *Aedes* (*Stegomyia*) *simpsoni*. Haddow examined a range of plants containing water-holding axils including *Colocasia*, pineapple, *Sanseveria* sp. and *Dracaena* spp. as well as species of banana. Teesdale's (1957) work was prompted by that of Haddow and took place in Kenya, both on the coast and a little inland in the Shimba Hills. Again the motivation for the work was the presence in the area of yellow-fever vectors which here included *A. aegypti* as well as *A. simpsoni*.

Both authors give some data on the frequency of occurrence of the habitats they studied. Both subdivided the *Musa* axils they examined by locality and by variety using local names which, although not identical, may be synonyms across the different languages involved. Accordingly it is difficult to combine their results by way of overview but Table 4.3 is an attempt so to do. The most important point arising from the results presented in Table 4.3 is that many axils were without water and, even in the best case, only 43% of those examined contained water. This varied from plant species to plant species and from variety to variety of banana. The results presented on volumetric capacity are discussed below.

Haddow (1948) carried this environmental overview one step further. For 'gonja' bananas and *C. esculentum* (only), he surveyed three different types of lowland country: an open, heavily cultivated area, an area of mixed cultivated/forested land, and a strip of forest edge close to undisturbed rainforest. Figure 4.4 is drawn from his results. Interestingly plants in the open country most frequently had both water (67%) and mosquito larvae (44%) present in their axils, whereas forest edge plants performed least well in this regard (water present – 54%, with mosquito larvae – 23%). Haddow presents the pattern of results without explanation. Possibly some combination of shelter from rainfall provided by canopy plants and the anthropogenic nature of mosquito populations may act in concert to produce the observed patterns.

Dimensions and volumetric capacity

The largest phytotelma I have seen was a water-filled log hole in an area of cool temperate rainforest in Tasmania. This had an horizontal opening about 1.2 m in length by 30–40 cm in width and a capacity (unmeasured) of many

Table 4.3. *The occurrence of water-filled axils in Kenya and Uganda*

Plant species	Location	Number of plants examined	Number of axils examined	Percentage with water	Average water content
Banana varieties					
Mkono wa tembo	Ganda, Kenya	215	2905	42.5	3.9
Mkono wa tembo	Kwale, Kenya	281	2645	6.8	3.2
Ya Kiume	Ganda, Kenya	574	5828	41	5.2
Ya Kiume	Kwale, Kenya	358	3382	6.3	3.7
Bokoboko	Ganda, Kenya	570	8329	33.6	4.1
Bokoboko	Kwale, Kenya	391	3443	5.7	4.2
Kibungala	Ganda, Kenya	219	1830	3.9	2.9
Kibungala	Kwale, Kenya	278	2689	20.3	3.1
Kisukari	Ganda, Kenya	201	1708	2.6	4.3
Kisukari	Kwale, Kenya	556	6856	25.9	3.5
Kiguruwe	Ganda, Kenya	530	6280	27.3	3.1
Kiguruwe	Kwale, Kenya	314	3052	5.6	5
Kipukusa	Ganda, Kenya	494	4141	4.7	4.4
Mboma	Kwale, Kenya	563	5733	27.9	4.5
Banana varieties					
Gonja	Bundibugyo, Uganda	100	670	35	4.5[a]
Menvu	Bundibugyo, Uganda	100	777	13	2.9[a]
Bitoke	Bundibugyo, Uganda	100	649	8	3.1[a]
Pineapple	Bundibugyo, Uganda	100	3188	16	4.2[a]
Colocasia	Bundibugyo, Uganda	100	480	45	14.6[a]
Dracaena	Bundibugyo, Uganda	100	5224	19	4.6[a]

[a] Estimated in a separate survey of 1000 axils of each plant species, when full (see Haddow 1948).
Data on Kenya from Teesdale (1957); on Uganda from Haddow (1948).

litres. In contrast I have measured the capacity of leaf axils at just a cubic centimetre or so. And yet both these habitat units and, of course, those of all possible intermediate shapes and sizes regularly contain thriving communities of animals. Size and capacity are among the most obvious and readily measured parameters of phytotelmata. Accordingly we have more information upon them although, again, it is perhaps surprising that more authors have not presented this very basic information.

I present here, on the one hand, a selection of results from phytotelmata

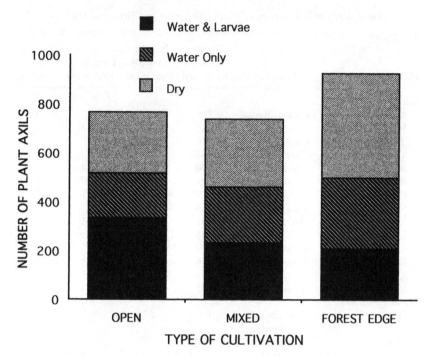

Fig. 4.4. The numbers of banana axils containing water and mosquito larvae in an open heavily cultivated area, mixed cultivated/forested area, and forest edge close to rainforest in Uganda (redrawn from Haddow 1948).

of different kinds and, on the other, a description of how these vary geographically for one example for which we have particularly extensive data – water-filled tree holes.

Table 4.4 contains information on the size and capacity of a range of phytotelmata. I present data on size and capacity of the bromeliad *Tillandsia utricularia* from Florida, calculated from the size, frequency and capacity data presented by Frank & Curtis (1981). In addition the table contains information from six species of *Nepenthes* from the lowland rainforests of Brunei (Clarke & Kitching 1993), from water-filled tree holes and bamboo internodes in northern New Guinea (Kitching 1990) and some previously unpublished results from the bract axils of inflorescences of the zingiberaceous plant *Tapeinocheilus ananassae* ('backscratcher ginger'), from the tropical north of Queensland. In addition the figures presented in Table 4.3 for water content of the Ugandan examples of axil waters (all, according to Haddow 1948, collected when the axils were full) may be set alongside those in Table 4.4 for comparative purposes. Other authors have presented more anecdotal

Table 4.4. *Physical characteristics of selected phytotelmata*

Habitat type	Location	Depth/length (cm)	Diameter (1) (cm)	Diameter (2) (cm)	Capacity (cc)
Bromeliads					
Tillandsia utricularia	Florida	25.3 ± 3.64	–	–	118.1 ± 43.56
Pitcher plants					
Nepenthes bicalcarata	Brunei	8.5 ± 1.91	6.1 ± 0.33	5.3 ± 0.25	148.7 ± 17.48
N. ampullaria	Brunei	5.3 ± 0.84	4.0 ± 0.11	3.3 ± 0.15	39.1 ± 3.4
N. mirabilis	Brunei	10.9 ± 2.90	3.3 ± 0.14	2.1 ± 0.13	39.6 ± 4.64
N. albomarginata	Brunei	11.6 ± 1.35	3.1 ± 0.18	2.9 ± 0.14	43.1 ± 4.05
N. rafflesiana	Brunei	8.7 ± 0.54	4.3 ± 0.17	4.1 ± 0.19	63.9 ± 9.32
N. gracilis	Brunei	8.3 ± 1.83	2.0 ± 0.06	2.1 ± 0.17	14.4 ± 1.17
Tree-holes	New Guinea	128.9 ± 11.76	188.3 ± 28.93	76.0 ± 8.37	–
Bamboo internodes	New Guinea	131.3 ± 15.87	49.5 ± 3.67	–	–
Bract axils					
Tapeinocheilos ananassae	Queensland	10.4 ± 0.82	–	–	38.7 ± 4.99

Bromeliad data from Frank & Curtis (1981); pitcher plant data from Clarke & Kitching (1993), tree hole and bamboo data from Kitching (1990).

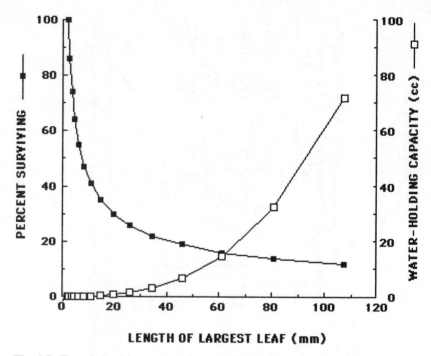

Fig. 4.5. The relationships amongst bromeliad tank size and percentage survival and water-holding capacity (drawn from data on *Tillandsia utricularia* in Frank & Curtis, 1981).

impressions of capacity. Laessle (1961), for instance, estimated the capacity of a large *Aechmea* bromeliad at about 2 litres and compared this, somewhat incredulously, with the estimate of 20 litres by Picado (1913) and, more realistically, with the 5 litres estimated by Pittendrigh (1948) as the capacity of very large specimens of *Hohenbergia*. Frank & Curtis (1981) presented particularly interesting data on how the capacity of the bromeliads studied changed with the size of those plants. In turn size was a clear indicator of plant age. This information (from their Table 1) is re-presented in Figure 4.5. As might be expected, as the size of the plant increases so its water-holding capacity increases. However, this increase is non-linear, presumably because the volumetric capacity is increasing by a cube rule whereas the leaf length (the measure of size used) is increasing linearly with age.

All these figures show the range of values presented by phytotelmata, with size and water-holding capacity varying widely over three orders of magnitude. Such variation must in turn have an effect on the community of animals which can form within each and this is an important factor to be considered when making across-site comparisons. An obvious conclusion, perhaps, is that valid comparisons may be made only within one phytotelm type. Comparisons across types are more problematical.

The large data set that we have accumulated over the years on the environment and fauna of water-filled tree holes in Australasia permits just this kind of comparison, and later chapters will explore this data in its various aspects. Table 4.5 begins this process by presenting information on the physical dimensions of tree holes in the sites we have investigated within the Australasian region to date. These stretch over almost 30 degrees of latitude and almost 1500 m of altitude.

In summary there is little statistical pattern which emerges from these data. There are weakly significant regressions between average hole depth and latitude ($r^2=0.35$, $F=5.32$, $pr(F)=0.04$) and between hole depth and altitude ($r^2=0.33$, $F=4.88$, $pr(F)=0.05$). These may reflect different growth forms of trees in forests at different latitudes and altitudes, but I am not convinced! In addition the average values of the two principal diameters of the tree holes (that is: the largest surface diameter and that at right angles to it) are also highly correlated ($r=0.83$) across regions.

Detritus dynamics

Almost all phytotelm communities are allochthonous and depend on two processes for obtaining basal energy and nutrients. They obtain most 'bulk' energy as detritus falling in from above, and in some instances (in particular in the cases of tree holes and axil waters) nutrients enter as run-off from the remainder of the plant in which they are contained. Accordingly patterns of entry of materials reflect processes external to the habitat unit: in particular, of litter fall and rainfall events. Processes of turnover within habitat units are further affected by decomposition processes within the habitat units.

The dynamics of detritus (and other nutrients) has received relatively little attention in phytotelm studies. The amount of detritus entering a phytotelm unit will reflect both the surface area of that habitat unit and the per unit area detritus fall in the neighbourhood of the phytotelmata concerned. Three sets of data which capture these two variables are presented, by way of example, in Figure 4.6. Two of these relate to studies of water-filled tree holes in England and Australia respectively (modified from Kitching 1983), the other is drawn from the work of Frank & Curtis (1981) and relates to leaf fall adjacent to *Tillandsia* bromeliads in subtropical Florida. As might be expected the patterns show different seasonalities reflecting their locations. The English data, from a deciduous beech woodland, show the greatest temporal heterogenity, the Florida data the least. It should be recalled, in interpreting this

Table 4.5. *Physical characteristics of water-filled tree holes from various Australasian sites*

Site	Latitude	Altitude (metres)	Number of holes	Depth (mm)	Diameter (1) (mm)	Diameter (2) (mm)
Madang, New Guinea	5° 00' S	c. 60	27	128.9 ± 11.76	188.3 ± 28.93	76.0 ± 8.37
Cape Tribulation	16° 05' S	c. 30	72	108.4 ± 8.06	193.5 ± 13.68	85.6 ± 5.33
Bellenden Ker	17° 25' S	120	18	140.0 ± 21.69	344.4 ± 47.46	140.8 ± 24.62
Atherton Tablelands	17° 13' S	c. 900	36	188.6 ± 15.63	207.1 ± 25.76	100.5 ± 10.48
Palmerston N. P.	17° 36' S	340	19	155.3 ± 24.17	210.0 ± 25.16	79.03 ± 10.97
Paluma	18° 57' S	850	29	169.5 ± 17.79	158.2 ± 19.07	88.6 ± 10.09
Eungella	21° 02' S	976	30	137.6 ± 12.23	194.8 ± 27.03	96.5 ± 14.17
Conandale Range	26° 42' S	790	13	156.2 ± 28.37	179.5 ± 37.05	74.2 ± 14.43
Lamington N. P.	28° 18' S	1100	10	158.0 ± 30.20	311.0 ± 35.90	153 ± 15.60
Dorrigo N. P. Site 1	30° 22' S	c. 150	9	178.7 ± 29.10	267.6 ± 56.96	105.9 ± 31.01
Dorrigo N. P. Site 2	30° 22' S	c. 200	6	196.7 ± 61.14	225.0 ± 32.02	102.5 ± 23.09
New England N. P.	30° 34' S	1200	10	169.3 ± 20.31	249.3 ± 37.31	85.6 ± 5.33

All sites but Madang are in Australia. Diameter 1 is the maximum diameter of the hole; Diameter 2 is the maximum diameter at right angles to Diameter 1.

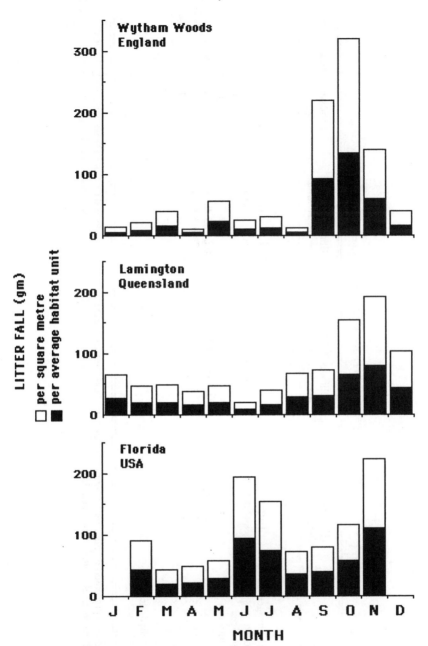

Fig. 4.6. Annual litterfall patterns adjacent to tree-hole study sites in Wytham Woods, UK (from Kitching 1969), and in Lamington National Park, Queensland, Australia (redrawn from Kitching 1983), and adjacent to a bromeliad study site in Florida, USA (from Frank & Curtis, 1981).

Table 4.6. *Chemical composition of stemflow water*

Measure	Illinois, USA *Fagus grandifolia*	Illinois, USA *Acer saccharum*	Illinois, USA *Quercus velutina*	Lancashire, England *Quercus petraea*
pH	6	5.8	5.6	3.7
Alkalinity	7.4	4	7	–
Ammoniacal nitrogen	1.12 ± 0.01	0.591 ± 0.015	3.43 ± 0.03	–
Nitrate nitrogen	3.44 ± 0.06	2.07 ± 0.19	10.5 ± 0.10	–
'Organic' nitrogen	–	–	–	0.21–1.2
'Inorganic' nitrogen	–	–	–	0.03–0.31
Phosphate	0.3 ± 0.03	0.338 ± 0.008	0.615 ± 0.018	0.02–0.18
Sulphate	5.22 ± 0.83	4.879 ± 0.03	29.0 ± 0.20	–
Potassium	–	–	–	1.7–9.9
Calcium	–	–	–	2.7–15.4
Sodium	–	–	–	8.0–40.6
Soluble carbohydrate	–	–	–	1.1–14.1
Polyphenols	–	–	–	2.0–9.0

All measures are presented as mg/l water (ppm). Alkalinity is calcium carbonate as mg/l. American data from Carpenter (1982), English data from Carlisle *et al.* (1966). American data are the mean ± 1 standard error, English data are presented as the range of observed values.

figure, that seasonality in Australia is the reverse of that in the other two sites and that the peak litter fall indicated is actually vernal, following new leaf flush on the evergreen trees of the rainforest, rather than the autumnal leaf fall of the deciduous woodland.

Little attention has been paid to the chemical qualities of leaf fall adjacent to phytotelmata (although there is a large literature on the composition of leaf litter in general – see, for example, Reichle 1970). I am unaware of any chemical analyses of the detritus pack *within* units of phytotelmata, although the chemical content of this component will be reflected in levels of water-borne nutrients (see below).

Many organic and inorganic nutrients enter phytotelmata as run-off. The magnitude of this input is hard to estimate although, as Carpenter (1982, 1983) convincingly demonstrates for the tree hole mosquito *Aedes triseriatus*, it may have a significant impact on growth, development and survivorship of larvae. Table 4.6 summarises the concentrations of selected nutrients in stem-flow water measured for three species of American deciduous tree by Carpenter (1982) and an additional species of oak from the United Kingdom by Carlisle *et al.* (1966). Turning such concentrations into nutrient loads entering tree holes, for example, requires information on the volume of run-off per unit diameter of tree trunk and the length of the trunk/tree hole interface. Such combinations of data are not available but Carlisle *et al.* (1966), for instance, estimates stem flow as between 1% and 16% of incident rainfall and records a maximum of 165 litres per tree per day during their two-year period of observation.

Water quantity and quality

As aquatic habitats, phytotelmata contain bodies of liquid, the quantity and quality of which may have major impacts upon well-being of the in-fauna and may well place limits on the complexity of the communities which they form. Many authors have measured simple fluctuations in water levels, fewer have examined easily measured chemical quantities such as pH and overall conductivity, and few indeed have carried out detailed chemical analyses on the water within habitat units.

Water levels

Data on water levels within phytotelmata and the way these may fluctuate through time make sense ecologically if expressed in one of two ways: either as measures of the proportion of habitat units dry as opposed to water-filled

74 *The phytotelm environment*

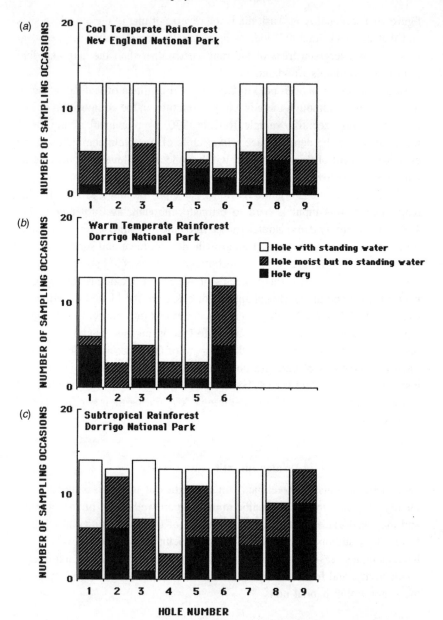

Fig. 4.7. The number of sampling occasions (over a thirteen-month period) that selected tree holes were dry, merely moist or with standing water, from three forest sites in northern New South Wales, Australia: (*a*) in cool temperate rainforest in New England National Park, (*b*) in warm temperate rainforest in Dorrigo National Park and (*c*) in subtropical rainforest in Dorrigo National Park.

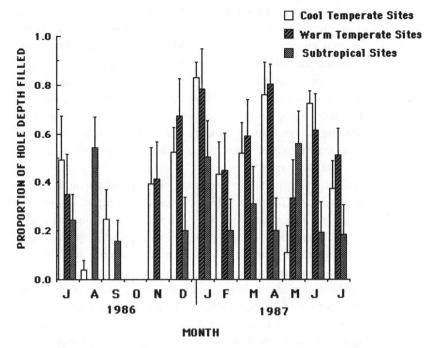

Fig. 4.8. The degree to which water-holding tree holes were filled with water over a twelve-month period at three Australian rainforest sites in northern New South Wales.

on each time interval, or, for a finer level of representation, when presented as proportions of the maximum capacity or depth for the habitat unit concerned. Needless to say these prescriptions have seldom been filled.

Figures 4.7 and 4.8 present data collected from tree holes in three sites in northern New South Wales during a thirteen-month period in 1986–87. These sites, which we encounter again in later chapters, were located in three different types of rainforest ranging from cool temperate, *Nothofagus*-dominated vegetation through more complex 'warm temperate' to subtropical rainforest. All three sites occur within a 40 km radius of each other, the different vegetation types resulting from differences in altitude, aspect and soil type. Within each site a set of tree holes was sampled monthly, during which time water depths were measured. Presence or absence of water was recorded by noting holes as being 'dry', 'moist' or 'with standing water'. Figure 4.7 presents these results on a habitat unit by habitat unit basis showing the considerable variation in the predictability of different tree holes as habitats for aquatic organisms, both within and between sites. Figure 4.8 presents a more quantitative analysis of these sites over the period of study. Each month for

each tree hole the water depth was measured and these data were converted
to a measure of 'proportion full': that is, the ratio of the measured water depth
to the known maximum water depth possible within each hole. The figure
presents the means and standard errors of these results for the sets of tree
holes studied in each of the three forest types. Across all three sets of holes
there is an indication of an overall seasonal pattern, presumably representing
a regional response to major weather patterns. That having been said, however, it is clear that the tree holes in some locations represent more predictable
aquatic habitats than in others – and this over a relatively short distance. I
return to this point in Chapter 9.

Less complete but nevertheless interesting data are available for other
classes of phytotelmata. Teesdale (1957) estimated the proportions of time
that leaf axils of *Musa* sp. were water filled during a seventeen-month period
in Kwale, Kenya. This information is presented as Figure 4.9, which I have
redrawn from Teesdale's paper. It is clear from the patterns that Teesdale
obtained that there is a short period of the year when most axils are dry. This
coincides with the end of the dry season in the January to March period. Once
filled, however, between 30% and 45% of axils contain water long after the
short wet season has ended. It is likely that this reflects two factors: first, that
the axil waters in bananas are more or less enclosed, reducing evaporative
loss and, second, that the large leaf area of the banana plants represent efficient collectors of moisture funnelling water into the watertight axils either
following dew formation or minor rainfall events.

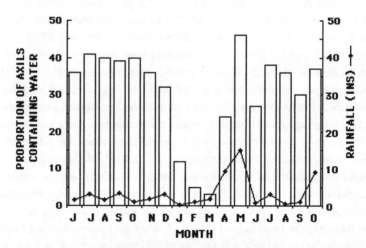

Fig. 4.9. Proportion of axils of *Musa* sp. which contained water estimated monthly
over a seventeen-month period, with associated rainfall data from Kwale, Kenya
(redrawn from Teesdale 1957).

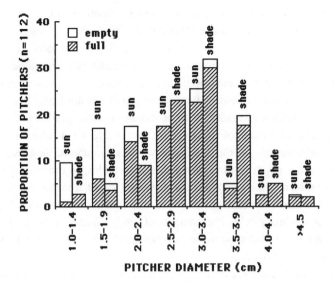

Fig. 4.10. Proportion of pitchers of *Sarracenia purpurea* in various size classes empty or water filled in both sun and shade, based on a survey of 112 pitchers in Michigan in June 1977 (redrawn from Kingsolver 1979).

Lastly, I present some data of Kingsolver's (1979) on the moisture regime in North American pitcher plants *Sarracenia purpurea*. In these plants the pitchers themselves are relatively short lived. Specialised leaves, the pitchers arise, grow, mature and die within one season. In other words, unlike a tree hole or the axils of a relatively long-lived banana plant, the plant container in this case is continuously changing in size. Kingsolver (1979) therefore was able to summarise the likelihood that pitchers of particular sizes would be wet or dry by taking a one-off 'horizontal' sample of plants and classifying them according to size. This is a viable alternative to making measurements on a marked set of habitat units through time. He presents the data collected in a survey of 112 pitchers made in June 1977 in a bog in Michigan. The results were partitioned into plants that were shaded and those in full sun and are presented in Figure 4.10. These show that the larger the pitchers, the more likely they are to contain water bodies. There is also a weak tendency for pitchers in full sun to be dry more frequently than those growing in the shade, and this is particularly so in smaller containers.

pH and conductivity

After any consideration of water quantity, the next most obvious environmental consideration in phytotelm studies is to specify some aspect of water quality. Measures of acidity (pH) are most frequently made partly because there is a long tradition of using levels of acidity to classify water bodies in the limnological literature, but also because such measurements are simply and cheaply made. In some instances, and for essentially the same reasons, measures of conductivity may be made. Such measures represent indices of the total concentration of electrolytes dissolved in a water body and, as such, are easily obtained substitutes for more extensive ionic analyses.

I have collated a range of these measures in Table 4.7. I have recalculated values from a number of the source documents, combining and rearranging data, sometimes taken from graphs rather than tables, in order to obtain the figures presented. Essentially phytotelm waters range from the very slightly alkaline through to extremely acid, with the overall mean within any phytotelm type distinctly acid. The contents of pitcher plants, in general, tend to be distinctly more acid than those of other plant containers. This is not surprising given their digestive functions.

Levels of acidity are responsive to changes brought about by dilution, runoff, litter entry and the activities of the biota living within them. These changes have been examined by a number of authors. Kurihara (1959) showed how acidity changed through the season in Japan (see Table 4.7). Fish & Hall (1978) measured pH levels in pitchers of *Sarracenia purpurea* of a variety of ages (see Table 4.7), and Laessle (1961) demonstrated that there was heterogeneity in acidity in different parts of the water bodies contained within a single bromeliad plant. In a parallel but very preliminary fashion we have been able to show that, within a single water-filled tree hole, pH gradients can occur both laterally across the water body and vertically, changing with depth.

A number of authors have estimated the interaction of pH levels with members of the in-fauna. Kurihara (1959) examined the interaction between the pH within bamboo containers and the pattern of oviposition of *Eristalis* species within the containers. He concluded that the egg-laying syrphids tended to avoid the most acid sites, preferring neutral or even alkaline sites when these were available. Viewing the problem from the opposite perspective, as it were, Benzing *et al.* (1972) were able to show that the generally acid waters contained within the tanks of *Aechmea* bromeliads rapidly became akaline when algal cultures were introduced, but the presence of insect larvae, in contrast, had little effect upon acidity.

Much less information is available on measures of conductivity. Tree holes into which mineral soil may enter as a result of run-off or rain splashing from the adjacent forest floor tend to have higher levels of electrolytes as measured by conductivity levels than do the 'cleaner' less permanent waters of axil plants. Interest in these aspects of water quality within phytotelms have led some authors to make much more extensive chemical analyses.

Specific chemical analyses

In limnological studies of water bodies larger than those customarily found in phytotelmata it is a commonplace that extensive chemical analyses of the water, generally focusing on key nutrient ions, are carried out as an adjunct to ecological studies. For a number of reasons, generally related to the availability of analytical facilities, students of phytotelmata have seldom made such detailed analyses. Information on the ionic concentrations of selected nutrients are available from water-filled tree holes and the tank waters of bromeliads.

Walker and his colleagues at Michigan State University have made extensive studies of the concentrations of selected ions from tree-hole water in Michigan and have related this to bacterial and mosquito productivity within particular tree holes (Walker & Merritt 1988, Walker *et al*. 1991). Figure 4.11 is taken from their 1991 paper (with permission) and indicates the fluctuations in levels of the cations nitrate, nitrite, sulphate and phosphate, over two seasons in 19 tree holes in 1985 and 17 in 1986. Much less complete results have been presented by early European workers, who used much less sophisticated analytical methods and, in one case at least, presented figures but not units (Karstens & Pavlovskiij 1927, Ramsden in Blacklock & Carter 1920, etc.). Walker *et al*. (1991) showed that nutrient levels in tree holes reflected inflow patterns, nutrient cycling processes within the habitat units and the excretion of nitrogenous waste by mosquito larvae. Rainfall-driven influxes of nutrients appeared to increase mosquito productivity partly, they argue, because of nutrient input but also because of the flushing of accumulated toxins from the microcosms. Bacterial numbers, on the other hand, appeared to be inversely related to mosquito density and rainfall flushing.

A data set for bromeliads focusing on selected anions and cations is presented by Torales *et al*. (1972) and is summarised in Table 4.8. Needless to say there is no overlap in the analyses selected by Torales *et al*. and those selected by Walker *et al*.! We await the basis for proper comparisons. Such comparisons would be of particular interest, not only because of their intrinsic merit, but also because of the close taxonomic similarities between the faunas of different types of phytotelmata. The different nutrient regimes would be an obvious avenue of investigation in tackling questions relating to habitat partitioning and the environmental limitations displayed by each of the taxa involved.

Table 4.7. *Acidity and conductivity estimates in phytotelmata*

Habitat type	Location	Author	pH	Conductivity
Bromeliads				
	Argentina	Torales *et al.* 1972	5.6 ± 0.15	–
	Jamaica	Laessle 1961	(4.0)–4.9–(5.8)	–
Pitchers				
Nepenthes bicalcarata	Brunei	Clarke & Kitching 1993	4.3 ± 0.14	–
N. ampullaria	Brunei	Clarke & Kitching 1993	3.7 ± 0.09	–
N. mirabilis	Brunei	Clarke & Kitching 1993	4.3 ± 0.11	–
N. albomarginata	Brunei	Clarke & Kitching 1993	3.8 ± 0.18	–
N. rafflesiana	Brunei	Clarke & Kitching 1993	2.6 ± 0.17	–
N. gracilis	Brunei	Clarke & Kitching 1993	2.4 ± 0.23	–
Sarracenia purpurea <60 d old	Eastern USA	Fish & Hall 1978	(5.0)–5.98–(6.3)	–
S. purpurea 30–60 d old	Eastern USA	Fish & Hall 1978	(3.6)–3.93–(4.5)	–
S. purpurea >30 d old	Eastern USA	Fish & Hall 1978	(3.4)–3.85–(3.8)	–
Tree holes				
Deciduous woodland	England	Kitching 1983	6.4 ± 0.15	339 ± 62.0
Cool Temperate rainforest	N. NSW	unpublished	5.2 ± 0.17	–
Subtropical rainforest	S. Queensland	Kitching 1983	5.9 ± 0.23	226.5 ± 75.5
Tropical rainforest	N. Queensland	unpublished	5.4 ± 0.054	–
Tropical rainforest	New Guinea	Kitching 1990	5.6 ± 0.12	159.5 ± 47.72
Bamboos				
Tropical Rainforest	New Guinea	Kitching 1990	5.5 ± 0.15	415 ± 146.8
Cultivated (August)	Japan	Kurihara 1959	(3.8)–6.4–(8.4)	–
Cultivated (September)	Japan	Kurihara 1959	(3.2)–6.9–(8.4)	–
Cultivated (October)	Japan	Kurihara 1959	(3.2)–7.3–(8.4)	–

Axils				
Heliconia spp.	Venezuela	Machado-Allison *et al.* 1983	7.23 ± 0.08	–
Curcuma australasica	New Guinea	Kitching 1990	5.4 ± 0.34	140 ± 27.3

Where an error term is shown this is ± 1 standard error; ranges about means are shown as (min.)-mean-(max.).

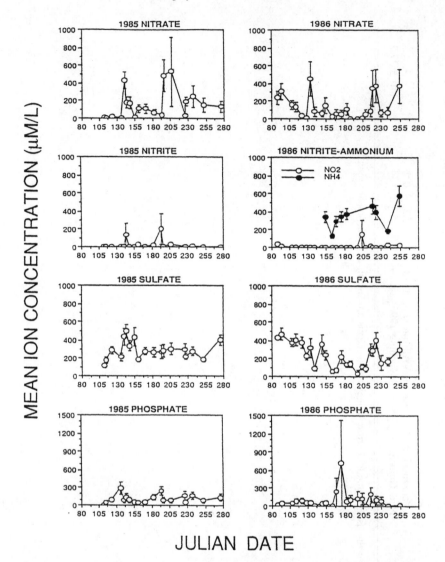

Fig. 4.11. Levels of nitrate, nitrite, sulphate and phosphate cations in 19 water-filled tree holes in North America (from Walker *et al.* 1991).

Temperature and dissolved oxygen regimes

For arthropods in particular and ectothermic organisms in general, ambient temperatures are a crucial aspect of the environment, principally because their rates of development are temperature dependent (see, for example, Kitching

Table 4.8. *Chemical analyses of the water content of the bromeliad* Aechmea distichantha *in Corrientes, Argentina*

Variable	Estimate (± 1 s.e.)
Water content (cc)	614.4 ± 85.18
Organic matter (ppm O_2 consumed)	43.1 ± 7.11
Ionic concentrations	
Hardness (ppm $CaCO_3$)	21.9 ± 3.49
Chloride (ppm)	6.5 ± 1.83
Calcium (ppm)	5.5 ± 0.54
Potassium (ppm)	5.2 ± 0.54
Magnesium (ppm)	1.6 ± 0.45
Sodium (ppm)	6.1 ± 1.21
Turbidity (log turbidity units)	188.5 ± 58.65

After Torales *et al.* (1972)

1977, Kingsolver 1979). For aquatic animals the medium itself may freeze or become intolerably heated. In an aquatic medium the need on the part of the arthropod inhabitants to obtain oxygen for respiratory purposes interacts with temperature in at least two important ways. Ambient temperature determines oxygen carrying capacity in the water body on the one hand and, on the other, has a direct impact on the rate of decomposition processes and microorganismal productivity which represent major competing demands for available oxygen. These impacts may be particularly acute in phytotelmata simply because their small size reduces the buffering capacity of the water bodies.

For these general reasons the temperature and dissolved oxygen regimes in phytotelmata have received attention from a number of authors. Again pitcher plants, bromeliads and tree holes have received the greater part of the attention directed to the particular environmental variables. Whereas water quality and other physico-chemical variables within plant containers may vary on a time scale of days or weeks, temperature and oxygen regimes will vary on a diel basis, within a longer term seasonal pattern in most cases. Accordingly, estimating the natural variability of these variables is no small task.

Tree holes in Czechoslovakia and Germany

In 1938 Karel Mayer published a detailed account of his studies on the temperature regimes in holes in beech trees. In a single hole Mayer measured

The phytotelm environment

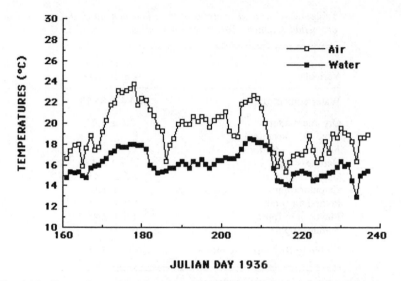

JULIAN DAY 1936

Fig. 4.12. Comparison of temperatures within the water body in a Czech beech tree hole and adjacent air temperatures over an eighty-day period in the summer of 1936 (redrawn from Mayer, 1938).

temperatures within the water body and in the adjacent air mass at 0800, 1200 and 1800 hours every day over a three-month period from June to August 1936.

I have used the daily means Mayer obtained to prepare Figure 4.12. What is immediately clear from these data is that although the water temperatures do track those of the air there is considerable insulation, resulting in lower water temperatures which exhibit less variability than those in the adjacent air. Condensing the whole sequence of data shown in the figure to time-based averages shows that at 16.1 °C the water temperature is about 3 °C cooler than that of the air (at 19.3 °C). More important, the variation about these mean values is considerably less in the case of the water temperatures (a coefficient of variation of 7.5 as opposed to 10.9 in the air). This pattern, writ small, is repeated on a daily basis as indicated in Figure 4.13 in which I have plotted a thirty-day segment of Mayer's data (from June 19th to July 19th 1936) collected at 0800, 1200 and 1800 hours separately.

Surprisingly few estimations of oxygen concentrations appear to have been made in water-filled tree holes. Certainly the estimation of levels of oxygen saturation within small water bodies presents technical difficulties, but these are by no means insuperable. Brandt (1934) examined selected environmental factors within a small series of tree holes in German beech trees. In some of these he measured oxygen levels using the titrimetric Winkler method. He

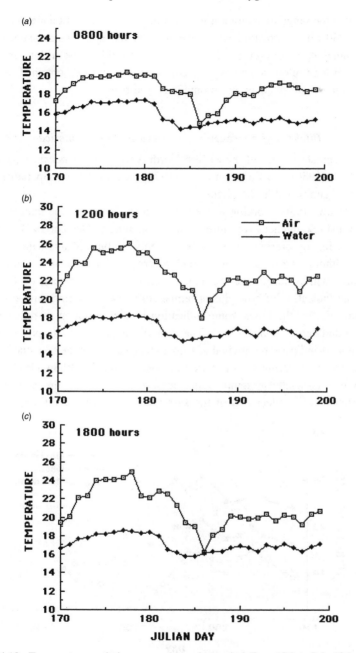

Fig. 4.13. Temperature variation over a thirty-day period (June 19th to July 19th 1936) in a Czech water-filled tree hole measured (*a*) at 0800 h, (*b*) at 1200 h and (*c*) at 1800 h (redrawn from Mayer 1938).

estimated the range of saturation levels at 3.0% to 29% about a mean of 15.1 (s.e. 4.230 after correction for temperature at the time of analysis). More determinations of oxygen levels in tree holes would be useful but Brandt's figures at least allow some basis for comparison with those obtained by others in *Sarracenia* pitchers and bromeliads (see below).

Pitchers *of* Sarracenia purpurea *in North America*

In two separate studies of the eastern North American *Sarracenia purpurea*, Bradshaw (1980) and Cameron *et al.* (1977) examined the temperature and oxygen regimes within the pitcher liquid.

Bradshaw (1980) working in New York State measured the temperatures within a single pitcher every three hours from June to October 1971. Unfortunately, for my present purposes, he presents the bulk of this data as daily means although he does include weekly averages of sunrise and sunset temperatures within the pitcher. So Figures 4.14 and 4.15 present, on the one hand, an indication of how pitcher temperatures varied during this period in general and, on the other, some indication of diel variation as reflected in sunrise and sunset temperatures. The general temperature data presented in Figure 4.14 indicate the gradual seasonal changes in temperature one would expect to occur during such a period of observation. Bradshaw (1980) does not give us associated air temperatures but it seems likely that the range within the pitchers will reflect that of the surrounding air, with additional radiant

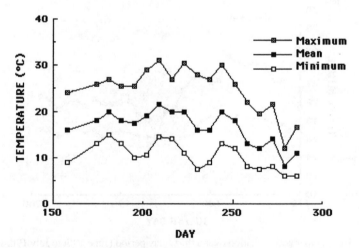

Fig. 4.14. Seasonal changes in daily maximum, mean and minimum temperatures within pitchers of *Sarracenia purpurea* from New York State, USA (redrawn from Bradshaw 1980).

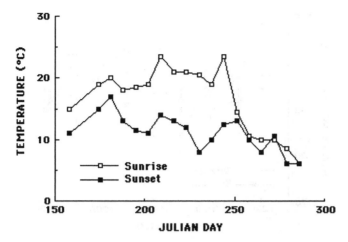

Fig. 4.15. Seasonal changes in sunrise and sunset temperatures within pitchers of *Sarracenia purpurea* from New York State, USA (redrawn from Bradshaw 1980).

heating effects, particularly in the most exposed pitchers. This is indicated in Figure 4.15, which shows a warming trend during each day with sunset temperatures from 4 °C to 10 °C higher than those at sunrise. This is entirely what one would expect in situations such as the exposed bogs in which *Sarracenia* plants occur. The convergence of sunrise and sunset temperatures late in the season when air temperatures are declining and, presumably, insolation occurring for shorter periods is consistent with this model.

Cameron *et al.* (1977) examined *Sarracenia* pitchers in the bogs of New Brunswick in eastern Canada. Their study focused on the dissolved oxygen regimes within pitchers and included measurements made in the water bodies of experimental plants in both October and November. These results are reproduced (and redrawn) in Figure 4.16. Levels, in general, are very high and point up the very great contrast between these phytotelmata and tree holes as an aerobic environment. On both occasions when measurements were made there was a daily rhythm in oxygen saturation with nocturnal nadirs and diurnal peaks. The peak for the October data is somewhat earlier in the day than that observed in November. Cameron *et al.* were unable to show statistically significant changes in dissolved oxygen concentration with temperature, date or even time of day. In addition these authors compared surface and near-bottom measurements within pitchers, finding that, in both cases, dissolved oxygen levels were near saturation (surface mean, 98.5%, range 86–106.5%; near-bottom, mean 95.2%, range 77–102.2%). It appears that in these pitchers oxygen concentration is maintained by diffusion through the pitcher walls

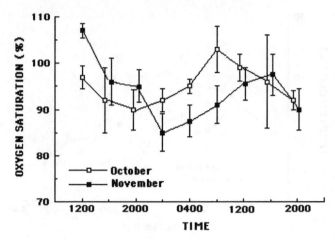

Fig. 4.16. Diel variation in dissolved oxygen levels within pitchers of *Sarracenia purpurea* in October and November, in New Brunswick, Canada (redrawn from the results of Cameron *et al.* 1977).

and is unaffected by photosynthetic activity, either of the pitcher plant itself or of any of the contained biota.

Bromeliads in Jamaica

Lastly, in this section, I present some results obtained by Laessle (1961) as part of his extensive studies of Jamaican bromeliads. In two separate studies Laessle presented results collected from three large plants of *Aechmea paniculigera*. Within each he measured both water temperatures and dissolved oxygen in the tanks. In each study he made a series of measurements of temperature or oxygen concentration in the central chamber and in each of the surrounding bract axils. I have summarised these results statistically in Table 4.9.

The temperatures he presents from his 'Plant 69' represent a late morning series when the plant was shaded and an afternoon series when the plant was fully insolated. The two sets of temperatures show highly significant differences indicating the importance of shade or lack of it in determining conditions within the water bodies. These results complement those of Pittendrigh's (1948) presented earlier in this chapter in which he classifies bromeliad species on the basis of their degree of exposure within the canopy. In other results Laessle (1961) comments that he observed a temperature range from 17.5 °C to 30 °C in more extensive studies but did not record night temperatures within high elevation bromeliads which, he anticipated, might be considerably lower.

Table 4.9. Summary statistics from temperature and dissolved oxygen measurements in Jamaican bromeliads

Plant and variable	Mean	Standard error	t-value (and comparison)	Probability
Plant 70: Full sun				
Dissolved oxygen, day	71.5	11.39	4.46	0.004
Dissolved oxygen, night	18	2.96	Plant 70 day vs night	
Plant 75: Shade				
Dissolved oxygen, day	23.3	3.81	−0.195	0.858
Dissolved oxygen, night	28	4.56	Plant 75 day vs night	
Plant 69				
Water temperature, shade	22.8	0.087	−12.639	<0.001
Water temperature, sun	28.6	0.447	Plant 69 sun vs shade	

Data calculated from Laessle (1961).
Dissolved oxygen is presented as percentage saturation, temperature in °C.

Laessle's (1961) results on oxygen concentrations (Table 4.9) are particularly interesting. He presents a set of results taken during the day and a second at night from both a fully exposed ('full sun') plant and one that was growing in the shade. For the full sun plant the oxygen levels collected during the day were significantly higher than those collected at night (72% as opposed to 23% saturation). In contrast, in the shaded plant the difference between the two sets of readings was not significant and resembled the night results from the full sun plant (day 23%, night 28%). Laessle surveyed both plants for the presence of algae and found that the central reservoir, in particular, in the 'full sun' plant had very high levels of algal growth, whereas, in the shade plant, he found virtually no algal cells. It seems very likely then, as Laessle concluded, that dissolved oxygen regimes in these bromeliads reflect the presence or absence of algae which, in turn reflects the position of the plant with respect to sun or shade. Needless to say the higher temperatures experienced by the 'sun' plants will also favour rapid algal growth.

Laessle's (1961) results with respect to oxygen concentrations in bromeliads are in marked contrast to those measured by Cameron *et al.* (1977) for *Sarracenia* pitchers. I presume that the much more robust leaves of the bromeliads represent a barrier to oxygen diffusion in a way that the much thinner walls of *Sarracenia* pitchers do not. Nevertheless the lowest levels observed in bromeliads are equal to or greater than the upper end of the range seen in water-filled tree holes in Brandt's (1934) studies.

Part II:

Methods and theories

I had sieved the tree-hole samples collected that morning from a set of holes perched in the buttress roots of trees deep in the forests near Baiteta in Madang Province of Papua New Guinea. Now in the cool (and, more important, mosquito-free zone) of the laboratory of the Christensen Research Institute I was sorting the detritus under the microscope, counting each of the organisms I encountered. I had carried out the same procedures in Australia, America and Europe and felt the fauna could hold no surprises. I pipetted out a series of mosquito larvae – the heavyweight wrigglers of species of Tripteroides, *the even larger predatory monsters that turn into Toxorhynchites. These joined my counts of the larvae of scirtid beetles, midges and so forth. And then, out of left field, into my view swam a paddleworm! This was a surprise: paddleworms, polychaetes, are marine animals which I associated with the rich rockpools of Robin Hood's Bay in North Yorkshire, or the rock shelf at Arrawarra, New South Wales, two of the formative places of my earlier biological education. But not so: it turns out there is just one small subfamily of freshwater polychaetes previously known from the water in tropical leaf axils – and now from water-filled tree holes. It was then perhaps less of a surprise, but no less satisfying, when in a similar circumstance while I was examining samples from water-filled tree holes collected in the lowland rainforests of the Daintree in North Queensland, a flatworm glided across the field of view of yet another sorting tray: a different class of annelid appearing to progress with no visible means of locomotion.*

From the steamy heat of the rainforests of New Guinea and northern Australia to the alpine zone in Tasmania is a large step climatically, and physiologically, but there, in 1993, I was asked to look into the water-filled axils of the amazing giant epacrid known as 'pandanni'. These massive spiky plants are reminiscent of the giant Lobelias of the equatorial African mountains (which possess yet other container habitats!) and their axils contained another zoological surprise. Washing out the debris and water from the leaf axils of five of these massive plants, I encountered the 'usual' range of animals plus an oddity. In each axil were several middle-sized mites each looking somewhat like a flattened dog-tick, sculling around on their eight limbs, quite unlike the other groups of mites that are familiar to those that peer into the groins of plants. These were anisitsiellids: belonging to a family that now occupy a number of Southern Hemisphere sites and whose ancestors were separated as the Gondwana supercontinent broke up during the Cretaceous period. Never before recorded from container habitats this discovery added a whole new family to the phytotelm bestiary.

In this section of the book I attempt two things. In Chapter 5 I discuss the ways in which phytotelm data are collected and how foodwebs are constructed from such information. The chapter concludes by a brief treatment of the descriptive statistics which can readily be extracted from foodwebs once they have been defined. Chapter 6 is one of the keys to this whole work dealing with available and new theory available to us to explain the patterns we observe in phytotelm foodwebs. It defines a series of ideas, hypotheses and predictions against which the patterns – to be described in Part 3 – can be compared. Towards the end of the book, in Chapter 13, I revisit these ideas to see how well they stand up to their juxtaposition with real data.

5

The construction and quantification of food webs
Drawing the webs

A food web in its simplest form is merely a diagrammatic or mathematical summary of the trophic interactions in a community: an indication of what eats what.

This having been said, it remains noteworthy how many different ways authors have chosen to represent what they perceived as this 'simple' concept. The purpose of this short chapter is to illustrate the different ways in which I have chosen to represent phytotelm food webs. I have found it appropriate to use a range of forms of representation of food webs to illustrate different points or test different hypotheses. To insist on a single standard is to misplace emphasis on method rather than question. Nevertheless it has not always been made abundantly clear in food-web descriptions what the ground rules of construction have been: and I have been as guilty of this misdemeanour as others. Accordingly I lay out here sets of rules for constructing webs that are used in ensuing chapters.

In addition the chapter introduces and defines a number of summarising statistics which may be extracted from food webs. Again I define only those quantities which I use subsequently: other authors have used other statistics. Parenthetically the chapter also introduces a food web from each class of phytotelmata as a tool for explaining web construction and the extraction of the statistics.

From field data to food webs

Figure 5.1 is a full representation of the metazoan food web, at the macro-organism level that I compiled for water-filled tree holes at Baiteta in lowland rainforest north-west of Madang in Papua New Guinea (Kitching 1990).

The web as represented in the figure compiles taxonomic information with trophic information in a binary fashion; that is, the presence of a trophic

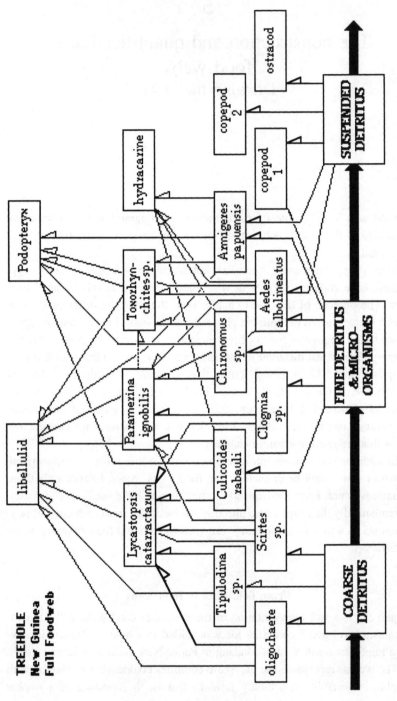

Fig. 5.1. A full representation of the metazoan food web from water-filled tree holes at Baiteta, Madang, Papua New Guinea.

interaction, however strong or weak, frequent or infrequent, is represented by an upwardly directed arrow between two species boxes, indicating flow of material in the direction of the arrow.

In practice a number of decisions have been made, implicitly or explicitly, in constructing the diagram.

The web was constructed during a six-week visit to the region during the wet season of 1987. During that time I searched the forest in the region for water-filled tree holes. Twenty-seven sites were encountered during this period. These sites were a combination of rot holes in tree stumps or broken branching points, or pans in the axils of branches and buttress roots. None was higher than four metres from the ground. Subsamples of the contents of each were obtained by dipping out the contents by hand, ladle and siphon. These were returned to the laboratory where they were sieved, sorted and counted. Unknown larvae were set up in rearing containers and emerging adults were collected, killed and mounted. Upon my return to Australia and over a period of years, these specimens were identified for me by a range of taxonomic specialists.

So much for the basic methodology: very similar methods are adopted for other phytotelm types. So, already there are a number of constraints upon the end product: the food web. In summary:

- Only a limited number of sites over a limited period of time were examined.
- Only low-level easily accessible sites were sampled.
- Only larvae which could be separated to the species level were ennumerated at that level. In other words 'chironomid midge larvae' might be counted but, after completion of the rearing process, turn out to be two or three species. There is then no way to go back and re-assign larvae to Species 1, 2 or 3. Various rules of thumb may be used but each carries further assumptions.
- Only larvae which are reared successfully can be identified further, and this may restrict the accuracy of the resulting food web.

Once a species list is drawn up then trophic positions must be assigned. This depends heavily on the sort of background knowledge set out in Chapter 3 and the concluding Annexe for each family of organisms encountered. It is seldom difficult to sort saprophages from predators. But then the much more subjective task of drawing links must be faced. In a long-term study the likelihood of any one species eating any other can be tested by a series of arranged encounters. This is particularly useful in identifying which species in an assemblage are, at least potentially, the top predators. In other cases links

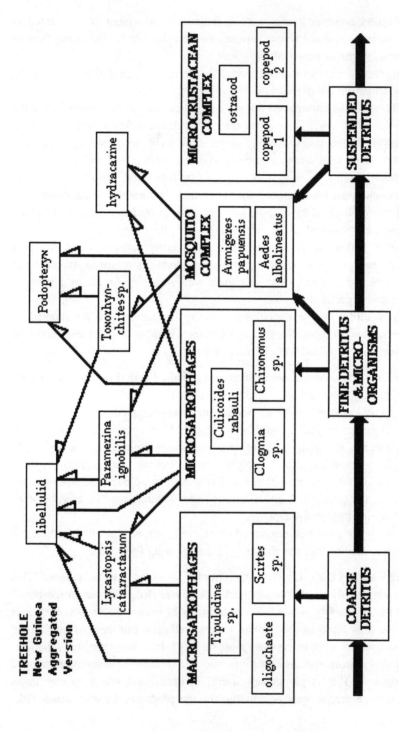

Fig. 5.2. The food web from water-filled tree holes in Papua New Guinea as in Fig. 5.1 but drawn in 'aggregated' form.

may be drawn on general principles of the relative sizes of the 'prey set' and the 'predator set', on the known prey range of members of the same genus, say, as the subject species, and sometimes on informed intuition. In phytotelm food webs, at least, this process is aided by the fact that in those situations which have been most thoroughly worked up it seems predators tend to the generalist rather than specialist, giving some added confidence to the webs that are drawn. At the end of this process of observation and decision-making a web of the sort illustrated in Figure 5.1 can be drawn. There will remain a further set of concerns:

- species assumed to behave trophically like close relatives may not, in fact, do so,
- trophic links assigned on the basis of arranged 'encounters' may not occur in field situations,
- assumed trophic links may in fact occur in the field but be so rare as to be irrelevant in any process of generalisation, and
- assumed trophic links may change during development of the species concerned.

As already discussed in the introductory chapter, the last two problems, of interaction strength and ontogenetic change, remain among the greatest challenges for further field research on food webs in phytotelmata.

Problems associated with the identification of species–species feeding links can be reduced significantly by drawing the web in the aggregated form shown in Figure 5.2. Such a simplified web contains almost as much usable information as Figure 5.1 and most of the derived statistics can still be calculated. The exception to this is connectance, which needs information on the full set of species interactions for its calculation (see below). It must also be added that the food web from New Guinea tree holes is one of the most complex webs so far constructed for any phytotelm community. Most of the webs encountered are much simpler and can be constructed with less information, fewer assumptions and more confidence.

One remaining problem arises at this stage of the investigative process. It has always been my experience that the occasional larva or other animal turns up in the sorting of organisms, particularly from larger phytotelmata such as tree holes, for which the question arises: 'Is this organism really to be considered part of the container food web, or is it merely an accidental?' Accidentals may occur, for instance, in drier tree holes where inhabitants of moist leaf litter become admixed with the sludge and water samples on which I base my food webs. To include them, without question, in food webs can make the webs unrealistically complex and can skew food-web statistics subsequently used in comparative analyses. Several considerations assist in making decisions as to whether to include a species in drawing up the food web, or not. Mere rareness, or even uniqueness within samples, is not in itself reason to exclude. For instance, as indicated in the preface to this part, in New Guinea two examples of the namanereine polychaete *Lycastopsis catar-ractarum* turned up in the samples. The first thought, that these were mere contaminants from neighbouring forest streams, was set aside after I learned that the only other known examples of this species were also from water-filled plant containers (see Annexe). More useful are some of the considerations discussed in the taxonomic accounts in Chapter 3 and the Annexe, in which some taxa can be considered as accidentals on the basis of the known biology of the groups concerned. For example, I occasionally find living Collembola in tree-hole and plant-axil samples. In general these are of terrestrial groups that occur commonly on tree bark and plant surfaces, in moss beds and leaf litter. I do not include these in the aquatic food webs that I construct. Ultimately though, there is a measure of judgement in decisions about inclusion and non-inclusion based on experience. The problems discussed here, of course, are diminished by larger sampler sizes, repeated visits to the same regions, and more intensive study of particular species – but these solutions are not always available.

Other representations

The two web representations so far discussed contain, within them, taxonomic designations and as such are necessarily complex and sometimes forbidding for the non-taxonomist. Often what is needed for a particular analysis is a summary of the structure of the web, indicating clearly the trophic linkages and the number of species at each trophic level but not concerning itself with the identity of the species themselves. Figure 5.3 is such a representation and is what I call a 'graphed' version of the web (because those who approach food-web analysis from the point of view of graph theory tend to draw their webs in this fashion – see, for example, Cohen 1978). A minor variation upon this form is to number the nodes and provide a key naming the species involved (as in Figures 5.7 and 5.9).

The graphed version of the tree-hole food web also provides the facility for representing a particular example of the food web which is characteristic of a particular area of forest, say, but which may not occur in every unit of the phytotelm type. So Figure 5.4 might represent the food web observed in a particular tree hole in northern New Guinea (in fact it is imaginary) which is a subset of the food web which is characteristic of the habitat type in the

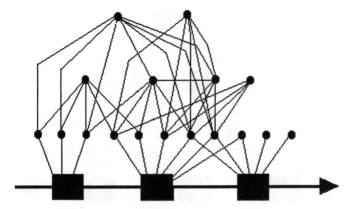

Fig. 5.3. A 'graphed' form of food-web representation of the Papua New Guinea tree-hole food web presented in Fig. 5.1.

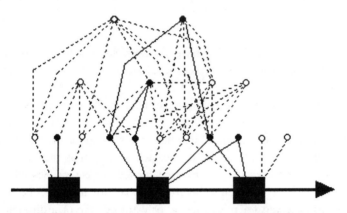

Fig. 5.4. A graphed food web from a single habitat unit with the complete regional web 'ghosted' in the background, based on the Guinea tree-hole web presented in Fig. 5.1.

area. This mode of represention is particularly useful for summarising local spatial and temporal variation in food web structure (e.g. Kitching 1987a, Kitching & Beaver 1990), often used in conjunction with the final mode of food-web representation that I shall describe here.

As I have mentioned already, a major shortcoming of the simpler representations of food webs described hereto is that they include no measure of the strength of the trophic interactions that are represented. Ideally some measure of the amount of material that flows along each trophic arrow would be included, especially when detailed analyses of spatial and temporal variation are required. New methods of labelling prey and measuring flow rates into

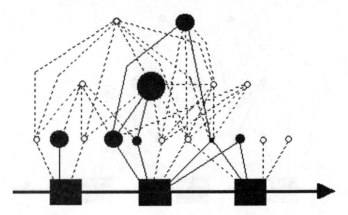

Fig. 5.5. A food web with the relative abundance of the participating organisms represented by different sized symbols. In this case the four sizes of solid symbols representing nodes indicate the presence of a particular species, the size of the symbol, the log abundance quartile for that species (across all habitat units in the region).

predators are now available but have not been applied, hereto, to any phytotelm webs (although these present ideal situations for the application of these methods).

Figure 5.5 represents a compromise solution to this problem in which the range of abundance values in the data set are divided into quartiles (usually after logarithmic transformation) and their presence in the food webs within particular phytotelm units is then represented by a symbol, the size of which reflects the abundance quartile observed for the subject species within that habitat unit. This circumvents the problem that, for instance, the range of abundances observed for midge larvae or ostracods are of a different order of magnitude than those observed for tipulids or odonates. The device of dividing observed abundance ranges into quartiles standardises across such organisms. Figure 5.6 is an example of webs constructed in this fashion for data from bamboo internodes collected in Sulawesi and previously presented in Kitching & Beaver (1990). Dividing the observed range of densities into quartiles is arbitrary. The range could be divided into fives or sixes or whatever ranges that the investigator chooses and that the inherent accuracy (or otherwise) of the data will bear.

This weighted graphical form of representation has the additional property that the web can be written as a numeric string in which numbers ranging from 0 to 4 (that is: species absent, species present at abundances levels in the first quartile, second quartile, etc.) can be used to represent a local food web. As such, the string may be used in the calculation of correlations across

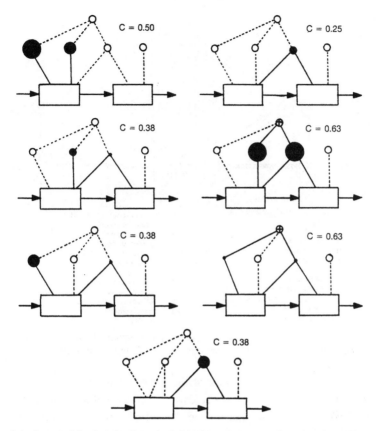

Fig. 5.6. A set of food webs from individual bamboo cups from northern Sulawesi, constructed in the manner of Fig. 5.5 (from Kitching & Beaver 1990).

habitat units, providing a useful index of web similarity for some purposes (see, for example, Kitching 1987a).

Some examples

Figures 5.7 to 5.10 represent examples of full metazoan webs from bromeliads, bamboo internodes, pitcher plants and axil waters. Figures 5.1 to 5.5, presented earlier, cover the case of water-filled tree holes in this context.

A bromeliad web

I have compiled Figure 5.7, with some difficulty and reinterpretation, from Laessle's (1961) excellent paper on the fauna of Jamaican bromeliads, using

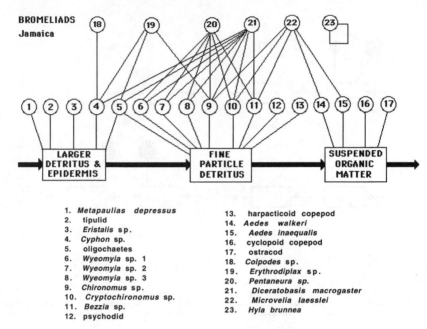

Fig. 5.7. A food web constructed to represent the aquatic fauna of water-holding bromeliads from Jamaica (constructed from information in Laessle 1961).

the 'labelled' graph form because of the great number of species involved. The faunistic aspects of Laessle's study revolved around the destructive sampling of twenty bromeliad plants in the Chestervale region of eastern Jamaica. Both micro- and macro-organisms were sampled and distributed for critical identification. The food web as I have drawn it is restricted to the metazoans, excluding rotifers, gastrotrichs and nematodes. Other than these I omitted a few very occasional or uncertain records mentioned by Laessle, and restricted myself to clearly aquatic species. In general I included those trophic links identified or inferred by Laessle (1961) himself but added a few more (such as that linking the anisopteran dragonfly larvae with the oligochaetes, for instance) on general principles (see taxon accounts).

A pitcher-plant web

Figure 5.8 represents the aquatic web found in pitchers of *N. albomarginata* from Penang Hill, Pulan Pinang, West Malaysia. This representation is based on that of the work of Beaver and his colleagues (Beaver 1979a, b, 1985; Kitching & Beaver 1990). The figure omits the araneid species *Misumenops* for present purposes because of its largely terrestrial habits.

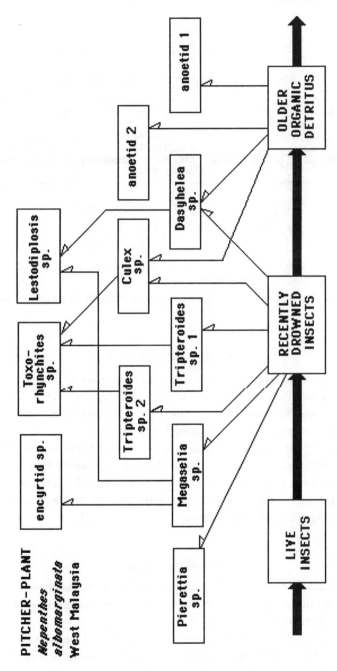

Fig. 5.8. The food web from pitchers of *Nepenthes albomarginata* in West Malaysia (after Beaver 1979a, 1985).

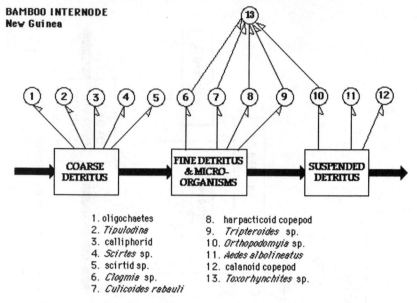

Fig. 5.9. The food web from bamboo internodes in Papua New Guinea (from
Kitching 1990).

A bamboo-internode web

Bamboo internodes present woody, if monocotyledonous, containers, the full
faunas of which have been seldom studied (see Table 2.4). Where appropriate comparisons have been made it emerges that bamboo food webs display
considerable overlap with co-occurring tree-hole webs.

Figure 5.9 summarises information based on my studies of the fauna in
northern New Guinea in the environs of Madang (Kitching 1990). The web
is based on surveys of the fauna of twenty-six water-filled internodes sampled by sawing off the bamboo stump below the last complete nodal plate
and emptying the contents completely.

An axil-water web

The fauna of water-filled axils formed by leaves or bracts has often been sampled but the full range, even of metazoan, organisms found within them has
seldom been examined (see Table 2.5).

Figure 5.10 presents the food web that occurs within the axils of the floral bracts of the zingiberaceaous plant *Curcuma australasica* in northern New
Guinea. Like the tree holes and bamboo webs presented in this chapter, the

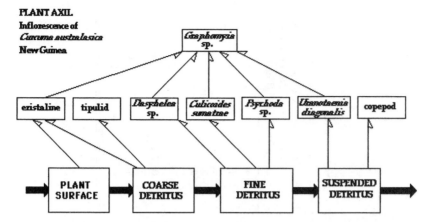

Fig. 5.10. The food web from the water-filled bract axils of *Curcuma australasica*, from Madang, Papua New Guinea (from Kitching 1990).

data on which the web is based were collected around the town of Madang during 1988. Twenty-six spikes of the plant were sampled by cutting below the water-holding section and inverting the whole into a plastic bag. Separate studies on the successional changes that occur in these food webs were made and are presented, *inter alia*, in Chapter 10.

Food-web statistics

There are a great number of ways in which food-web data of the sort described above can be summarised and rearranged for comparative purposes, some simple and some complex. I have always used the simpler measures and I present here the statistics I have used and use in this work. I include even the most obvious, not as a slur on the reader's intelligence, but by way of laying unequivocal ground rules for what follows. It is surprising how frequently, in the literature, different statistics have been presented under the guise of equivalence (see, for example, Cohen 1978, Jeffries & Lawton 1985, Closs 1991).

The various statistics I allude to are summarised in Tables 5.1 and 5.2. Table 5.1 presents most of these as they may be extracted from the five webs illustrated earlier in this chapter. Table 5.2 contains the 'summary statistics' (see below) associated with the web from water-filled tree holes in New Guinea.

Table 5.1. *Basic and derived food-web statistics from examples of five types of phytotelmata*

Statistic	Tree hole New Guinea (Figure 5.1)	Bromeliads Jamaica (Figure 5.7)	Pitcher plant Malaysia (Figure 5.8)	Bamboo internode New Guinea (Figure 5.9)	*Curcuma* axils New Guinea (Figure 5.10)
Number of species	17	23	11	13	8
Number of predators	6	5(6)	3	1	1
Number of top predators	2	0	0	0	0
Number of macrosaprophages	3	5	2	5	2
Number of feeding links	43	42	16	17	13
Number of trophic levels	3	2	2	2	2
Predator:prey	0.55	0.35	0.38	0.08	0.14
Mean prey per predator	4.8	4	5	5	2
Mean food-chain length	2.44	1.72	1.73	1.42	1.86
Connectance	0.27	0.41	0.4	0.35	0.5

Simple counts

Comparisons of food webs in space and time are most clearly made using very simple statistics extracted from the webs and the information underlying their construction. Such counts may be made from the food-web diagrams and, where many habitat units have been surveyed, may have variance terms attached to them. The first six entries in Table 5.1 present some of the most obvious and useful. I have adopted a number of conventions for resolving practical difficulties with respect to some of these quantities:

- I include both first-order and top predators in counts of 'predators'.
- Any predator acting as a top predator is so counted regardless of whether or not it also feeds at other trophic levels.
- The 'number of feeding links', an important count involved in the calculation of connectance (see below), includes the links between the saprophages and components of the detritus base.
- The 'number of trophic levels' is the number of levels of organisms actually engaged in feeding. In detritus-based systems such as phytotelmata this includes the saprophages and the various orders of predators that feed upon them. In autochthonous systems the photosynthetic organisms add an additional layer. It is a moot point how the situation involving some minor herbivory at the base of phytotelm food webs, as in some pitcher plants and axil waters, should be assessed in this regard (see for instance the food web in *Curcuma* (Figure 5.10) in which eristaline larvae may scrape away epithelial cells from the walls of the container). Where the question has arisen I have generally ignored the containing plant in trophic level counts and have regarded the whole as a fundamentally allochthonous community.

These simple rules are, of course, merely conventions and alternative usages no doubt could be justified.

Descriptive statistics

In addition to the counts which follow from construction of the food web I also draw upon the summarising statistics reflecting the abundance levels of each of the participating species. Table 5.2 presents an example of this analysis calculated for bamboo webs in New Guinea and modified from Kitching (1990). Of course such statistics can only be calculated and used if access to basic data on numbers of organisms per unit volume of sample is available. Hence, although it is often possible to construct a food web from the published

Table 5.2. *Summarising statistics for abundance levels of organisms in bamboo internodes in New Guinea*

Species	Statistic	Percentage occurrence	Mean	Maximum	Minimum	Standard deviation	Coefficient of variation
Oligochaetes		50	13.1	50.9	1.31	13.36	101.98
Copepods		23	–	–	–	–	–
Culicoides rabauli		46	19	37.9	0.3	14.19	74.68
Toxorhynchites sp.		35	1.3	2.2	0.7	0.56	43.08
Other culicids		85	19.3	67.5	0.4	19.09	98.91
Tipulodina sp.		15	3.7	7.3	0.8	2.71	73.24
Clogmia sp.		4	10.1	na	na	na	na
Calliphorids		15	3	3.7	1.7	1.13	37.67
Scirtids		11	16.5	29.1	2.6	13.3	80.61

–, abundances not estimated. na, not applicable.
Abundance levels are based on numbers per 100 ml of sample; recalculated from Kitching (1990).

accounts of the faunistics of particular phytotelmata (as in the case of the bromeliad web presented in Figure 5.7) abundance statistics cannot be deduced from the original accounts.

As is clear in Table 5.2, abundance statistics are constrained in a number of ways. First, when animals are counted as immatures and subsequently reared, the statistics can only be calculated at the resolution allowed by the level of identification of the larvae. So in the table all culicids other than *Toxorhynchites* were lumped at time of sorting. Rearing showed this group to represent three species but, of course, the larval counts cannot be accurately partitioned among the three species. Again if the study site can be revisited following the initial analysis then larvae may be able to be separated more finely. Second, the range of summarising statistics included represents a personal choice from a wide range of options. The percentage level of occurrence is included here for convenience: it is not strictly a summary statistic reflecting presence or absence rather than abundance, but does indicate the frequency of occurrence of a species across habitat units examined. I include the coefficient of variation ([standard deviation × 100]/mean) in my summaries as a useful way of weighting the variance measure used to reflect the general levels of abundance characteristic of each species.

Derived statistics

In addition to the simple counts and the statistics summarising levels of abundance, a number of indices are commonly used to capture important variation within and across food webs. The following measures are used elsewhere in this work.

Predator:prey ratio

The ratio of the number of predatory species to the number of prey species represents an important emergent property of any community. It is readily calculated from simple species counts but does require formalisation when top predators are present. The inherent problem revolves around the question: should one include predatory species within the count of 'prey' when these species are used as food by top predators? If the answer is 'yes' then a primary predatory species that is also a prey item for a top predator will be counted in on both sides of the ratio and will have no net effect on the calculated quantity. For this reason I calculate predator:prey ratios as:

$$\frac{\text{number of species of predators (including top predators)}}{\text{number of species of primary consumers}}$$

Mean prey per predator

A useful measure for cross-web comparisons is arrived at by averaging the number of prey species that are used by each of the predators. This is a measure of the degree of specialization (or otherwise) of the predators forming part of the web. It may usefully be weighted sometimes by the total number of species in the web. The mean value will also, of course, have an associated variance term which may be useful in some cases.

Average food-chain length

Each of the potential 'vertical' paths through a food web represents a food chain. Accordingly for any web we may calculate the mean food-chain length by averaging the number of links present in all chains present in the web. This is clearly illustrated in Figure 5.11 for a number of different hypothet-

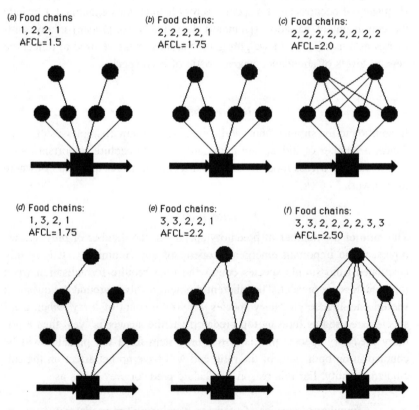

(a) Food chains
    1, 2, 2, 1
    AFCL=1.5

(b) Food chains:
    2, 2, 2, 2, 1
    AFCL=1.75

(c) Food chains:
    2, 2, 2, 2, 2, 2, 2, 2
    AFCL=2.0

(d) Food chains:
    1, 3, 2, 1
    AFCL=1.75

(e) Food chains:
    3, 3, 2, 2, 1
    AFCL=2.2

(f) Food chains:
    3, 3, 2, 2, 2, 2, 3, 3
    AFCL=2.50

Fig. 5.11. A series of imaginary food webs showing the calculation of the statistic 'average food-chain length' (AFCL).

ical web structures made up of six nodes at two trophic levels (Figure 5.11a–c) or seven nodes at three trophic levels (Figure 5.11d–f). Within each set of three the left-hand webs are minimally connected (Figure 5.11a, d) and the right-hand webs are maximally connected (Figure 5.11c, f). The remaining two webs represent intermediate situations. There is a complex, non-linear relationship between the value of the average length of food chains calculated in this fashion and the three variables: number of species, number of trophic levels and number of trophic links. The exact relationship need not concern us here; suffice it to say that the statistic provides a useful summary of food-chain length. Of course like any average it will have an associated variance term.

Connectance

Finally among statistics derived from food webs the commonly used quantity connectance may be calculated. Pimm (1982) defines this simply as 'the actual divided by the possible number of interspecific interactions' within a web. I have assumed that 'possible' in this definition assumes the qualifiers: 'given a particular distribution of organisms among trophic levels, and that top predators may act as predators within any web'. This common measure of web complexity is readily understood graphically. Consider Figure 5.12 in which three hypothetical webs are presented. Each has an identical number of nodes at a fixed number of trophic levels, but each displays a different degree of connectance. The web in Figure 5.12a is minimally connected: any fewer links and at least one member of the community would drop out. In contrast the web in Figure 5.12c is maximally connected and, hence, has a connectance value of unity. The web in Figure 5.12b is deliberately drawn to present an intermediate value.

Turning to Table 5.1 it is clear that the level of connectance within a web reflects firstly the overall size and 'height' (that is: the number of trophic levels) of the web and, second, the extent to which the participants are generalists or specialists. Thus the complex tree-hole web (Figure 5.1) contains potentially many possible trophic links reflecting its complexity, but each of the predators utilises only a subset of the potential prey items. In contrast the much simpler plant-axil web (Figure 5.10) which contains but a single, generalist predator and no top predators displays the highest connectance value of the five webs summarised in the table.

Diversity measures

In general I have not used measures of α-diversity in analyses of food webs. Where considerations of species richness and evenness (the component parts

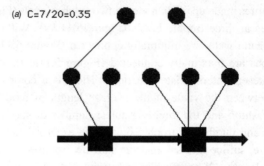

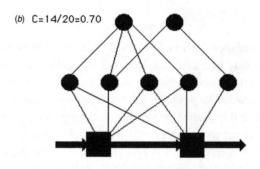

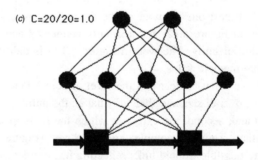

Fig. 5.12. Three imaginary food webs with the same species richness but contrasting levels of connectance.

of most diversity indices) have been important I have used these explicitly rather than in indicial form.

Whereas considerations of α-diversity require measures of community composition within a sampling unit (however defined) indices of β-diversity calculate the similarity (or difference) between the composition of any two samples, A and B. As in the case of α-diversity these may be calculated simply using numbers of species, or by taking account of the relative abundance of each species within the samples. In the present work I use only the simplest of these, based solely on the co-occurrence or otherwise of particular species across pairs of food webs. For this purpose I have used the familiar Sørensen's Index (C_S) calculated as:

$$C_S = 2j/(a + b)$$

where a is the total number of species found in sample A, b the total number in sample B, and j the number of species shared between the two samples. This metric varies between 0 and 1 in value with the unit value representing exact coincidence in the species occurring in the two samples that are being compared.

6

Processes structuring food webs
Explanatory theories and other ideas

Examine with me the fauna inhabiting a tree hole in the subtropical rainforest of south-east Queensland. First we must find a suitable tree: in general, in this forest, water-filled holes are commonest among the interdigitating buttress roots of any one of half a dozen species of tree. We should seek mature canopy trees to maximise our chances of finding such sites. The younger trees are not subject to the same forces of stress due to wind sheer and so do not have buttresses or water-filled hollows. Trees that are senescent have often had such holes during their mature years but now are old and in partial decay. The integrity of their bark is violated and most hollows formed are no longer watertight.

Having found a water-filled tree hole, we remove its contents and sort them into their various kinds and so construct a food web following the principles set out in Chapter 5. We can then calculate the various statistics that may characterise the community in this tree hole at the moment of its sampling. Of course the adjacent tree holes, although within the same patch of forest, are unlikely to possess exactly the same faunal components at the same levels of abundance. If we return to our selected tree hole at regular intervals we will see the species composition of the community change from time to time and the relative abundances of each faunal component will also vary. To restate the obvious, the food-web structure within the tree holes in this patch of forest varies in space and time.

A simple study such as that described here may stretch over a few hectares of forest and a small number of years of study. But other spatial and temporal scales are of interest to us as community ecologists.

Thinking of the distribution of organisms in space we may ask a number of questions. How do the tree-hole food webs in this patch of forest compare with those in adjacent forest patches? How do the food webs we identify in south-east Queensland rainforest compare with those in rainforests in Tas-

114

mania, northern New South Wales, central or far northern Queensland? How do the Australian webs compare with those in other parts of Australasia, say, or other biogeographical regions of the world? In practical terms spatial questions are attractive because we can actually go out and observe the food-web structures at different places, and observations of this kind underpin many of the discussions in ensuing chapters.

On the temporal scale we may enquire how the community has changed over the tens of thousands of years of post-glacial climate change as rainforests of one sort or another have spread and retracted amoeba-like across the face of the southern continent; and, we may ask how does the presence or absence of particular species within tree-hole communities reflect the recent evolutionary history of the group to which they belong. From a practical point of view the range of temporal scales gives us greater pause as field ecologists. We may measure changes that occur over time scales within that determined by human longevity, but we must rely upon inference and 'thought experiments' in attempting to identify changes occurring on greater time scales.

The essence of ecological inquiry has often been identified as comprising two parts: pattern and process. We measure pattern, and we infer process from the patterns. We proceed, where we can, to establish experimental tests of the importance of the processes we have identified as relevant in generating the patterns we observe. The information we collect on the variation in food-web structure in phytotelmata across the range of spatial and temporal scales to which I have alluded already is the *pattern*. The challenge then is to identify and test the underlying *processes*.

This chapter has a number of interconnected, even overlapping aims. It pretends to be a set of theoretical statements providing the context for all subsequent chapters. In this chapter, I attempt to identify the potential behavioural, populational, ecosystematic, biogeographic and evolutionary processes, both deterministic and stochastic, that may be involved in structuring food webs in phytotelmata across the full range of relevant spatial and temporal scales. The processes involved include some commonly associated with 'community theory' such as competition and invasion, alongside others that have received less attention in this context such as chance encounter or environmental predictability.

What few have attempted (with good reason, as may become evident) is a synthesis of processes across the many magnitudes of scale involved. And yet only such a synthesis will provide the proper context for interpreting any particular observation, and avoid the errors of assuming generality for a favourite process or set of processes out of their spatio-temporal context. In addition, data sets collected from phytotelmata are almost unique in providing comparative data across these many scales against which such a synthesis can be measured. Of course much remains to be clarified, but the combination of theory and data that I present also identifies key gaps in current knowledge and will act as a guide for future data collection.

In order to simplify what is a very complex situation I join other commentators in selecting a set of points in space and time for special attention. These represent artificially defined discrete points along the two continuous axes of space and time. I attempt to identify particular processes which structure the pattern we observe in food webs at particular scales in space or time. Sometimes there is a clear relationship between a particular scale selected and a particular ecological or evolutionary process: more commonly, particular processes have a spatial and/or temporal domain of operation which spans more than one of the points selected along the space or time axes. To complicate matters further some processes will act pre-eminently at a particular scale but will still play a role, albeit a less significant role, at one or more adjacent scales. Accordingly, to summarise the set of candidate processes is a challenging task!

Food-web theory: pattern, process and progress?

The theoretical aspects of food-web structure have been among the most active and vigorous aspects of modern ecology. This modern interest in the subject stemmed initially from the work of Gilberto Gallopin (1972), elaborated extensively by Joel Cohen (initially in 1978). During this period of a revival in interest in food webs key contributors were few. Robert May and Stuart Pimm were pre-eminent. Pimm's early work (e.g. 1979, 1980) and key works in collaboration with John Lawton (Pimm & Lawton 1977, 1978, 1980) culminated in the seminal book *Food Webs* (Pimm 1982). In parallel to, and interdigitating with, these works May presented a series of results related, initially, to the complexity/stability dynamic in ecosystems (May 1972, 1973, 1974) proceeding to a more explicit involvement with food-web properties in later work (1983).

Interest has grown out of these pioneering works to the point where, currently, a dozen or so authors contribute repeatedly to the field; many others participate from time to time. Most who write on the theoretical aspects of the subject use data collected by others but collated by them into a data base which has grown to contain a hundred or so webs which are reliably documented and appropriately verified. Early episodes in the development of the field revealed the shortcomings of many sets of data which, in mitigation, had generally been collected in quite different scientific contexts (see, for example, Martinez 1993). Currently accepted webs are generally available through data-sharing agreements and mechanisms of one sort or another (Cohen *et al.* 1993, Martinez 1993).

What I attempt here is not a comprehensive review of the field. Others have performed this service from time to time (Pimm 1982, 1992; Pimm *et al.* 1991; DeAngelis 1992) and their works should be consulted for more complete treatments. Instead I make comments upon some aspects of the body of current food-web theory which bears upon my subsequent remarks and analyses. Accordingly many worthy contributions are omitted: this in no way represents a dismissal of their importance or historical role within the development of the field.

Much effort has been spent on the identification and championing of emergent properties of the data sets which derive from studies of food webs. Among others these so-called scale invariant properties have included constant predator:prey ratios (Cohen 1978, Briand & Cohen 1984), constant ratios of the number of feeding links and the number of species in the web (Cohen 1978, Martinez 1992), and statements on the generalisation of predators and the vulnerability of prey (Havens 1993, Schoener 1989). Schoenly *et al.* (1991) added a new dimension to the debate by analysing insect-based food webs according to habitat type. In addition generalisations have been made concerning the role of omnivores (Pimm & Lawton 1978) and the relative physical size of species occupying different trophic positions within webs (Southwood 1977, Colinvaux 1978). Some have concerned themselves with aspects of web topography (Sugihara 1984) and stability properties of mathematical constructs proposed as models of food webs (Pimm 1979).

In each of the cases identified above, and others that I do not discuss, the validity of the proposed generalisation has usually been supported by analysis of some subset of the available published data. In almost all cases the particular generalisation has been disputed and debated vigorously, as is appropriate in an active field of science. Sometimes this debate has touched on the vitriolic but in most cases progress has been made, generally as more and better data sets have become available. Seldom have data been collected specifically to test pre-stated hypothetical generalisations.

Several of the most robust of these empirical generalisations have been the subject of a more integrated approach in which simple assumptions of general trophic patterns have been used to construct models of food webs, in turn to generate emergent properties. The most exciting and extensive example of this in recent years has been the so-called Cascade Model of Cohen and his colleagues (Cohen & Newman 1985; Cohen *et al.* 1985, 1986; Newman &

Cohen 1986) which generates two empirical 'general laws' on the basis of a third such 'law'. So the Cascade Model allows the derivation of the relative proportions of species at different trophic levels ('the species scaling law') and the proportions of trophic links at each trophic level ('the link scaling law') from the ratio of links to species within the web ('the link-species scaling law').

These generalisations concern themselves with *pattern* only. If their existence can be sustained, they join a distinguished set of comparably derived empirical patterns from Boyle's Law to Cope's Rule (with many stops in between), all of which play a major role in their respective sciences (physics and biogeography in the two exemplars presented). They have the potential, once established, to allow us to predict community properties from, for example, a simple estimate of species richness and this may well prove to be a valuable tool in biodiversity management for conservation (Havens 1994). But the identification of such general patterns is only part of the body of knowledge which must form the theory of any part of science. As indicated earlier (and it certainly bears repetition), any pattern is generated by underlying *processes*, and general statements concerning these processes must complement those made with respect to pattern. Of course a particular pattern may be generated by more than one process, either separately or in combination. In many cases only painstaking experiment will differentiate among candidate processes. The potential production by a single process of more than one pattern outcome (as in the branching processes of chaos theory – see May 1975, Gleick 1987) is a greater challenge to field ecologist and statistician.

Perhaps curiously, recent treatments of process within the literature on food-web theory have been much less common than those relating to pattern (although, admittedly, the distinction between the two can easily become blurred). An appropriate theoretical literature does exist but, for the most part it sits within theoretical population ecology rather than within community ecology.

So, in an enormous literature, considerations of competition and predation, resource limitation, nutrient dynamics and ecological energetics have been invoked (e.g. Hutchinson 1959, Watt 1968, MacArthur 1972, Schoener 1974, Whittaker 1975, Cousins 1987, DeAngelis 1992) to explain that phenomenon which we know as the ecological niche of any organism. More rarely this approach has been integrated to consider multiple co-occurring organisms as problems in species 'packing' along particular resource dimensions (Pianka 1975, MacArthur 1972, Price 1975 and references therein). Such approaches seldom contemplate more than two trophic levels at any one time.

A growing number of writers, most notably Stuart Pimm, have examined the emergent properties of food-web processes at the level of the community itself, considering the integrated and integrating mechanisms which may produce observed food-web structures (Pimm & Lawton 1977, Pimm 1980, 1982, 1992; Briand & Cohen 1987). Even in these cases only a subset of potentially important processes has been considered. So, in a spatial context, processes acting at, say, grand biogeographic levels or extremely local levels have not generally been considered. And yet, as we shall see, continental drift or chance encounter may have as much to do with the food-web structure we observe in a particular case as do multi-species population dynamics. On the temporal scale, changes across millennia such as local species radiations, seral changes within habitats such as serial modification of the resource base, or changes within days such as successful discovery and oviposition, may be as important as the 'few-to-many generations' time scale of population dynamics.

So, there is an imbalance in food-web studies, in my view, between the current investment in theoretical investigation of pattern and that of process, in favour of the former. I am by no means the first to make this observation and the sympathies I have expressed above are essentially coincident with or complementary to those expressed by Cousins (1987), Lawton & Warren (1988), Paine (1988) and Polis (1994). For me, as for others, 'good' theory must produce predictions which can be tested in manipulative or 'natural' experiments; these must be established or sought *after* the development of the theory on which the predictions are based. Food-web studies are only slowly entering the experimental phase. We have shown that such manipulations can be done using phytotelmata as experimental systems (Pimm & Kitching 1987, Jenkins *et al.* 1992); others are making comparable experiments at much grander scales (e.g. Carpenter *et al.* 1987, 1992, 1994).

In all cases known to me, actual manipulative experiments have been restricted to local or regional spatial scales, on short to medium time scales. There are very good reasons for this: it is very difficult to carry out manipulations on global scales or in evolutionary time frames. For these last purposes we must seek the grander experiments carried out by nature and live with the uncontrolled variation that we strive to eliminate in manipulative experiments. Of course, such 'natural' experiments are also a vital part of our analyses even in those cases where we can envisage manipulations.

So what I attempt to establish in what follows is, on the one hand, an identification of processes which will structure food webs at a variety of scales (remainder of this chapter) and, on the other, sets of phytotelm data which make some progress in the testing of the ideas that emerge (remainder of the book).

Structuring mechanisms
Appropriate scales

Before we can launch into a review of processes operating at various scales we must first select the scales. However one does this, the choice will be a personal one and I make no apologies for the choices made. Nevertheless I have included among my choices a subset of spatial and temporal scales which have been used, implicitly or explicitly, by other authors.

Any spatial analysis includes considerations of how organisms, singly or in sets, get from place to place. Such processes may be passive or active and may engage time scales from the immediate to the evolutionary – in other words from weeks and months to millennia.

Spatially I focus on phytotelmata at the *global* scale (10^3–10^4 kilometres, the *continental* scale (10^2–10^4 kilometres), the *regional* scale (10^1–10^2 kilometres) and at the *local* scale (10^{-2}–10^1 kilometres) (Figure 6.1).

Temporal scales essentially invite thought about why a particular species or set of species is here now as opposed to some time in the past (or projected future). It demands consideration of change and a consideration of phytotelm communities as dynamic objects that increase or decrease in complexity through time.

Following biological precedents I address temporal change at the level of

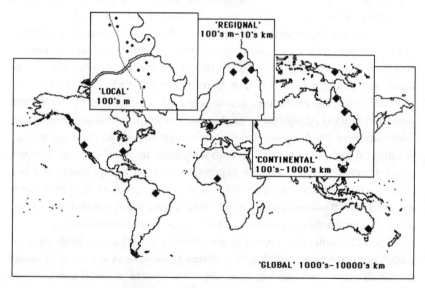

Fig. 6.1. World map illustrating the spatial scales designated in the text. The 'sampling' points shown are notional only.

the *evolutionary* (10^4–10^6 years), the *ecological* (10^{-1}–10^3 years depending on the habitat type), and the *behavioural* ($<10^{-1}$ years). The missing time scale (10^3–10^4 years) is, traditionally anyway, ignored by biologists and yet it is the time scale at which climate change becomes apparent. Perhaps, with tongue-in-cheek, we may call this a *millennial* time frame: although giving it a name does not help us either understand or cope with change at this scale.

The global scale

The animal species that occur in any phytotelm food web, regardless of their trophic position, are drawn from the biogeographically determined set of available biota. It is not likely that we will find organisms representative of South American families, for instance, occupying food webs in any habitat in, say, South-east Asia. These biogeographic sets impose broad limits on what may or may not be present within a particular food web, and processes acting to produce food-web patterns on smaller scales of both space and time do so only within the limits set by the composition of these biogeographic sets. This truism nevertheless turns our attention to any or all of the large-scale processes of biogeography which result in the presence of particular taxa in any location.

Continental drift

The most widely discussed of such processes is undoubtedly continental drift. The serial break-up of the Cretaceous supercontinent Pangaea divided the fauna of the time into a number of segments which, subsequently isolated, have undergone long periods of more or less separate evolution. This process has produced a set of endemic faunas associated with the various land masses which have come into existence since the initial break-up of the supercontinent. So we talk freely of Gondwanic biotas, or Neotropical endemics, or Malagasy specialists.

The fundamental limitations on the compositions of faunas resulting from their tectonic history is translated into situations where specialised families occur within phytotelmata but only over restricted geographic regions. Examples include the neurochaetid flies of the Pacific region and the peculiar distribution of the genus *Metriocnemus* within the Chironomidae: Orthocladiinae. Among vertebrates the mantellid frogs are highly characteristic of axil waters in Madagascar but are restricted to that island. Such restriction presumably reflects the geological history of the island of Madagascar, one of the first blocks of land to break free from the Gondwanan continent in the late Cretaceous.

Plant distribution patterns

As described in Chapter 2 some classes of phytotelm are restricted to particular plant taxa: some may be defined by the plants in which they occur (in
fact, all except tree holes are more or less restricted in this fashion). Hence
the biogeographic history of the container plants themselves will play a role
in determining food-web structure. In a trivial sense a phytotelm food web
cannot be present if the container plant is absent. So, for instance, there are
no pitcher plants found on the continent of Africa, and only very restrictive
occurrence of pitcher plants within Australia. Bromeliads are essentially
American plants and their special fauna is not to be sought elsewhere –
although in this case the proclivities of gardeners to move plants around the
world, and that of mosquitoes and their kin to exploit any available water
body, enabled Thienemann (1934), for example, to talk about bromeliad faunas in the Bogor Botanic Gardens.

Metahabitats

In addition to the presence of a particular habitat type within an ecosystem,
the richness of a particular phytotelm food web may reflect, among other
things, the set of other phytotelmata in the same region. An incomplete overlap does occur between the faunas of tree holes and pitcher plants, bromeliads and axil waters and so on. The set of different phytotelmata in any one
of them forms what we may call a metahabitat (by analogy with the commonly used term 'metapopulation'). The larger the metahabitat the greater
the range of phytotelm species that will occur in any location and the more
complex the food web in any one of them is expected to be.

Adaptive radiation

The fundamental patterns of distribution of organisms, plants and their associated faunas are generated by continental drift but, more often than not, these
patterns are modified by the action of other biogeographic processes, particularly at the edges of regions. So, for example, as the Australasian tectonic
plate approached, and ultimately collided with the complex of plates which
are South-east Asia, so the possibility of faunal exchange arose by the ecological process of dispersal. Accordingly we may expect any pattern produced
by continental drift to be 'fuzzy' at the edges. Such pattern may be further
complicated by more or less extensive adaptive radiation producing clouds
of closely related species at locations other than that at which the clade itself
originated.

Within certain taxa accelerated speciation events may have occurred over millions of years leading to exceptional diversity of the closely related descendent taxa within particular regions. These speciation events, although occurring independently of plate tectonics, will nevertheless increase the species pool from which the fauna of particular communities within phytotelmata may be drawn. For example, within the Old World tropics the filter-feeding mosquito larvae of the genus *Tripteroides* have been the product of evolutionary radiation to produce a diverse and widespread group of organisms. Many of these organisms occur in tree holes, pitcher plants and other container habitats in these regions. The genus is absent from Western Europe and the North American continent and hence, not surprisingly, is unavailable to play any role in the structure of food webs in these situations.

Latitude, synoptic climate and productivity

At the global scale, latitude may be expected to be a good predictor of species richness and food-web complexity. Latitude itself, of course, is unlikely to be the causal factor involved but is a correlate of synoptic climate and total productivity. Aspects of these phenomena we may reasonably expect to be related causally to food-web structure.

Synoptic climate, summarised generally in terms of the temperature and rainfall regimes, may have both direct and indirect effects on phytotelm organisms.

The temperature regime will inevitably affect the length of the growing season for any poikilotherms – and all phytotelm animals fall into this class – and that of the container plants themselves. Short seasons will tend to restrict the fauna to either short-lived species which can complete at least one generation in a short time or longer-lived species which can overwinter in a more or less quiescent state. In either case the strictures imposed by such an environment will act as a selective filter excluding a range of taxa which have not evolved such adaptations. Temperature along with rainfall and soil type will also have a direct impact on ecosystem productivity in the environment surrounding any particular phytotelma. This will be reflected in detritus production and, hence, in the magnitude of the allochthonous energy base available within the habitat unit. A number of authors have discussed the potential impact of such productivity on food-chain length and complexity (Hutchinson 1959, Pimm 1980, 1982, Kitching & Pimm 1985, Pimm & Kitching 1987, Briand & Cohen 1987). In general we would expect to encounter longer food chains in more productive environments simply because more basal energy is available within the food webs of which the chains are part. That this is not a clear-cut pattern within food webs (Pimm 1982, Kitching & Pimm 1985) is more an indicator of the presence in ecosystems of further, modifying processes which mask the simple energetic effect than it is evidence of the invalidity of the generalisation.

In the vast majority of cases the water body within phytotelmata is the direct result of the input of rainfall either directly or as run-off. In the few cases where this is not the case, the water body is partly at least the product of ground water drawn up by the plants and secreted into the phytotelm cavity. In both cases the ambient rainfall regime has the potential to affect directly the environmental quality within the habitat unit. Indeed water-filled plant parts are simply not present in exceedingly dry environments: in others, they may be highly seasonal.

Rainfall of course will also affect environmental productivity, as does ambient temperature (above) and indirectly the energetic base within phytotelm units that occur in the vicinity.

To this point I have discussed the potential impact of climate in terms of the magnitude of the incident heat and rainfall on the food web within a plant container. The actual patterns of temperature and rainfall may have an impact separate from that produced by the mean levels of heat and water input alone. In other words environmental predictability may be a factor in itself. Pimm & Lawton (1977) postulated that environmental predictability may have a direct effect on the equilibrium structure of food webs determining the 'permissable' complexity in any particular place. The temporal pattern of primary productivity around habitat units, determined, as already suggested, by rainfall and temperature, will also contribute to the integrated variable 'environmental predictability' through the variable pattern of energy availability within habitat units throughout one or more seasons. Pimm & Lawton (1977) demonstrated that for model food webs the time taken to return to equilibrium increased with increasing complexity. They argued that if natural food webs behave in a similar way then we would expect complex food webs with long food chains only in the more predictable environments. The longer return times associated with these complex food webs would preclude the long-term persistence of such communities in environments in which the average period between perturbations was greater than this return time.

Lastly, the synoptic climate at different latitudes in conjunction with aspect, substrate type and altitude (see below) plays a major role in determining the predominant vegetation type in any place. This in turn will determine the availability of any particular phytotelm type and, perhaps more important, the frequency of occurrence and range of such habitats in any location. The fact that most phytotelms occur in plant species associated with mesic forest environments points up the importance of whatever factors permit the existence of forest in particular locations.

Box 6.1 summarises the predictions which follow from identification of these processes operating at the global scale.

Box 6.1 Processes influencing food-web structure in phytotelmata at the global spatial scale: hypotheses and predictions.

Note: some of these hypotheses may be be mutually incompatible – where this is the case they may be regarded as alternatives.

Continental drift

H1. Hypothesis: Elements of the fauna of phytotelmata in a particular geographic location will reflect the tectonic history of the location.

P1. Prediction: Some faunal elements will be unique to particular locations at levels higher than the species.

H2. Hypothesis: The fauna of phytotelmata will include faunal elements from adjacent biogeographic areas due to dispersal.

P2. Prediction: Examples of particular phytotelm types close to the boundary between two biogeographic regions will contain elements not found further from the boundary.

Host plant biogeography and metahabitats

H3. Hypothesis: The foodwebs found within particular plant groups will reflect in part the biogeographic history of the container plants.

P3. Prediction: Where a particular plant taxon has a well-defined range, the contained food web may vary from the middle to the edge of the range of the taxon.

H4. Hypothesis: The richness and complexity of a food web within any one phytotelm type will be increased if a range of other phytotelm habitats is also present.

P4a. Prediction (1): Food webs within a particular phytotelm type will be more speciose and complex when a range of other phytotelm types is also present.

P4b. Prediction (2): Where a set of phytotelm types occur in a particular place there will be at least a partial overlap in species composition across phytotelm types.

Box 6.1 (cont.)

Adaptive radiation

H5. Hypothesis: Regardless of other patterns, phytotelm communities in general will have a 'local' flavour, reflecting the occurrence of some taxa particularly associated with the land mass or region on which they occur.

P5. Prediction: The most speciose clades within the higher taxa (in general) occurring in phytotelmata will generally be those best represented in phytotelm communities, and these will differ from biogeographical region to region.

Latitude, synoptic climate and productivity

H6. Hypothesis: Food webs respond in richness and complexity to the environmental gradients correlated with latitude.

P6. Prediction: There will be clear gradients in food-web statistics with latitude, higher latitudes having simpler webs.

H7. Hypothesis: Production is correlated with latitude, and higher levels of productivity allow for the existence of more complex webs, containing longer food chains.

P7. Prediction: Foodwebs in more productive areas will be more complex with greater average food-chain lengths.

H8. Hypothesis: More predictable environments allow more complex food webs to occur, containing longer food chains.

P8. Prediction: Environments with climate or other environmental variables that are less variable should contain more complex webs and longer food chains.

The continental scale

The processes which act to determine food-web structure at the continental scale will include several of those already identified as relevant to the global scale: latitudinal changes in climate, and in environmental predictability will, in general, be in evidence at the continental scale, as will changes in ecosystem productivity. Further, where a particular group of plants defines a class of phytotelms, the food web that occurs within them may well show features which reflect the within-continent distribution and history of the plants concerned. Further, the richness and complexity of food webs will be related to the number and range of phytotelm habitat units in particular places.

The finer grain of continental patterns

Global processes will act at the continental scale, but superimposed on the resulting patterns will be variation reflecting the finer grain of variation within continental boundaries. Higher taxa representing the end products of within-continent radiations will be 'over'-represented in food webs within the continent. In addition the topography of the continent will produce 'special' distributions of particular biomes across the landscape which, inevitably, will affect the distribution of phytotelms and their predictability as a habitat for animal species. Further, environmental variation and its various correlates will exist along more subtle axes related to, for instance, altitude, rain shadows and distance to coastlines, which complicate the simple latitudinal model. In particular, factors determining the distribution of rainforests on the continental scale will have a substantial effect on patterns of phytotelm occurrence and the nature of phytotelm food webs simply because plant containers are so well represented in rainforest ecosystems.

Climate change and 'species' pumps

The interaction of evolutionary radiation and climatic change over a time scale of a few to a moderate number of millions of years have been called upon to provide an explanation for why the moist forests of the world – in the northern Neotropics, West and Central Africa, and the Indo-Australian region – are so highly diverse. As cyclical climate change occurs over many millions of years so patches of tropical forest advance and retreat across the landscape. At periods of low temperature, as during the Ice Ages, they remain as islands in a landscape of other vegetation. This island situation isolates sets of organisms which then evolve in isolation, potentially allowing for the evolutionary divergence of species that formed a single panmictic species prior to being isolated. Then, as the climatic cycle moves on to restore more equable, warm and moist conditions, so rainforest patches spread out and regain contact one with the other. The flora and fauna of the previously isolated areas now mix, producing a highly diverse and speciose combined biota. Periods of expansion and contraction of the forest, together with processes of speciation occurring within isolated patches, have been described as a 'species-pump' mechanism – just one of a number theories available to account for the very high species diversity that we observe in rainforest. Whatever the reason this is also one of the explanatory tools available to us for accounting for the increased speciosity and complexity of organisms in food webs in phytotelmata from tropical rainforests compared with subtropical or temperate forests.

Patterns of habitat segregation and inter-specific interactions

Within taxa sharing broadly similar biologies, it is at the continental scale that we observe broad patterns of habitat segregation. The evolved behavioural activities which produce the observed patterns of habitat segregation, of course, occur at a much more local scale. Nevertheless I suppose that the evolution of such patterns reflects processes which have occurred on a much broader spatial scale.

It is also at the continental level that we may expect to see the impact or lack of impact reflecting the presence of particular predators. So returning to the situation with tree-hole mosquitoes in North America discussed in Chapter 12, it is clear that the presence or absence of larvae of the keystone predator *Toxorhynchites* have a major impact on food-web characteristics within tree holes. But the distribution pattern of *Toxorhynchites* reflects climatic and topographic changes on the continental scale. The genus of mosquito is absent from northern and western North America and so it is again a truism to point out that whatever formative effect it may have on tree-hole food-web structure cannot occur in those regions.

All this having been said, food-web variability at the continental scale is still likely to be explicable in terms of biogeographic and ecosystematic processes interacting with physiological factors at a 'coarse' scale.

The set of additional hypotheses and associated predictions that emerge from considerations at the continental scale are summarised in Box 6.2.

***Box 6.2* Processes influencing food-web structure in phytotelmata at the continental spatial scale: hypotheses and predictions**

Hypotheses already noted in Box 6.1, which relate to the global scale of organisation as well as the continental scale, are *not* repeated here.

Finer grain of continental patterns

H9. Hypothesis: Long-distance latitudinal patterns will be overlain and complicated by other environmental gradients such as rainfall patterns, altitude, distance to coastline.

Box 6.2 (cont.)

P9. Prediction: Significant proportions of the variance associated with the values of food-web statistics along continental gradients will be accounted for by variables other than latitude.

Climate change and species pumps

H10. Hypothesis: Phytotelm communities in ecosystems which have been subjected to pulsed climate change, which nevertheless allowed patches of mesic forest to persist, will be richer than those in areas where climate change has been either less periodic, or more severe.

P10. Prediction: Tropical rainforests should contain the richest and most complex of food webs within particular types of phytotelmata.

Habitat segregation and interspecific interactions

H11. Hypothesis: Parallel evolution of sets of ecologically similar species within a single land mass will lead to clear patterns of habitat preference and consequent segregation.

P11. Prediction: Where a particular, ecologically uniform taxon is diverse within a phytotelm type we expect to see marked differences in behavioural patterns in adult insects leading to (at least partial) segregation in space within particular ecosystems.

H12. Hypothesis: The presence or absence of 'keystone' predators within phytotelmata will have major qualitative and quantitative impacts on the food webs they contain.

H12. Prediction: Dramatic differences in food-web structure will occur across the boundary of the geographical ranges of keystone predators when these are not uniformly dispersed over an entire continent.

The regional scale

The regional scale of variation is the one at which I expect population dynamic and island biogeographic processes to be most significant in determining food-web structure. In particular the metapopulation features of the participating species will be of great importance.

Metapopulation dynamics

All populations of phytotelm organism will occur in a patchy fashion across the landscape. This patchiness will reflect the local availability of habitat units as well as the history of success or otherwise in invading the region. Over a period of time populations of any species will come and go as the variety of processes which cause local extinction change in importance in that particular segment of the landscape. As with all other populations the long-term health of the species is reflected in the dynamic interplay between extinction and re-establishment of populations within the more widely spread metapopulation which exists at the level of the landscape. The concept of 'metapopulation' is discussed further in Chapter 9.

Island biogeographic processes

Even when a selected location is within the full biogeographic range of a species, the actual occurrence of that species at the location will be a response to a set of processes different from those that define the geographical range itself. The presence of individuals of a species at any location at the regional scale depends on the presence or absence of a population of that species at the location, which in turn reflects the outcome of the processes of invasion, survivorship and local extinction which come under the purview of island biogeographic theory (MacArthur & Wilson 1966).

The likelihood of establishment of an invading species, in turn, will reflect the availability of suitable habitat units or even the variety of habitats available in a particular location. This will be affected by a range of variables including the size and age of the patch of, say, forest, in which the habitat units occur.

Interspecific interactions

Processes which operate on the behavioural time scale will also be important at this regional spatial scale. Species–species interactions such as competition, predation and mutualism may well play a role in structuring phytotelm food webs (see Chapters 8, 9 and 12). Two-species interactions demand the presence of both species in the pair. And the occurrence of both species in any particular place and time will reflect not only the chance processes of encounter or otherwise of a particular habitat unit but also the coincidence of healthy populations of each of the players in the two-species interaction in that place.

Hypotheses and associated predictions based on this discussion of processes at the regional scale are summarised in Box 6.3.

Box 6.3 **Processes influencing food-web structure in phytotelmata at the regional spatial scale: hypotheses and predictions.**

Interspecific interactions are covered in Box 6.2 (H11, H12).

Metapopulation dynamics

H13. Hypothesis: Populations of phytotelm organisms and the food webs within which they occur will vary at the scale of the landscape, reflecting population-level processes of extinction and establishment.

P13. Prediction: Populations of selected species will occur in some, spatially defined, subsets of phytotelm units and not in others: local extinctions may occur and these patterns may be more marked in rarer species such as predators.

Island biogeographic processes

H14 Hypothesis: The presence of particular phytotelm animals within a patch of forest (say) will reflect the island biogeographic processes of invasion and extinction.

P14. Prediction: Phytotelms of particular types occurring in larger, less isolated ecosystem patches should have richer faunas and consequent food webs that those occurring in smaller and/or more isolated patches of the same ecosystem type.

The local scale

It is at the local spatial scale (of metres rather than kilometres) that we observe food-web variations from unit to unit of a particular phytotelm habitat: we need to think about processes which produce the difference we observe from tree hole to tree hole, bromeliad tank to bromeliad tank, or pitcher to pitcher. Three sets of processes operating on different time scales and/or in different classes of phytotelms will be especially important: succession, seasonality and the set of processes subsumed under the heading stochastic discovery and survival.

Succession

Succession within phytotelm units occurs over the ecological segment of the general time scale; that is, over a period of months to a small number of years. It will be of particular importance in relatively short-lived phytotelms such as pitchers or leaf axils in which new habitat units are continually forming, maturing and senescing. Accordingly, at any point in time a study of the food webs within these habitat types would almost inevitably lead to encounters with all stages in any successional sequence occurring in food-web structure and composition. In contrast, long-lived habitat units such as water-filled tree holes will exhibit successional changes less frequently: either following some catastrophic set-back to a pre-existing community within the habitat unit or, even more rarely, during the period following the formation of a new unit of habitat. In either case structuring processes will occur as the food web builds up within the habitat unit.

The 'traditional' models of succesion in the literature relate to patterns of vegetation change following the opening up of a new bare area of soil for colonisation. The different models which have been proposed for this set of processes have been reviewed by Connell & Slatyer (1977). I also base this brief account on the very clear review present in the text by Begon *et al.* (1990). Essentially these authors compare, contrast and, ultimately, combine the processes of facilitation, inhibition and tolerance, each of which has featured as the driving force in earlier models. The oldest of these relies on the process of facilitation (Clements 1916). Under this model the presence of particular species within the community opens the way for other species to invade and become established later in the successional sequence. Such modification of the environment may be as simple as the establishment of prey species being a necessary precursor to successful establishment by specialist predators, or it could be the much more subtle chemical or physical modification of the environment by the feeding or other habits of the precursor species. In contrast, inhibition-driven succession envisages an exclusion process in which early invaders prevent, or substantially reduce, invasion by later slower growing species (Sousa 1984, Bazzaz 1979). Lastly the tolerance model relies on the ability of later invaders to tolerate the presence of earlier invaders, eventually shading them out (Connell & Slatyer 1977).

All of this theoretical background and the very extensive literature (see, for example the volume edited by Glenn-Lewin *et al.* 1992) which has grown up around it does little more than provide a vocabulary for dealing with, on the one hand, animals in succession (which may or may not follow the patterns established by the underlying vegetation) and, on the other, the rather different environments presented by habitats such as phytotelms. Depending on the longevity of the phytotelm unit, we may observe a sequence of states in the biological assemblage which reflect external changes in the resource base (as it matures, senesces and dies perhaps) or a real succession in which there is some serial connections between earlier and later stages in the development of the community.

In phytotelm habitats no food web can become established until there is a resource base available for the earliest invaders. In habitat units such as exposed tree holes or open pitchers this resource base is established by the entry of dead and dying organic matter allochthonously. Presumably such matter carries with it the micro-organi•:ns which will begin the process of decay and release of resources. Once such a resource base is established, detritivorous organisms can invade. In one sense this is *facilitated* by the entry of the dead and dying organic matter, and/or the micro-organisms involved in their break-down. Equally though the interaction between micro-organisms and invading saprophages could be one of mere *tolerance*. Once established there is evidence of a coarse level of resource partitioning among the saprophages based roughly on their relative sizes. It is likely that once a saprophagous fauna is established then the establishment of predators, characteristic of a later seral stage in the succession, is facilitated. Given the small size of many phytotelm units there is distinct possibility for *inhibitory* processes to occur, by which the first invading generalist predatory species (or even individuals) will preclude the entry of others – possibly by the simple expedient of eating the propagules of any latecomers. It is also likely that where there are more complex webs containing several species of predator, then patterns of specialisation will be observed and the number of voracious generalist species (such as *Toxorhynchites*) will be limited.

Seasonality

Seasonality at the level of the ecological population or community is, simply stated, the occurrence of regular patterns reflecting responses to annual cycles in climate and associated physical and biological phenomena. When we consider the role of seasonality in the structuring of phytotelm communities, two situations need to be recognised. First, seasonality within the food webs themselves can only occur in long-lived types of phytotelm. Second, in less long-lived containers, there may be a seasonality associated with the life cycle of the container plant itself which determines the presence or absence of the particular phytotelm food web at all. This second case is probably worthy of little further attention except to say that I would expect a preponderance of generalist aquatic fauna in such situations.

Returning to situations such as tree holes which will persist for many years. I would expect to observe seasonal patterns in food-web structures only in those habitats in which there is a clear externally driven seasonal cycle in environmental conditions. In the perhumid tropics of Borneo, for example, clear seasonality is unlikely to be manifested. Even in those situations in which there is clear external seasonality, as in temperate forest, I would expect the appearance of seasonality to be damped by the nutrient-poor nature of the allochthonous detritus resource which, once established, provides a continuous, if poor, source of food for the saprophages in the food web. Of course developmental rates of insect larvae and other poikilotherms will vary, in particular with temperature, and this will lead to seasonal changes in relative abundance and age structure. However, the presence or absence of species (and in consequence many structural features of the food webs they comprise) will be less seasonally dependent especially in situations where life cycles are extended to more than a single season, presumably because of the poor quality of the resource base. Clearer seasonality may occur in the predatory cohort within the food web for one or both of two reasons. First the resource quality is higher: it is the detritivorous prey species that have had to cope with poor resource quality; they themselves represent a relatively high quality food resource to predators. Second, some predators in phytotelm food webs represent the relatively short-lived aquatic stages of species that spend most of their lives elsewhere. Species of this kind, such as the larvae of frogs, for instance, may well have a clear-cut seasonality in their life cycle driven by events which impact upon other life stages in environments external to the phytotelms themselves.

Stochastic discovery and survival

Lastly there is food-web variation that can be observed at the local scale which reflects behavioural activities of the component species (Chapters 9, 13). So major differences in food-web structure from habitat unit to habitat unit are observed which do not occur as part of a longer term change in community structure such as we might see during succession or seasonality, merely reflecting the chance processes of discovery: that is, whether or not a particular mature habitat unit is found over a particular period of time by a particular species of organism.

In any habitat unit this will be a reflection of whether or not egg-carrying females or other reproductive propagules of a species of organism find that habitat unit at a certain window in time. Further changes in the overall food web will reflect whether or not such invaders are able to establish at this very local level. The processes of successful establishment or otherwise within a habitat unit may be in contrast to those which may or may not occur at the regional level (see preceding sections). Survivorship within a habitat unit may reflect local conditions of food availability and microclimate, they may reflect the presence or absence of prey or predatory species with which the invad-

Box 6.4 **Processes influencing food-web structure in phytotelmata at the local spatial scale: hypotheses and predictions.**

Succession

H16. Hypothesis: Successional processes will occur on a habitat unit to habitat unit basis reflecting the history of that particular unit.

P16. Prediction: Distinct changes in community structure should occur through time following either the appearance of a new habitat unit *or* catastrophic 'clearing' of a pre-existing one.

H17. Hypothesis: Given the simple nature of phytotelm food webs, a facilitation model of succession within habitat units seems most likely.

P17. Prediction: Predators and other higher trophic levels will appear in developing foodwebs only after a set of saprophages ('prey') is well established.

Seasonality

H18. Hypothesis: Seasonal patterns within phytotelms are only possible in long-lived containers: even in these habitats seasonality is unlikely to be obvious at the level of the food web, although it may be apparent in the population dynamics of particular species within the communities.

P18. Prediction: Values of food-web statistics sampled regularly should show little variation which can be correlated with change of season, although within more complex webs the relative abundance of some species may show such patterns.

Stochastic discovery and survival

H19. Hypothesis: At a single point in time the presence or absence of particular species and consequent web structures within individual habitat units will reflect the chance processes of discovery and subsequent survival by the dispersing propagules of that species.

P19a. Prediction: Individual habitat units will contain random subsets of the species that make up the regional foodweb.

P19b. Prediction: No individual habitat unit will contain all the specialist organisms which make up the regional food web for that phytotelm type.

ing species may interact, and they may reflect the amount of time during which a particular habitat unit may present an equable environment to that species. At this level the discovery of a particular habitat unit by a single female of a predatory species such as *Toxorhynchites* may lead to the local extinction of a prey species, with significant consequences for the whole food web. Similarly at this local level the degree of exposure or otherwise of a particular habitat unit may cause it to be flushed clean during a storm event, or not as the case may be.

Hypotheses and predictions relating to processes and phenomena at the local spatial scale are presented in Box 6.4.

Connections

A set of processes which, potentially, affects food-web structure in phytotelmata has been identified in this chapter. In turn this has led to a number of hypotheses and predictions which should follow (see Boxes). The remainder of the book describes appropriate data sets from phytotelmata and the ways in which these can be used to test one or more of most of these hypotheses. Chapters 13 and 14 revisit and synthesise these theoretical ideas and spatial and temporal data from a variety of phytotelmata.

Part III:

Patterns in phytotelm food webs

The Nothofagus *beech forests of northern New South Wales may not have been the models for Tolkien's Fangorn but the similarities are striking. The gnarled and convoluted trunks of these survivors of the great forests of Gondwana are covered in Spanish moss and wreathed in mists on many a morning. Perched in the high country and experiencing a changing, sometimes severe and often moist climate, these extraordinary rainforests are home to the superb lyrebird, the olive whistler and, even, the strange bristle bird – itself a survivor, perhaps, of the earliest passerines originating on this ancient land mass. The lyrebird reveals itself, more often than not, by piercing mimicry of other species – often at a volume that the original models could never match! Other inhabitants are less obvious in their presence: a grey kangaroo looms out of the mist by the side of the track, a much smaller swamp wallaby scuttles into the undergrowth as we approach.*

Presenting the utmost contrast, the lowland rainforests of the Daintree north of Cairns in Australia's wet tropics tumble down to golden beaches in an effulgence of luxuriant vegetation. The wet season here is warm, and the field worker may often experience that extraordinary sensation of watery continuity as the warm monsoonal rain buckets down from above to join the standing water on the forest floor and the temporarily intercepted 'drip layer' in between. The almost monospecific canopy of the beech forests is replaced by a bewildering admixture of up to 200 species of trees within a single hectare, their canopies intermingling and emerging to form a wildly heterogeneous matrix for other life. No lyrebirds here, but the raucous and piercing call of the yellow-footed scrub fowl as it stalks the forest floor fills the noise gap in quantity if not quality. Long-tailed paradise kingfishers, enormous cassowaries, migrant glossy starlings, sparkling sunbirds and others complement the scrub fowl to produce a rich and exciting bird fauna. Mammals are less obvious here but the fan palms at night just might produce a

137

skunk-like striped possum for the spot lighter, while the enormous white-tailed rat, one of Australia's many native rodents, shows no hesitation in rifling through food stores, clothes, papers and biological samples left injudiciously within reach.

But both of these forest types are exciting and rich sources of water-filled treeholes and represent two of the twelve or so sites we have visited within Australasia, building up a more and more comprehensive picture about how the community of animals within treeholes varies on the continental scale. Add to this information from New Guinea, Sulawesi and Borneo and our appreciation of variation over a huge scale begins to emerge.

Part III of this work deals with the patterns observed in the food webs of container habitats and the variation we observe in space and time at a range of scales from the global to the local, the recent to the geological.

7

Food-web variation across geographical regions
Spatial patterns: the global scale

The subtropical rainforest of south-east Queensland is a far cry from the deciduous woodland of southern England. Walking south from O'Reilly's Lodge towards the New South Wales border I pass the giant liane-draped trees of the rainforest. The strange Gondwanan hoop pines are interspersed among red carrabeen and black booyong, strangler fig and brush box. Here one can pass sixty or seventy species of tree in a kilometre's walk, many of them carrying crowns of crows' nests and orchids, stagshorns and mistletoes. Among the trees and epiphytes are the homes of mound-building scrub turkeys and the rifle bird, Australia's bird-of-paradise. Ring-tailed possums and red-necked pademelons bring a mammalian presence to the forest, complicated by the occasional dingo and marsupial native cat. And here too one may stumble upon python and black snake, forest dragon and leaf-tailed gecko. The list of plants and vertebrates alone is hundreds of entries long. In fact this is a rainforest, not as rich as the ones further north in Queensland and beyond but, nevertheless, a place of great biological excitement and richness.

In contrast, the biological riches of Wytham Woods in Berkshire are altogether more understated, although nonetheless significant in their way. Strolling along the Singing Way on a summer's day thirty years ago, I recollect spreading beech trees, planted almost 200 years earlier along the old London to Oxford stage road, now reduced to a muddy track. Beeches are interspersed with oaks and ashes, horse chestnuts and limes, but the list is modest. Perhaps a dozen species of tree are common here. As in the rainforest the trees have their denizens: the great tits and willow warblers, the bullfinches and woodpeckers. Mammals too may be sensed, if seldom seen: badgers and fallow deer, foxes and weasels. But this deciduous woodland is altogether tamer and more garden-like than the subtropical rainforest, reflecting not only climate and latitude but the heavy hand of humans for over 2000 years. The Australian rainforests, where they were inhabited at all, were lived in by humans who used them for hunting and gathering, ceremony and shade, but not as the raw material for a different landscape that they set about creating with axe and plough.

But the contrast is not absolute. In the spreading massive buttresses of the beech trees in Wytham Woods are water-filled tree holes superficially indistinguishable from those in the plank buttresses of the black booyong trees of

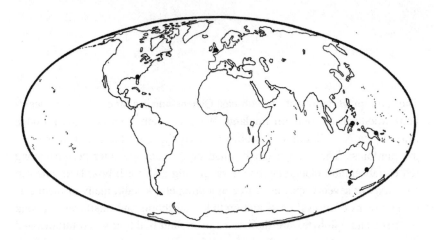

Fig. 7.1. Map of the world indicating (solid circles) the sites from which data are available on the food webs in water-filled tree holes and which have been used in the analyses.

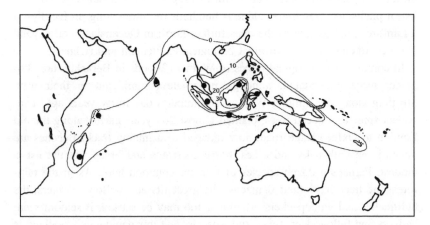

Fig. 7.2. Map of the Eastern Hemisphere indicating (solid circles) the sites from which data is available on the food webs in *Nepenthes* pitchers and which have been used in the analyses. The isopleths indicate the species diversity of the genus *Nepenthes* throughout its range (from Kurata 1976).

Lamington National Park. This similarity of habitat units across vast geographical distances is one of the outstanding reasons why phytotelmata have proved such useful objects of study in recent years.

In this first chapter dealing with the pattern of spatial variation in the structure of phytotelm food webs I collate information on the largest of scales available to us. I focus on two classes of habitat in this account: water-filled tree holes and tropical *Nepenthes* pitcher plants. Tree holes occur in moist forests from the temperate zone to the perhumid tropics, and in all continents but Antarctica. Information on food webs within tree holes is available from four continents and a variety of forest types. *Nepenthes* pitcher plants occur throughout the Old-World tropics (see Chapter 2) and we have information from many species within this range. Figures 7.1 and 7.2 map the sites which acted as data sources for my analyses.

Tree holes in Britain, USA, Sulawesi and Australasia

It is clear from the account presented in Chapter 3 that tree holes as habitats for animals have been investigated in many places. There are records of frogs from the West African rainforests, giant *Mecistogaster* damselflies from Panamanian tree holes, tipulid larvae from rot holes in Paraguay and so forth. Water-filled tree holes are a feature of wet forests worldwide. But to construct a food web, even when deliberately omitting the micro-organisms as in these accounts, requires more than the taxonomically restricted records in which the literature abounds, fascinating as that is. I have described in Chapter 5 the procedures my co-workers and I have used to obtain sufficiently complete faunal information for the construction of a food web and the extraction of appropriate statistics from it for further analysis.

Surprisingly, very few workers have provided the information necessary for this purpose. A number of European workers have analysed the contents of water-filled tree holes in Britain, Germany, Czechoslovakia and Russia (refer to Table 2.3). In North America Fashing (1975) provides a faunal list for tree holes in Virginia (collected during his studies of the biology and taxonomy of tree-hole mites), Snow (1958) has presented perhaps the most detailed studies available for North America, although based upon observations of but one site in Arkansas, and the combination of Bradshaw & Holzapfel's (1983) work on the mosquito assemblage and later work of mine on other elements of the fauna allows us to construct a web for sites in northern Florida. Elsewhere in North America we could perhaps construct a tentative food web for, say, the western seasonally wet forests by combining the records of individual taxa originating from many different authors, but we

Table 7.1. *Location data and food-web statistics from water-filled tree holes from selected locations around the world*

	Wytham Woods, England	Tall Timbers, Florida, USA	Lamington, SE Queensland, Australia	Mt Field, Tasmania, Australia	Dumoga-Bone, Sulawesi, Indonesia	Baiteta, Madang, New Guinea
Location data						
Latitude	51°45' N	30°40' N	28°50' S	42°00' S	1°00' N	5°00' N
Mean monthly rainfall (mm)	54.3	129.6	97.2	48.5	270.8	297.6
Rainfall coefficient of variation	15.28	36.24	47.84	13.46	na	35
Mean monthly days of rain	12.4	10.2	6.5	12.3	na	17.2
Highest monthly maximum temp. (°C)	21.7	32.6	25.5	na	na	30.2
Lowest monthly maximum temp. (°C)	6.5	17.9	16	na	na	28.6
Highest monthly minimum temp. (°C)	12.4	22.1	16	na	na	23.2
Lowest monthly minimum temp. (°C)	1	5	9	na	na	22.8
Statistic						
Number of species	8	16	12	7	13	17
Number of predators	0	2	4	0	5	6
Number of top predators	0	0	1	0	1	2
Number of macrosaprophages	3	3	1	1	3	2
Number of feeding links	10	26	25	9	33	43
Number of trophic levels	1	2	3	1	3	3
Predator:prey	na	0.13	0.33	na	0.38	0.35
Mean prey per predator	na	3.67	3.75	na	4.8	4.8
Mean food-chain length	1	0.68	1.63	1	2.61	2.44
Connectance	0.42	0.37	0.58	0.43	0.49	0.27

Climatic data based on the nearest meteorological station, where available.

would have no guarantee at all that this represented a whole assemblage at any defined place and time.

For central and South America, although naturalists do examine water-filled tree holes, as evidenced by published records of selected taxa, I know of no published list (as of 1997) available for the whole metazoan fauna of a tree hole site anywhere in the region. The situation is even more acute in the great rainforests of West Africa from where there is a dearth even of the singular records and monospecific studies that are available from the Neotropics.

It is from the Indo-Australasian region that we have most information on tree-hole communities. During this time we have accumulated many site-specific faunal lists from Australia, from Tasmania to Cape York, from New Guinea, from Sulawesi in Indonesia and from the lowland dipterocarp forests of northern Borneo. There still remain enormous gaps in our knowledge, particularly with respect to the South-east Asian rainforests of Indo-China, Malaysia and Indonesia, and the southern Asian rainforests of India and Sri Lanka, but at least we can construct some food webs for the region. Some of the data we have accumulated from tree holes on the Australian continent is included in the global comparative analysis presented here. A more complete treatment of the eleven sites we have studied within continental Australia is the subject of Chapter 8 and a detailed analysis of regional and local food-web variation at selected Australian sites is in Chapter 9.

For the present analyses I have selected six data sets: two from Australia (from subtropical rainforest in Queensland and from cool temperate rainforest in Tasmania), one from New Guinea, one from Sulawesi, and one each from western Europe and the southern USA (Figure 7.1). In each case I have deliberately chosen the sites from which we have collected original data ourselves. This will not only make generally available many unpublished observations but also means that I am clearly aware of the shortcomings of the database in each case.

The locations

Table 7.1 contains such climatic information as I have been able to collect on the sites used in this analysis. The accounts below summarise briefly the vegetation at each location.

Wytham Woods, England

The Wytham Woods estate comprises just over 400 ha of managed woodland situated in a loop of the river Thames, north-west of the city of Oxford. Its history, geology and botany are reviewed in detail by Elton (1966). The woodland itself is largely artificial, having been created by plantings on agricultural land between 1814 and 1872. The planting sequence and other aspects of the estate's history are described by Grayson & Jones (1955). A number of areas of the woodland, including those from which the tree-hole data were collected, comprise a continuous overstorey of beech trees (*Fagus sylvatica*) with a sparse herbaceous undergrowth dominated by *Mecurialis perennis* L. and *Rubus* agg. Adjacent areas of woodland contain mixed stands of oak (*Quercus robur, Q. petraea*), ash (*Fraxinus excelsior*), sycamore (*Acer pseudoplatanus*) and hawthorn (*Crataegus monogyna*).

The data used in the analyses were collected between 1966 and 1969. Details of the sampling methods are presented by Kitching (1969, 1971).

Tall Timbers Research Station, Florida, USA

The Tall Timbers Research Station is located on the southern edge of Lake Iamonia in northern Florida. This account of its vegetation is based closely on that presented by Bradshaw and Holzapfel (1983). The Research Station comprises 1200 ha of mixed hardwood forests in a series of hammocks and on Hall, Gay's and Sheep Islands in the Lake (now actually linked with causeways). The most abundant tree species are sweet gum (*Liquidambar styraciflua*), live oak (*Quercus virginiana*), water oak (*Q. nigra*), beech (*Fagus grandifolia*), magnolia (*Magnolia grandifolia*), ironwood (*Ostrya virginiana*), tupelo gum (*Nyssa sylvatica*), holly (*Ilex opaca*) and pignut hickory (*Carya glabra*). Interspersed among the hardwoods are plantings of *Pinus elliottii* and *P. taeda*. The entire Research Station is subjected to controlled burns each year in the late winter or early spring.

The data on the basis of which the food webs used in these analyses were drawn were collected by W. Bradshaw & C. Holzapfel between October 1977 and July 1978 (Bradshaw & Holzapfel 1983) augmented by my own observations in July and August 1988.

Dumoga-Bone National Park, Sulawesi, Indonesia

The Dumoga-Bone National Park in the Indonesian province of Sulawesi Utara stretches from Kotamabagu in the east to Gorontala in the west. The eastern part of the Park, particularly the Tumpah and Toraut catchments in the upper Dumoga valley, were the site for the Royal Entomological Society's centenary expedition, 'Project Wallace', in 1985.

The area in which tree holes were studied comprises lowland rainforest at about 300–350 m altitude. The composition of the forest has not been studied in any detail but is relatively open, often having an understorey dominated by palms (especially *Calamus* spp.). A general account of the vegetation of Sulawesi may be found in Whitten *et al.* (1987). The park has an equatorial monsoon climate characterised with constant high temperatures, high and evenly distributed rainfall and high humidity. The annual rainfall in the northern peninsula is between 2000 mm and 3000 mm, with mean temperatures and relative humidity more or less constant at approximately 36 °C and 71% respectively.

A total of sixteen rot holes and pans was sampled in January and February 1985. A preliminary account of these results is included in Kitching (1987b).

Baiteta, Madang Province, New Guinea

The region around the village of Baiteta, some 15 kilometres north of Madang in Papua New Guinea, comprises so-called 'lowland hill' forest (Saunders 1976). It is a high-diversity rainforest with frequent occurrence of trees such as *Intsia bijuga, Terminalia* spp., *Spondias cytheraea* and *Mastixiodendron pachyclados*. The region is moderately populated and the forest is considerably modified in places due to clearing, shifting agriculture and human habitation. More complete accounts of the vegetation are presented by Saunders (1976) and Robbins *et al.* (1976).

The tree-hole data included in these analyses were presented in Kitching (1990) and the details of the collection process are described in a more general context, in Chapter 5.

Lamington National Park, Queensland, Australia

Lamington National Park is an area of total protection covering some 18 800 ha just north of the Queensland/New South Wales border in eastern Australia. The flora of the region has been described by McDonald & Whiteman (1979). The forest is a complex notophyll vine forest (after Webb 1959) classified by Floyd (1990) as 'subtropical rainforest. *Agyrodendron actinophyllum* alliance, suballiance 11 (*Caldcluvia paniculata–Cryptocarya erythroxylon–Orites–Melicope octandra–Acmena ingens*)'. The areas in which tree-hole studies where carried out had a wide range of tree species including (in addition to those listed already) *Lophostemon conferta, Ficus watkinsiana, Araucaria cunninghamii,* and *Geossois benthamina*.

The information on which the food webs used in these analyses is based was collected during the period 1979 to 1986, although the principal data from eleven repeatedly sampled holes were collected from September 1982 to August 1983.

Mt Field National Park, Tasmania, Australia

Mt Field National Park is located in the south-west of Tasmania. Ogden & Powell (1979) describe the vegetation of the area along an altitudinal gradient from 158 m above sea level to the treeline at 1220 m. Treeholes were located within 'microphyll fern forest' dominated by *Nothofagus cunninghamii* at about 670 m. This forest formation is co-dominated by *N. cunninghamii and Atherosperma moschatum*.

The patterns

The food web obtained from water-filled tree holes from the Baiteta, New Guinea, site has already been presented in Figure 5.1 and the associated statistics in column 1 of Table 5.1. The webs obtained from the other five sites used in the 'global' analysis are illustrated in Figure 7.3*a–e*. The associated food-web statistics are summarised in Table 7.1.

Wytham Woods, UK and Mt Field, Tasmania

The simplest webs of all are those found in English and the Tasmanian sites (Figures 7.3*a* and 7.3*c*). In each case our studies failed to find any metazoan predators and wider faunistic studies (see Chapter 3) confirm that such additional elements in these food webs are unlikely. Skidmore (1985) does record *Graphomya maculata* (Muscidae) as a predator of syrphid larvae in rot holes in England and Tate (1935) notes *Phaonia exoleta* (Muscidae) feeding on mosquito larvae in similar places, but in the three years of close study that resulted in the web in Figure 7.3*a* I did not encounter these species, neither of which is a specialist associate of water-filled tree holes.

The sets of saprophagous species found in England and Tasmania are almost congruent in some respects. Both webs contain oligochaetes, eristaline syrphids, scirtids, chironomids, ceratopogonids and mosquitoes. The single species of mosquito of the Tasmanian webs is 'matched' by two species in the English webs; the Tasmanian webs contained copepods, the particular English webs under study did not. However, in other nearby tree holes in England, the copepod *Moraria arboricola* was abundant. Very few further Tasmanian sites were examined and other species of mosquito may well be found in a more extensive study.

Tall Timbers Reserve, Florida

The web obtained in northern Florida (Figure 1.3*b*) contained not only a much wider range of saprophagous species but two predatory larvae in addition. The predatory species *Toxorhynchites rutilus* and *Corethrella appendiculata* together make the food webs from this site the most complex recorded to date in North America. The *Toxorhynchites* species has a widespread distribution in the eastern two-thirds of the continent but that of *Corethrella* is much more restricted.

The saprophages in the Floridan webs contain elements equivalent to those in the English and Tasmanian webs. There are eristaline syrphids, scirtids, ceratopogonids, non-predatory mosquitoes and copepods. Psychodid larvae also occur within this community as they do in those in continental Europe (but not in the English examples studied to date). However, there are two species of endemic mites (absent entirely from Europe as far as we are aware), and there are ostracods – recorded from European webs but not from those studied here. Both the mosquito and the ceratopogonid assemblage are more speciose than those recorded in Europe and Tasmania. There have been no chironomids recorded from North American tree holes.

Lamington National Park, Australia

The food webs recorded from Lamington National Park in south-east Queensland (Figure 7.3*d*) contain about thirteen species of metazoans, less than the sixteen in the Floridan web. However, a species of top predator (tadpoles of the frog *Lechriodus fletcheri*) is a significant structural addition, adding a trophic layer to the web additional to that seen in the North American web.

Of the three predatory species confirmed in this community, two are species of mites, including the endemic hydracarine *Arrhenurus kitchingi*. The most abundant predators in these situations, however, are the larvae of the tany-podine chironomid *Anatopynnia pennipes*. Other tanypod predators occur in tree holes elsewhere in Australasia and South-east Asia but they have not been recorded from American or European systems. Species of *Toxorhyn-chites* occur in the region but were not encountered in our studies. They occurred frequently in tree holes in the Australian tropics (see Chapter 8).

The saprophagous assemblage in these Queensland tree holes is more or less 'typical' of tree holes elsewhere – scirtids, chironomids, ceratopogonids, psychodids, mosquitoes, mites and ostracods. Eristaline larvae have been recorded from the area but were not encountered in our studies.

Dumoga-Bone National Park, Sulawesi

The web presented in Figure 7.3*e* contains but 13 species, albeit organised into three trophic levels. It displays some notable absences and, for reasons that will be discussed in due course, is considerably simpler than might be expected for a near equatorial web. Indeed it is markedly simpler than the web recorded from New Guinea and even from tropical Australia.

The top predator in this ecosystem is an odonate, *Lyriothemis cleis*, and the predatory layer comprises three species of mosquito and a tanypodine chironomid, ecologically similar to the species found in Australia and New

(a)

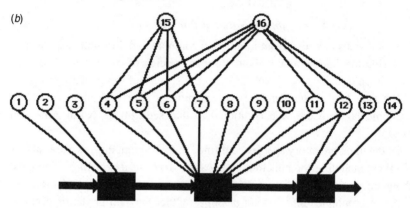

1. *Myiatropa florea*
2. *Prionocyphon serricornis*
3. oligochaete
4. *Dasyhelea dufouri*

5. *Metriocnemus cavicola*
6. *Aedes geniculatus*
7. *Anopheles plumbeus*
8. copepod

(b)

1. tipulid
2. *Mallota cimbiciformis*
3. *Prionocyphon discoideus*
4. *Culicoides guttipennis*
5. *Culicoides nanus*
6. *Dasyhelea oppressa*

7. psychodid
8. *Naiadacarus arboricola*
9. *Hormosianoetus mallotae*
10. harpacticoid copepod
11. *Orthopodomyia alba*
12. *Aedes triseriatus*

13. *Anopheles barberi*
14. ostracod
15. *Corethrella appendiculata*
16. *Toxorhynchites rutilus*

(c)

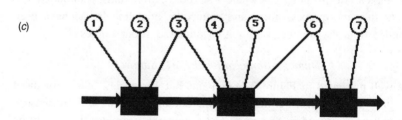

1. eristaline
2. *Prionocyphon niger*
3. oligochaete
4. ceratopogonid

5. chironomine
6. *Aedes* sp.
7. copepod

(d)

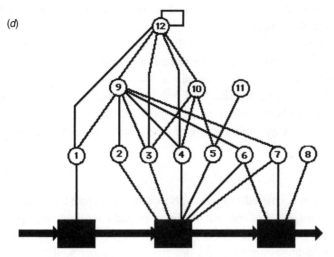

1. *Prionocyphon niger*
2. *Clogmia* sp.
3. *Culicoides angularis*
4. *Polypedilum* sp.

5. algophagid mite
6. *Aedes candidoscutellum*
7. *Aedes* sp.
8. ostracod

9. *Anatopynnia niger*
10. *Arrhenurus kitchingi*
11. *Cheiroseius* sp.
12. *Lechriodus fletcheri*

(e)

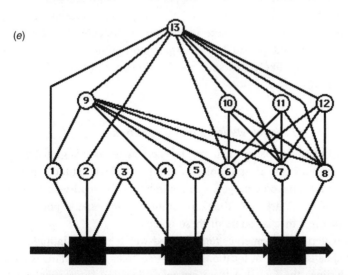

1. scirtid
2. *Tipulodina* sp.
3. *Gyraulus* sp.
4. *Dasyhelea* sp.
5. *Megaselia* sp.

6. *Tripteroides* sp.
7. *Culex ('Culiciomyia)* sp.
8. *Aedes ('Stegomyia)* sp.
9. *Anatopynnia* sp.
10. *Toxorhychites* sp. 1

11. *Toxorhynchites* sp. 2.
12. *Culex ('Lutzia)* sp.
13. *Lyriothemis cleis*

Fig. 7.3. Tree-hole food webs from (*a*) southern England, (*b*) northern Florida, (*c*) Tasmania (*d*) south-east Queensland and (*e*) Sulawesi. Sources of information are given in the text.

Guinea. In an intensive study we recorded no frogs, or mites (predatory or otherwise), from the community.

At the saprophagous level the 'usual' scirtids, mosquitoes and cerato-pogonids were joined by a species of tipulid, a phorid and, unique among food webs studied to date, a small gastropod. No non-predatory chironomids were recorded from these situations and only three species of mosquito larvae were found. In both cases I suspect more extensive sampling would change the situation, expanding the size of the saprophage grade within these webs.

Baiteta, New Guinea

The food webs I encountered in northern New Guinea (Figure 5.2) contained at least seventeen species of metazoan animals spread over three trophic levels. Two species of top predator, both odonates, were found together with a four-species predator assemblage. As described in Chapter 5, this predator assemblage contained the extraordinary freshwater polychaete *Lycastopsis catarractarum*, in addition to familiar tree-hole species such as *Toxorhynchites*, a species of *Arrhenurus* and a tanypodine chironomid.

The saprophages in this web comprised eleven species, representing many of the groups encountered elsewhere in tree holes. Notable absences were saprophagous mites and phorid larvae. Again no syrphids were recorded although it is *a priori* likely that such species do occur in tree holes in the region, albeit rarely.

Food-web statistics and global trends

The most obvious sort of pattern that may be encountered in this global analysis of food-web variation in tree holes is one that reflects latitude. Indeed, significant relationships do emerge when a number of the food-web statistics presented in Table 7.1 are plotted against latitude. Two general points need to be made before I discuss some of these results.

First, any pattern which emerges relating food-web features to latitude must be interpreted as inferring 'at locations where tree holes occur'. This is a truism but underlines the fact that there are many locations at particular latitudes, even on the same land masses, at which moist forests and hence tree holes do not occur. Any causal relations between latitude and food webs that may be imputed need to be qualified in this fashion.

Second, and of more substance, is the observation that many of the food-web statistics used in the analyses are far from independent of one another. In particular, high values for simple linear correlation coefficients exist among the values for the number of species, number of predators, number of top predators, numbers of feeding links, number of trophic levels and average food-chain length (Table 7.2). This is not surprising as the first three of these variables are serially nested subsets of each other and the number of feeding links and trophic levels are, in a sense, simple transformations of the more fundamental values. It is a little more difficult to understand the implied relationship between average food-chain length and four of the other variables (and why this average length is not correlated with the values for total number of species).

Average food-chain length, as explained in Chapter 5, is computed by enumerating the lengths of all possible food chains within a web and averaging the values so obtained. Such enumerations are obtained by selecting the end-point of each food chain and counting the number of links back to its base. When predators and top predators are present they represent the ends of the food chains in that web in a disproportionately high number of cases. In addition, frequently, each predator represents the end of several alternative food chains. So as the number of predators and top predators in a web increases so the contribution each makes to the average food-chain length will increase disproportionately, and the average will increase accordingly.

The existence of high correlations among pairs of variables implies some computational connection between them. It does not mean that the relationship between them will necessarily be linear. In addition there may be significant non-linear relationships among pairs of variables which do not generate a high linear correlation value. Figure 7.4 illustrates these points.

The relationships between the number of predators and the number of species in the food web (Figure 7.4a) and that between the number of feeding links and the number of species in the food web (Figure 7.4b) are apparently linear implying a more or less constant value for the ratios between these pairs of variables. The first part of Table 7.3 summarises a regression analysis of these variables and confirms what is obvious from the figures, that the relationship between the second pair of variables is 'better' than that between the first pair.

The third relationship shown in the figure, that between the number of trophic levels and the number of species in the food webs represents a situation where a non-linear relationship provides a better representation of the inter-relationship than does a linear one, even though these two variables generated a high linear correlation coefficient (0.74, Table 7.2). A simple logarithmic transform of the independent variable produces the curvilinear fit shown in Figure 7.4c and the regression summarised in Table 7.3.

The last case illustrated in the figure represents a situation where there

Table 7.2. *Linear correlations among selected food-web statistics from water-filled tree holes*

	Number of species	Number of predators	Number of top predators	Number of feeding links	Number of trophic levels	Average food-chain length
Number of species	1					
Number of predators	0.77	1				
Number of top predators	0.62	0.92	1			
Number of feeding links	0.91	0.96	0.87	1		
Number of trophic levels	0.74	0.97	0.83	0.89	1	
Average food-chain length	0.68	0.96	0.87	0.86	0.99	1

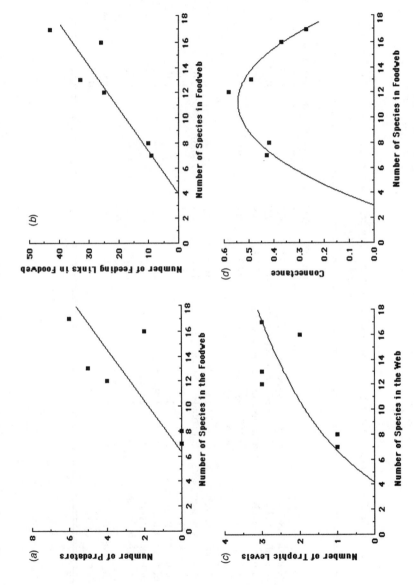

Fig. 7.4. Relationships among food-web statistics derived from the tree-hole webs of Fig. 7.3.

Table 7.3. *Regression analyses among selected food-web statistics from tree holes, and between food-web statistics from tree holes and latitude*

Dependent variable	Independent variable(s)	Intercept	Slope (1)	Slope (2)	Coefficient of determination	F-value	p|F
Number of predators	Number of species in web	−3.07	0.48	–	0.59	5.83	0.073
Number of feeding links	Number of species in web	−11.31	2.93	–	0.82	18.18	0.013
Number of trophic levels	Ln(Number of species in web)	−3.09	4.95	–	0.63	6.6	0.062
Connectance	Number of species in web, Number of species in web**2	−0.47	0.18	0.008	0.91	14.75	0.028
Number of species in web	Latitude	16.05	−0.15	–	0.65	4.36	0.104
Number of species in web	Latitude, Latitude**2	14.65	0.1	−0.01	0.65	2.73	0.211
Number of predators	Latitude	5.99	−0.12	–	0.87	26.33	0.007
Number of macrosaprophages	Latitude	–	–	–	0.02	0.08	0.709
Number of feeding links	Latitude	40.33	−0.6	–	0.84	21.53	0.01
Number of trophic levels	Latitude	3.29	−0.04	–	0.75	12.04	0.026
Average food-chain length	Latitude	2.74	−0.03	–	0.75	12.15	0.025
Predator:prey	Latitude	0.41	0.0079	–	0.81	16.7	0.015
Prey per predator	Latitude	5.5	−0.01	–	0.8	16.24	0.016
Connectance	Latitude	–	–	–	0.03	0.11	0.752

appears to be an important non-linear relationship between two variables which was not at all apparent from the simple correlation analysis. The connectance values and the total number of species in a food web are linked by a second order polynomial relationship (Figure 7.4d) which generates a coefficient of determination of 0.74 and a significant F-value (see Table 7.3). The inverted U-shape of the relationship is a consequence of the fact that connectance values are low when the number of species in the web is low, and rise to a peak and then decline again as numbers increase. The low connectance values when species numbers are low reflect the fact that in the absence of predators (the usual case when species numbers are at the low end) connections in the web are merely between saprophages and particular components of the detritus base. Detritus components are defined by particle size and the different saprophages specialise on different-sized components. When just a few predators are present in webs they tend to be generalists and increase connectance dramatically. In yet more speciose webs top predators appear and some first-level predators can 'afford' to be specialised, leading to a reduction in connectance values. In addition the number of potential trophic links (the denominator in connectance calculations) in speciose webs with three trophic levels becomes very large, decreasing the likelihood of a saturation situation and, accordingly, the calculated connectance values.

I have discussed at length the inter-relatedness of food-web statistics because such interconnections will occur within each tree-hole and pitcher-plant data set that I shall examine in this chapter and in Chapters 8 through 11. Elsewhere I will illustrate and present analyses of these but will not rehearse the above arguments.

Turning, then, to the important relationships among food-web features and latitude within this data set, I refer to Figure 7.5 and the greater part of Table 7.3. There are five clearly significant relationships which emerge from the regression analyses presented in Table 7.3. Three of these can be subsumed under a simple predator/latitude rubric: the relationships of latitude and number of feeding links, and of latitude with number of trophic levels, are clearly derivative. The high significance of the predator/latitude regression is noteworthy and is illustrated in Figure 7.5b. I have illustrated this as a curvilinear relationship in the figure although a linear regression generates the more significant F-value (see Table 7.3). Whatever the precise nature of the relationship it is clear that the number of predators within the food webs declines steeply with increasing latitude. This declining relationship is forced by the total lack of predators in the most temperate webs and their, relative, abundance in the equatorial ones. The existence of intermediate values, however, indicates a gradual decline rather than a simple lots/none contrast. The decline in number of

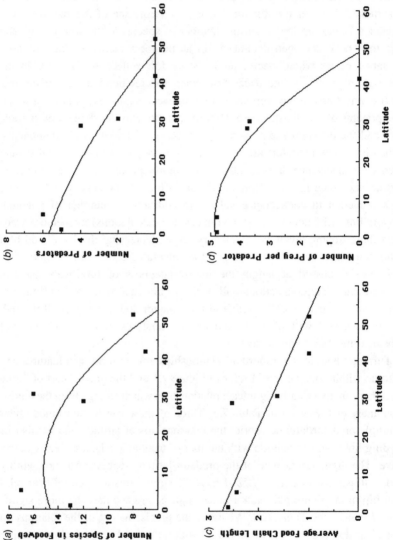

Fig. 7.5. Relationships between selected food-web statistics from water-filled tree holes and latitude.

predators mirrors the decline in total numbers of species which also follows increasing latitude (Figure 7.5*a*). In this more general case, however, the data hold greater variance and the regression relationships between numbers and latitude (both linear and curvilinear) are not conventionally significant.

The relationship between average food-chain length and latitude is shown in Figure 7.5*c*. There is a clear and significant relationship between the variables with average length declining with increasing latitude, reflecting the progressive simplification of the community which accompanies the decline in the predator cohort with latitude. One point, that for Lamington National Park, Queensland, stands out as an outlier above the trend line. Indeed, without this point the significance of the regression increases by an order of magnitude (r^2=0.97, F=115.41, p=0.002). This outlying point is caused not so much by the increased number of predators found at this site but largely because one of them is a generalist top predator, thereby contributing a large number of chains of length three to the computation of the average.

I make this point not to undervalue the particular datum but to stress two points. First, it is clear that this particular statistic is highly sensitive to a single, strategically placed addition to a food web. Second, these observations underline the importance, within the general comparative approach that I take in these analyses, of assembling as large a number of data sets as possible in order that global trends may be identified even where particular data may deviate from this trend.

The last relationship illustrated in the figure (Figure 7.5*d*), that between the average number of prey items per predator and latitude, is somewhat weaker than the others discussed. This relationship is typical of a set of such pairwise relationships which occur in my analyses which involve ratios – and, in the absence of predators, take zero values. The significant decline observed in the plot shown in the figure is forced by the existence of two zero values. The intermediate values for the Floridan and south-east Queensland webs would not in themselves produce a significant relationship, although indicate the same declining trend. Nevertheless the zero values represent 'real' if negative data and there is no statistical argument for omitting them.

Nepenthes pitcher plants in the Old-World tropics

The general distribution of the described species of *Nepenthes* pitcher plants has been discussed in Chapter 2. Figure 7.2 summarises this information drawn from data presented by Kurata (1976). In addition the map given in the figure indicates the locations from which I have assembled the data analysed in this section.

Data sources and study sites

The data base has been assembled partly as a result of primary work by my students and myself (Borneo, Sulawesi), partly from the observations of Beaver on West Malaysian species (Beaver 1979a,b, 1980), from Thienemann's (1932) observations in Sumatra and Java, and in the remaining cases (Madagascar, Seychelles, Sri Lanka, Australia) by compiling records from a variety of authors each writing on restricted sites of taxa. For the first three of these four 'compiled' cases I follow Beaver's (1985) lead and have interpreted obser- vations in the same fashion that he did. Unlike Beaver's (1979a, 1985) webs, however, I omit non-aquatic organisms, including the spider *Misumenops nepenthicola*, although I am aware of recent discussion about the occasional aquatic forays of these spiders and the consequent impact on food-web sta- tistics (Clarke 1998). I have selected the most complex web available for each land mass where a choice was available (i.e. West Malaysia and Borneo).

In few of these cases are details of the study sites presented. In general, *Nepenthes* species are rosette plants in the immature form, becoming scram- blers or climbers as they mature. In general the lower altitude species (which category includes all those in this analysis) are plants of disturbed situations such as road verges or light gaps within tropical forest. *Nepenthes maxima* in Sulawesi occurred on a floating sphagnum mat on the margin of a lake. Such details as are available on the locations included may be found by con- sulting the original data sources given earlier.

The patterns

The food web that occurs within the pitchers of *Nepenthes albomarginata* in West Malaysia was studied and defined by Beaver (1979a,b). This web is illustrated as Figure 5.9 and key direct and derived statistics that relate to it are contained in Table 5.1. The other food webs used in these analyses are illustrated in graphed versions in Figure 7.6[1]. Table 7.4 contains information on each site and the food-web statistics relating to each of them. The site information included has been selected so that a range of the hypotheses pro- posed in Chapter 6 can be evaluated. I have provided the latitude and longi- tude and the area of the land mass on which it is located (except for Aus- tralia, where the range of *Nepenthes mirabilis* is restricted to the tropics of

[1] At the time of carrying out this analysis I was unaware of the work of Kato *et al.* (1993) on the fauna of ten species of *Nepenthes* in Sumatra. Their subjects did not include *tobaica* – the species studied by Thienemann (1932) – but clearly will provide invaluable data for future geo- graphical analyses.

the north-west). This land area is included because so many of the sites are on oceanic or continental islands. In addition I include two variables specifically related to the genus *Nepenthes*. I list the distance from the sampled site to the current point of maximum diversity of the genus. On the basis of current knowledge this is in and around Mt Kinabalu, Sabah. In addition the number of co-occurring species of *Nepenthes* on each land mass provides a measure of 'metahabitat' availability.

Madagascar

The food web from Madagascar contains four species, each feeding directly on the detritus base. A histiostigmatid mite, *Creutzeria tobaica*, and two species of the culicine mosquito, *Uranotaenia*, were identified by Paulian (1961), who also recorded an unidentified chloropid larvae as occurring at the same sites. Paulian (1961), elsewhere in his extensive monograph on the zoogeography of Madagascar, notes an additional mosquito species, *Ficalbia pollucibilis* from other sites where it also breeds in the axil plant *Ravenala*. I have followed Beaver (1985) in not including this third species of mosquito in the compiled food web.

Seychelles

The food web recorded from pitchers of *Nepenthes pervillei* in the Seychelles are the simplest of those investigated to date. One species of histiostigmatid mite, *Creutzeria seychellensi*, and one of *Uranotaenia* occupy the pitchers according to Mattingly & Brown (1955) and Nesbitt (1979). Both are saprophagous.

Sri Lanka

The fauna of *Nepenthes distillatoria* in Sri Lanka is more complex than those occurring further west in the Indian Ocean. Two species of mosquito occur, both from characteristic phytotelm genera (*Tripteroides* and *Uranotaenia*), a histiostigmatid mite and a phorid belonging to the genus *Megaselia*. I have not included the psychid moth *Nepenthophilus tigrinus* recorded from these pitchers by Günther (1913) in the food-web compendium. I excluded this species for two reasons: firstly so my usage should be consistent with that of Beaver (1985) and secondly because of the doubts expressed by Erber (1979) as to the host specificity of the species.

Penang, Malaysia

The web collected by Beaver (1979a,b) from pitchers of *Nepenthes albomarginata* from Penang, Malaysia (Figure 5.9) contained eleven species

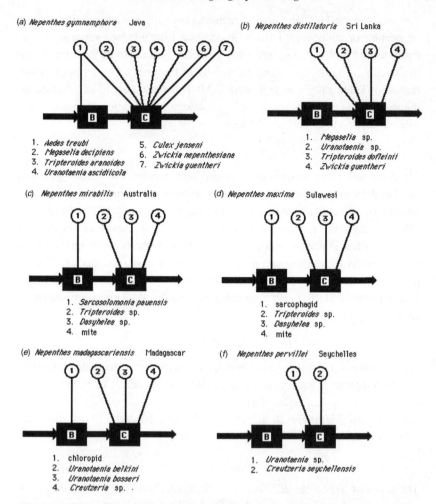

Fig. 7.6. Food webs from the pitchers of *Nepenthes* from (*a*) Java, (*b*) Sri Lanka, (*c*) northern Australia, (*d*) northern Sulawesi, (*e*) Madagascar, (*f*) the Seychelles, (*g*) Sumatra and (*h*) northern Borneo. The species of *Nepenthes* are indicated on the figure. Sources of data are discussed in the text.

organised at two trophic levels. The saprophagous layer included the syrphid species *Pierettia* and the phorid *Megaselia* (presented by Beaver as *Endonepenthia* – see Annexe). Three species of mosquitoes, one of ceratopogonid and two of mites, complete the saprophagous assemblage. Three first-order predators complete the web, including an encyrtid parasitoid and a cecidomyiid fly larva.

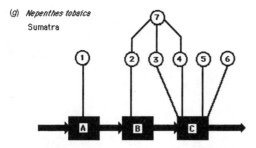

(g) *Nepenthes tobaica*
Sumatra

1. *Phyllocnistis nepenthae*
2. *Syritta capitata*
3. *Megaselia* sp.
4. *Dasyhelea confinis*
5. *Tripteroides aranoides*
6. *Uranotaenia gigantea*
7. *Lestodiplosis syringophora*

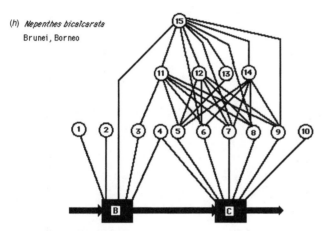

(h) *Nepenthes bicalcarata*
Brunei, Borneo

1. *Wilhelmina nepenthicola*
2. *Nepenthosyrphus* sp.
3. *Dasyhelea* sp.
4. *Polypedilum* sp.
5. *Tripteroides* sp.
6. *Armigeres* sp.
7. *Uranotaenia* sp.
8. *Culex* sp. 1
9. *Culex* sp. 2
10. *Zwickia* sp.
11. *Corethrella* sp.
12. *Toxorhynchites* sp. 1
13. *Toxorhynchites* sp. 2
14. *Toxorhynchites* sp. 3
15. *Campanotus schmitzi*

Sumatra

The food web constructed for *Nepenthes tobaica* from Sumatra, compiled
from Thienemann (1932), contains seven species. It is unusual in a number
of respects. First the lepidopteran *Phyllocnistis nepenthae* feeds subaqueously
on the inner surface of the pitchers and, second, Thienemann (1932) recorded
neither mites nor the predatory larvae of *Toxorhynchites* from these systems.
Thienemann did record mites from *N. gymnamphora* in Java and *Toxorhynchites* from other phytotelmata (Thienemann 1934) so he was well acquainted

Table 7.4. *Basic and derived food-web statistics from Nepenthes pitcher plants*

	N. madagascariensis Madagascar	*N. pervillei* Seychelles	*N. distillatoria* Sri Lanka	*N. albomarginata* West Malaysia	*N. tobaica* Sumatra	*N. gymnamphora* Java	*N. bicalcarata* Borneo	*N. maxima* Sulawesi	*N. mirabilis* Australia
Location data									
Latitude	22°30′ S	4°30′ S	7°30′ N	5°00′ N	4°00′ S	7°00′ S	4°30′ N	2°00′ N	12°30′ S
Land mass area	587 041	443	65 610	128 666	473 610	134 980	746 337	189 026	[a]
Distance from Sabah	8116	7496	3881	1654	1546	1622	187	1107	3587
Number of co-occurring species	0	0	0	11	21	2	30	6	0
Statistic									
Number of species	4	2	4	11	7	7	15	4	4
Number of predators	0	0	0	2	1	0	5	0	0
Number of top predators	0	0	0	0	0	0	1	0	0
Number of macrosaprophages	1	0	0	2	1	1	4	1	1
Number of feeding links	4	2	4	15	9	8	41	4	4
Number of trophic levels	1	1	1	2	2	1	3	1	1
Predator:prey	–	–	–	0.25	0.17	–	0.5	–	[a]
Mean prey per predator	–	–	–	2.5	3	–	6	–	[a]
Mean food-chain length	1	1	1	1.9	1.5	1	2.5	1	1
Connectance	0.5	0.5	0.5	0.47	0.31	0.57	0.6	0.5	0.5

[a] Not relevant because of very restricted distribution of *N. mirabilis*.
– Not applicable.

with both taxa. The single predator recorded from *N. tobaica* was a species of the cecidomyiid *Lestodiplosis* such as Beaver (1979a) encountered in West Malaysia in *N. albomarginata* and Clarke & Kitching (1993) found in *N. mirabilis* and *N. gracilis* in Borneo (see Chapter 8).

Java

Pitchers of *Nepenthes gymnamphora* in Java contained a simpler food web than that encountered in the neighbouring but larger island of Sumatra. Seven species of saprophages were found by Thienemann (1932) in this *Nepenthes*. Four species of culicid larvae belonging to three genera were complemented by a species of phorid, and two of mites.

Brunei

Nepenthes bicalcarata is a robust plant with large long-lived pitchers. It is restricted to the northern coastal regions of Borneo and contains a relatively complex food web made up of fifteen species spread across three trophic levels. The saprophages include the curious calliphorid *Wilhelmina nepenthicola* (Figure A.27) and a specialist syrphid *Nepenthosyrphus* sp. together with a 'usual' assemblage of mosquitoes, ceratopogonids, chironomids and mites. A subset of this saprophagous assemblage is preyed upon by a suite of three *Toxorhynchites* spp. and a single species of chaoborid, a species of *Corethrella*. Finally the curious 'swimming' ant *Camponotus* (*Colobopsis*) *schmitzi* (Clarke & Kitching 1995, Clarke 1997) acts as a generalist predator feeding, *inter alia*, on the larvae of *Toxorhynchites*.

Lake Mo-oat, Sulawesi, Indonesia

Pitchers of *Nepenthes maxima* have a modest fauna of only four species, all of which are saprophages. Two of the species involved are unidentified but belong to the families Sarcophagidae (Diptera) and Histiostigmatidae (Acari) respectively. *Tripteroides* mosquito larvae and larvae of the ceratopogonid genus *Dasyhelea* compete the food web.

Cape York, Australia

Nepenthes mirabilis has a very extensive distribution which takes in the tip of Cape York in Australia. There are also one or two outlying populations further south in the eastern tropics of that continent. The food web collated for this species is structurally identical to that recorded from Lake Mo-oat, Sulawesi. The four saprophagous species present include species of *Tripteroides* and *Dasyhelea*, an unidentified histiostigmatid mite and the sarcophagid *Sarcosolomonia pauensis* described by Yeates *et al.* (1989).

Food-web statistics and global trends

In considering patterns of variation across food webs at this global scale in water-filled tree holes I carried out a dual analysis: first examining the statistical relationships within the set of food-web measures and, then, relating these measures to latitude. In that analysis the only obvious independent variable to use was latitude and this, as we saw earlier, may be taken as a covariate of a range of latitude-related environmental factors (I return to this point later). In considering the larger data set available from *Nepenthes* pitchers I follow a comparable pattern: looking for internal correlations and relationships, then attempting to relate selected food-web variables to one or more independent variables. Unlike the tree hole case, however, there are four candidate variables and I analyse each of these in turn and then, where appropriate, together. Two of these variables, the number of co-occurring *Nepenthes* species and the distance to the centre of the plants' distribution, are negatively correlated (Figure 7.7) but the relative importance of their roles can be readily distinguished. The selected variables are latitude, the number of co-occurring species of *Nepenthes*, the distance to the centre of distribution of *Nepenthes*, and the size of the land mass on which the sampled species of *Nepenthes* occurs (Figure 7.7).

Internal relationships

If anything the various statistics derived from *Nepenthes* food webs are even more closely correlated one to the other than was the case within the tree-hole data set. Table 7.5 shows the correlation matrix for all of the variables used except connectance, which was uncorrelated with the other quantities. Table 7.6 summarises regression analyses which use the total number of species as the independent variable and each of the other variables in turn. Highly significant linear relationships exist with the number of predators, the number of macrosaprophages, the number of feeding links, the number of trophic levels, the predator:prey ratio and the numbers of prey species per predator. Only in the case of the number of feeding links can this relationship be improved (in terms of F-value) by use of a non-linear regression (see row 4 of Table 7.6). The regression between total species in the web and the average food-chain length is also significant ($p=0.014$). As already indicated, the connectance values are not significantly related to the other variables.

    The very 'tight' level of inter-relatedness implied by these analyses suggest to me that communities in *Nepenthes* pitchers are much more structured than those in tree holes, perhaps because of the simpler and more circumscribed physical environment that they present to their inhabitants. The con-

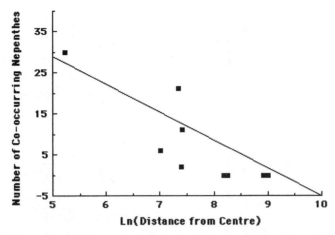

Fig. 7.7. The relationship between the number of co-occurring species of *Nepenthes* and the distance to the estimated centre of the distribution of the genus *Nepenthes*.

stant fractions of predators and macrosaprophages implied by these results suggest perhaps a misleading predictability. I reiterate that I deliberately chose the most complex food web available, when a choice was available (i.e. from West Malaysia and Borneo), and in some cases other co-occurring species of *Nepenthes* contain less fully 'packed' webs. I return to this point in Chapter 8.

Relationships with latitude

I carried out regression analyses using latitude as the independent variable and each of the food-web statistics in turn as the dependent variable. Not surprisingly in no case was there a remotely significant relationship. *Nepenthes* as a whole is a tropical plant and the species included in these analyses all occur in rainforests and swamp forests of various kinds. It would be interesting to repeat the analysis with food webs from within a region but along an altitudinal gradient. I would predict, in this case, that relationships not unlike those seen between tree-hole web statistics and latitude would be apparent. Unfortunately no such data set exists currently.

Relationships with the number of co-occurring species of Nepenthes

The number of co-occurring species of the genus *Nepenthes* on the land masses from which the present data originate, ranges from zero to thirty. A co-occurring set of thirty species of pitcher plants, each offering similar but not identical living conditions for potential inhabitants, represents a much 'safer' and more predictable environment in terms of the sheer number of habitat units, the availability of suboptimal 'fall-back' habitats, and the dimensions

Table 7.5. *Linear correlations among selected food-web statistics from Nepenthes pitcher plants*

	Number of species	Number of predators	Number of macro-saprophages	Number of feeding links	Number of trophic levels	Predator: prey ratio	Number of prey per predator
Number of species	1						
Number of predators	0.93	1					
Number of macrosaprophages	0.93	0.94	1				
Number of feeding links	0.94	0.99	0.95	1			
Number of trophic levels	0.92	0.98	0.86	0.92	1		
Predator:prey ratio	0.94	0.99	0.91	0.96	0.99	1	
Number of prey per predator	0.91	0.96	0.87	0.93	0.99	0.98	1
Average food-chain length	0.78	0.91	0.83	0.92	0.87	0.88	0.92

Table 7.6. *Regression analyses among selected food-web statistics from Nepenthes pitcher plants*

Dependent variable	Independent variable(s)	Intercept	Slope (1)	Slope (2)	Coefficient of determination	F-value	p>F
Number of predators	Number of species in web	-1.56	0.38	—	0.87	46.6	<0.001
Number of macrosaprophages	Number of species in web	0.63	0.67	—	0.88	49.54	<0.001
Number of feeding links	Number of species in web	-7.63	2.73	—	0.87	48.5	<0.001
Number of feeding links[a]	Number of species in web	6.21	-1.7	0.26	0.98	126.7	<0.001
Number of trophic levels	Number of species in web	0.41	0.16	—	0.85	35.5	<0.001
Predator:prey	Number of species in web	-0.15	0.04	—	0.89	55.7	<0.001
Prey per predator	Number of species in web	-1.72	0.47	—	0.82	32	0.001
Average food-chain length	Number of species in web	0.61	0.09	—	0.61	10.72	0.014
Connectance	Number of species in web	—	—	—	0.06	0.44	0.527

[a] Non-linear analysis.

of the available environmental 'space' (with some species of *Nepenthes* offering opportunities not presented by others) than does a set of but one species.
Accordingly we may expect that the number of co-occurring species of
Nepenthes may well be a powerful variable for explaining the observed variation in size and complexity of the communities which inhabit pitcher plants.

Table 7.7 summarises the regressions carried out using the number of co-
occurring species of *Nepenthes* as the independent variable. Four of the significant linear relationships obtained are graphed in Figure 7.7. Essentially
there were relationships significant at least at the 0.01 level (generally much
better) for all the food-web variables used except connectance, which
appeared to be unrelated to the independent variable. Of course the tight intercorrelation among the food-web variables described above means that if any
one such variable is significantly related to a particular independent variable,
then it is highly likely that the others will also be so related. I graph these
particular relationships for a number of reasons:

- The number of species in each web I take as the 'fundamental' statistic
  underpinning all the others. It summarises the whole set of processes of
  co-adaptation that have occurred between the phytotelm inhabitants and
  the habitat presented by the pitcher plants.
- The two statistics, the number of predators and the number of macrosaprophages, are each plotted to demonstrate that the significant
  relationship between total species and the number of co-occurring species
  of *Nepenthes* is not merely the result of a very close relationship between
  one guild within the fauna and the independent variable. It also shows
  that although the two relationships are both statistically significant, that
  involving the number of predators ($F=24.87$, $p=0.002$) is a much 'tighter'
  one than that involving the number of macrosaprophages ($F=13.58$,
  $p=0.008$). This relationship also points up a contrast with respect to the
  tree-hole data described above, within which no pattern involving the
  macrosaprophage guild could be identified.
- The average food-chain length is plotted because it explicitly reflects the
  two-dimensional structure of each food web. In addition I use this statistic for testing some of the hypotheses discussed in Chapter 6.
  (Hypotheses and results on spatial patterns and processes are brought
  together in Chapter 12.)

Relationships with the distance from the centre of richness of Nepenthes
As intimated earlier (Figure 7.2) the genus *Nepenthes* peaks in species richness on or about Mt Kinabalu, Sabah. The nine sites sampled range from

187 km from Mt Kinabalu (Brunei) to 8116 km (southern Madagascar). Because of this wide range I used a natural logarithmic transformation of the data before carrying out the regressions summarised in Table 7.6.

As in the case of the analyses against the number of co-occurring species of *Nepenthes*, almost all the regressions carried out gave results that were formally significant. For the dependent variables, number of species in the food webs and number of macrosaprophages, these regressions were more significant than previously; in other cases either equally or less so. Again there was no significant relationship between the connectance values and the independent variable. Three of the relationships are plotted in Figure 7.8. Again I have chosen to plot only a selection of the relationships identified and have chosen three of the four given in Figure 7.8 both for the reasons set out previously and for ease of comparison between the two figures. In all three cases set out in Figure 7.9 the strength of the relationship depends upon the point representing the results from *Nepenthes bicalcarata* from Brunei (at the top left of each graph). Without this data point, although the trend lines remain, the significances of the relationship are much reduced. Of course, there is no *a priori* reason to ignore this key data point.

Relationships with the size of the land mass

For this analysis, for reasons given earlier, the data point from northern Australia was ignored. Setting this aside, the sizes of the land masses used range from 443 km^2 for the Seychelles to 746 337 km^2 for the island of Borneo.

Table 7.7 indicates that, with the exception of the total number of species in the food web and connectance, each of the food-web statistics showed a weakly significant relationship with the size of the land mass (just under the 0.10 probability level in each case). The best relationship identified was that between the number of macrosaprophages and land-mass size. I repeated all the analyses after having transformed the land-mass data logarithmically, but in no case did this increase the strength of the relationship.

Inter-relationships among the independent variables

As may be expected, the independent variables used in these four sets of analyses (latitude, number of co-occurring species, distance from the centre of *Nepenthes* richness, land-mass size) are not themselves wholly uncorrelated. Table 7.8 indicates the values of the correlation coefficients involved. In particular the values for the number of co-occurring species are highly correlated with the distance measure. Again this is not surprising because the centre of richness of *Nepenthes*, on which the distance measures are based, is in turn based upon the isopleths of species richness shown in Figure 7.2.

Table 7.7. Regression analyses of food-web statistics from Nepenthes pitcher plants against geographical variables

Dependent variable	Intercept	Slope	Coefficient of determination	F-value	p>F
(1) Independent variable: Number of co-occurring Nepenthes species					
Number of species in food web	3.97	0.32	0.7	16.52	0.005
Number of predators	-0.17	0.14	0.78	24.87	0.002
Number of macrosaprophages	0.53	0.09	0.66	13.58	0.008
Number of feeding links	2.67	0.96	0.73	19.05	0.003
Number of trophic levels	0.95	0.06	0.92	77.34	<0.001
Predator:prey	-0.01	0.01	0.85	38.77	<0.001
Number of prey per predator	-0.19	0.19	0.94	103.23	<0.001
Average food-chain length	-0.96	0.04	0.85	38.87	<0.001
Connectance	–	–	0.01	0.09	0.775
(2) Independent variable: Ln distance from centre of Nepenthes distribution					
Number of species in food web	29.75	-3.05	0.71	17.42	0.004
Number of predators	10.13	-1.21	0.68	14.63	0.007
Number of macrosaprophages	7.98	-0.88	0.72	17.64	0.004
Number of feeding links	79.89	-9.14	0.74	19.16	0.003
Number of trophic levels	5.3	-0.5	0.64	12.4	0.01
Predator:prey	1.05	-0.12	0.66	13.46	0.008
Number of prey per predator	12.77	-1.5	0.66	13.3	0.008
Average food-chain length	4.14	-0.37	0.66	13.69	0.008
Connectance	–	–	0.009	0.01	0.945
(3) Independent variable: Area of land mass (excluding Australia)					
Number of species in food web	–	–	0.3	2.56	0.152
Number of predators	-0.18	4.1E–06[a]	0.4	3.93	0.093
Number of macrosaprophages	0.3	3.3E–05	0.49	5.78	0.053

Number of feeding links	2.04	3E–05	0.42	4.33	0.081
Number of trophic levels	0.98	1.8E–06	0.41	4.24	0.083
Predator:prey	–0.01	4.2E–07	0.39	3.77	0.098
Number of prey per predator	–0.16	5.5E–06	0.45	5	0.067
Average food-chain length	0.98	1.3E–06	0.39	3.85	0.095
Connectance	–	–	0.009	0.01	0.946

[a] Exponential format.

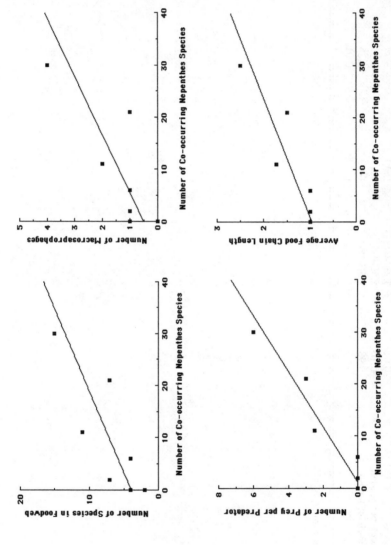

Fig. 7.8. The relationship between selected food-web statistics from *Nepenthes* pitchers and the number of co-occurring species of *Nepenthes* on the same land mass.

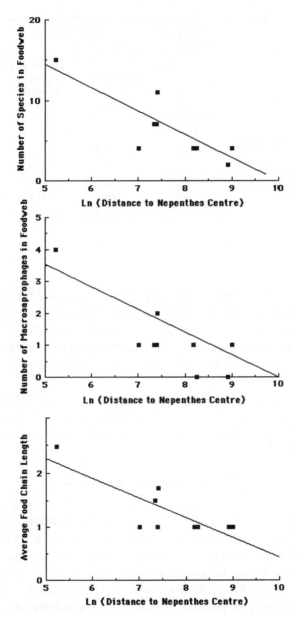

Fig. 7.9. The relationship between selected food-web statistics from *Nepenthes* pitchers and the distance to the estimated centre of the distribution of the genus *Nepenthes*.

Table 7.8. *Correlation matrix for the independent variables used in the regression analysis of* Nepenthes *food webs (see text)*

	Latitude	Number of co-occurring species	Ln distance to centre of *Nepenthes* richness	Area of land mass
Latitude	1	–	–	–
Number of co-occurring species	–0.44	1	–	–
Ln distance to centre of *Nepenthes* richness	0.55	–0.84	1	–
Area of land mass	0.35	0.69	–0.48	1

Multivariate relationships

The two sets of univariate relationships involving the number of co-occurring species of *Nepenthes* and the distance of any site from the centre of *Nepenthes* richness, respectively, provide adequate statistical power for both descriptive and predictive purposes. I tested whether a better prediction might result from analyses including both these independent variables simultaneously.

In three cases the use of both variables gave a better result than using the distance variable alone. In no case was the F-value generated using both variables together better than that using the number of co-occurring species of *Nepenthes* alone.

Summary of Nepenthes *analysis*

The tightly interconnected set of relationships that I have identified in this analysis are in part at least the result of covariability among the variables involved. There remains an underlying set of relationships in which the variable 'distance from the centre of *Nepenthes* richness' is the predictor of the various food-web statistics that are apparent within pitchers, as well as of the number of species of *Nepenthes* that occur in any one area. On a purely statistical basis, the best predictor throughout is the number of co-occuring species of *Nepenthes*. The size of the land mass is no doubt a contributing factor in some cases, especially when that size is small (as in the case of the Seychelles) but the magnitude of this area effect is negligible compared with that accounted for by the distance variable or its covariate, the number of co-occurring *Nepenthes* species.

Connections

The results presented in this chapter for both tree holes and *Nepenthes* are discussed in light of the hypotheses erected in Chapter 6 in the general discussion presented as Chapter 13. More detailed analyses, focusing either on smaller spatial scales or on changes through time, for the Australian tree hole results and some of those from *Nepenthes albomarginata*, are presented in the next four chapters.

8

Food-web variation within a continent: the communities of tree holes from Tasmania to Cape Tribulation
Spatial pattern: intermediate scale

The eastern seaboard of Australia at the time of the European invasion 200 years ago presented a continuum of forest from southern Tasmania to the tip of Cape York (White 1986). These forests presented then, as they do now, a complex mosaic. Fire-maintained, eucalypt-dominated woodlands and forests, in one form or another, occupied the drier segments of the landscape. In the moister, monsoonal North, in the cool wet uplands and in a series of smaller pockets on the more protected wetter side of hills, in gorges and valleys, highly diverse 'closed' forests occurred: the rainforests. Since the European invasion the overall extent of each of these forest types has changed. Vast areas of euclaypt woodland and forest have been cleared for timber and to produce grazing land. The rainforests were an early target, containing as they did valuable 'cabinet' timbers – red cedar, hoop pine, satin ash, silky oak – the common names all reflect the high value placed, then and now, on these precious timbers. About half of Australia's rainforests were cleared between 1788 and the present day, but patches remain from Tasmania in the south to the tip of Cape York.

All these rainforests have structural properties in common: they frequently form closed canopies; they are dominated by a mix of evergreen species; they contain a rich flora of epiphytes, vines and lianes; and they contain the continent's richest faunal assemblages. In addition they all produce water-filled tree holes as rot holes, water-tight hollows in buttress roots, and pools in branch axils.

Certainly water-filled tree holes do occur in eucalypts (see, for example, English *et al.* 1957) but only rarely, and only as temporary aquatic habitats. It is in the rainforests that they may be particularly abundant and long lived. Over the last 15 years we have sampled sets of tree holes at many rainforest sites along a 3000-kilometre axis from Mt Field in Tasmania to Cape Tribulation in far northern Queensland.

This chapter presents a comparative analysis of these results, in order to show how food-web structure in otherwise very similar habitat units can vary within a continent at a range of latitudes, altitudes and prevailing local climates.

Rainforest tree holes in Australia

It would be scientifically impressive to report that this comparative analysis has been based on a carefully designed study with evenly spaced sampling locations using strictly comparable efforts and protocols. Sadly this has not been the case as I have had to use data as and when they were available – reflecting the manifold vagaries of student projects, field expeditions and available funding.

The major data collections have been described, briefly, in Chapter 1. The map that is Figure 8.1 shows collecting locations used in these analyses together with the current extent of rainforests on the continent. I have based this latter element of the map on the distribution indicated by Adam (1992). Table 8.1 augments the mapped information with data on the positions of our sampling sites, their altitudes and approximate annual rainfall. The table also contains details of the collecting intensity and periods of our studies of tree holes at these locations.

In general, at each site, tree holes have been sought and sampled during single, sometimes extended field trips. The actual number of holes sampled reflects the frequency of occurrence of habitat units in particular forest patches as well as sampling effort. The sites in Tasmania, the Conondale Ranges, Eungella, Paluma, Bellenden Ker, Palmerston National Park, the Atherton Tablelands and Cape Tribulation were all sampled during one-off field trips during the wet season in each case (although not in the same year in all instances). In each case as many holes as possible were sampled. Sufficient sites were examined to produce a satisfactory 'snapshot' of the food web in tree holes, except in the Tasmanian case where extensive searching over the course of a week produced only three tree holes (although one of these was massive). The data from these eight sites were augmented with that from two studies in which a smaller set of holes was visited repeatedly over the course of a year.

In each case a sample of detritus and water from the tree holes was sieved and its fauna sorted. Larval insects were reared to adulthood. Foodwebs have been constructed based on previous knowledge of the groups concerned (reviewed in Chapter 3 and the Annexe) and intensive study of a few sites. The results from the study at Lamington National Park have been presented

Table 8.1. *Australian sites at which tree holes were sampled between 1979 and 1994*

Site	Approx. position	Altitude (m)	Period of study	Number of holes	Average annual rainfall	Forest type[a]
Mt Field, Tasmania	43° S 147° E	600	Jan–Feb 93	3	1000	MFF
New England NP, NSW	30° S 152° E	1320	Jun 86–Aug 87	9	2199	MFF
Dorrigo NP, NSW	30° S 153° E	760	Jun 86–Aug 87	8	2017	CNVF
Lamington NP, Qld	28° S 153° E	1050	Sept 82–Aug 83	11	2990	CNVF
Conondale Range, Qld	26° S 143° E	790	Mar–93	13	c. 1500	CNVF
Eungella, Qld	21° S 148° E	980	Jan–93	30	1789	CNVF
Paluma, Qld	19° S 146° E	800	Mar–93	27	2600	CNVF
Bellenden Ker, Qld	17° S 146° E	110	Mar–93	18	2000	CMVF, Type 1a
Palmerston NP, Qld	18° S 136° E	350	Mar–93	19	3228	CMVF, Type 1a
Atherton Tablelands	17° S 145° E	750	Mar–93	36	1419	CNVF, Type 5a
Cape Tribulation, Qld	16° S 145° E	25	Jan–Feb 88,89	76	3827	CMVF

[a]Key: CMVF, complex mesophyll vine forest; CNVF, complex notophyll vine forest; MFF, microphyll fern forest. Subdivided types follow Tracey (1982).

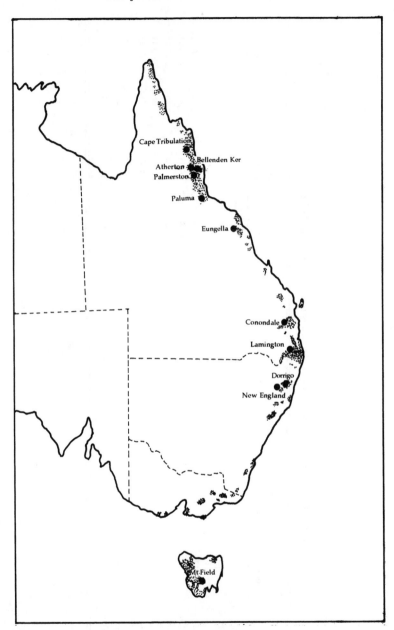

Fig. 8.1. Map of the eastern half of the Australian continent showing sampling locations for rainforest tree-hole studies. The stippled areas represent the approximate distribution of current rainforest (based on Adam 1992).

Table 8.2. *Food-web statistics from the Australian sites listed in Table 8.1*

Site	Number of species	Number of predators	Number of macrosaprophages	Number of feeding links	Number of trophic levels	Predator: prey ratio	Mean prey: predator ratio	Mean food-chain length	Connectance
Mt Field, Tasmania	7	0	1	9	1	–	–	1.00	0.43
New England NP, NSW	10	2	3	22	3	0.25	6.00	3.35	0.54
Dorrigo NP, NSW	9	2	2	20	2	0.25	4.50	1.92	0.50
Lamington NP, Qld	12	4	1	25	3	0.33	3.75	1.63	0.58
Conondale Range, Qld	8	1	1	15	2	0.14	6.00	1.89	0.53
Eungella, Qld	16	4	3	1	3	0.33	5.80	2.26	0.46
Paluma, Qld	12	5	0	33	3	0.71	5.00	2.66	0.53
Bellenden Ker, Qld	13	3	1	24	3	0.30	4.33	1.75	0.46
Palmerston NP, Qld	19	3	2	43	3	0.17	8.33	2.26	0.43
Atherton Tablelands	19	3	2	47	2	0.19	9.67	1.88	0.49
Cape Tribulation, Qld	20	5	2	66	3	0.27	11.50	2.56	0.61

by Kitching & Callaghan (1982) and these results together with those obtained from our Tasmanian studies have been included in the analyses described in Chapter 7.

Although all eleven sites from which results have been obtained were located in rainforests, the vegetation of the sites differed in type. The southernmost sites, in Tasmania and at New England National Park, were in cool temperate rainforest in which the overstorey is much simpler than in rainforests from warmer sites. In both cases the canopy was dominated by species of *Nothofagus: cunninghamii* in New England and *moorei* in Tasmania. These forests, although very moist, have a harsh, at times extreme climate, which includes considerable periods of frost. Our more northern sites were all more diverse, broad-leaved forests, ranging from subtropical rainforest at the Dorrigo, Lamington and Conondale sites to both high and low altitude tropical forests further north. The Eungella site at 21° S and 980 m altitude is generally considered more or less intermediate between the 'subtropical' and 'tropical' forest types. In botanical terms, and using the classification of Tracey (1982), the forest types we examined were microphyll fern forests (the 'cool temperate sites), complex notophyll vine forests (the subtropical sites and the higher altitude tropical ones) and complex mesophyll vine forest (the lower altitude tropical sites).

In quantitative terms (Table 8.1) the sites examined ranged over 27° of latitude from 16° to 43° south, and over a 1300 m altitudinal range from 25 m to 1320 m AMSL. Annual rainfall from these sites all lie in the high range (1000 mm per annum plus) but still present a variation from 1000 mm to 3827 mm. Although all sites experience a wet and a dry season, at some of the wetter and more northern sites this seasonality in rainfall is much less marked.

The food webs

The food webs generated from the eleven sites sampled are summarised in Table 8.2. The webs from Tasmania and the Lamington site have been illustrated in Figure 7.3c and *d*.

The number of metazoan species involved in the webs varied from 7 to 20, with the fewest species in tree holes from the harsh environment of the cool temperate rainforest of Tasmania, and the most in the ever-wet everwarm lowland forests of Cape Tribulation in the far north. As in all tree-hole webs the species in the webs are either saprophages of one sort or another or predators (including top predators).

The numbers of predatory species we found in the webs varied from none

Table 8.3. *Linear correlations among selected food-web statistics from water-filled tree holes within Australia*

	Number of species	Number of predators	Number of feeding links	Predator: prey ratio	Number of prey per predator	Number of trophic levels	Average food-chain length	Connectance
Number of species	1							
Number of predators	0.67	1						
Number of feeding links	0.94	0.75	1					
Predator:prey ratio	0.124	0.76	0.24	1				
Number of prey per predator	0.83	0.53	0.87	0.09	1			
Number of trophic levels	0.5	0.79	0.5	0.63	0.44	1		
Average food-chain length	0.5	0.71	0.64	0.65	0.68	0.73	1	
Connectance	0.072	0.47	0.31	0.35	0.38	0.36	0.41	1

Note comments in the text on the lack of independence among some of these pairs.

to five, again reflecting the south to north enrichment found for the fauna as a whole. The most widespread predatory species in this regard is the tanypodine chironomid *Anatopynnia pennnipes*, which occurred commonly from New England National Park northwards. The zygopteran *Podopteryx selysi* and an undescribed species of planarian occur from Eungella northwards, and in the lowland northern forests these are joined by an anisopteran (as yet unknown in the adult stage) and dytiscid beetles. Treeholes from the *Nothofagus*-dominated forests of New England National Park contained larvae of a hydrophilid beetle not encountered elsewhere and those from Lamington contained two species of predatory mite. Although they were known to occur in tree holes from northern New South Wales northwards, we did not encounter any larvae of the culicine predator *Toxorhynchites* spp. in our samples.

Other food-web statistics are presented in Table 8.2. Table 8.3 is the correlation matrix generated for eight of the nine food-web statistics from Table 8.2 (I omitted the number of macrosaprophages as these show little pattern across sites). Strong correlations clearly exist between the number of species and the number of predators ($r=0.67$), the number of feeding links (0.94) and the numbers of prey per predatory species (0.83). The number of predators in the webs, in addition, correlate with the number of feeding links ($r=0.75$), the predator:prey ratio ($r=0.76$), the number of trophic levels (0.79) and the average food-chain length (0.71). The number of feeding links is significantly correlated with the average food-chain length ($r=0.64$) and the number of prey per predatory species (0.87) in addition to the associations already noted. The predator:prey ratio associates with the number of predators, as already mentioned, the number of trophic levels ($r=0.63$) and the average food-chain length (0.65). The number of prey per species of predator is correlated with the number of species, the number of feeding links and the average food-chain length ($r=0.68$). Lastly, in addition to those relationships already mentioned, the number of trophic levels is highly correlated with the average food-chain length ($r=0.73$). As also noted in Chapter 7 there are no strong correlations between the connectance values and those for any other food-web statistic.

As discussed in the preceding chapter, the existence of these correlations is no surprise, given the way in which each of them is derived from the same set of food webs, no elements of which are truly independent of each other. The interpretation of any relationships obtained between the values of selected food-web statistics and environmental variables may be confounded by the inter-relatedness among variables, and due care must be exercised in this regard.

I have examined these data for patterns which may indicate relationships

Table 8.4. *Linear correlations among latitude, altitude and annual rainfall for tree-hole sites within Australia*

	Latitude	Altitude	Annual rainfall
Latitude	1		
Altitude	0.15	1	
Annual rainfall	−0.5	−0.2	1

between food-web values and one or more of three environmental variables: latitude, altitude and annual rainfall. As it happens, these three environmental variables, perhaps surprisingly, are *not* themselves intercorrelated (Table 8.4) and so each is used, separately or in combination, in the regression analyses which follow.

Table 8.5 presents the results of regression analyses carried out using the range of food-web variables already identified as dependent variables, and the three environmental quantities as independent variables (in various combinations). In the case of the first dependent variable selected (the number of species in the food web), I present the results of all the analyses carried out regardless of the significance or otherwise of their outcomes. For the other dependent variables I have presented only the 'significant' results, and even in those instances only those that add insight to the overall analysis. There are problems involved in the interpretation of these results (as in all situations in which data structure has been investigated using many regression analyses). First, as already discussed, there is covariance among the dependent variables and, second, with approximately fifty separate analyses involved, there is a non-negligible possibility of Type 2 errors if higher significance cut-offs are recognised

For the dependent variable 'numbers of species in the web', there is a significant relationship with latitude (Figure 8.2*a*) which, alone, accounts for 64% of the variance in the dependent variable. No other single variable regressions are significant for this dependent variable. The overall variance accounted for is increased usefully by the bivariate regression on both latitude and rainfall (rising to 67%). The incorporation of altitude into bivariate and trivariate analysis does not improve the power of the regression.

The variable 'numbers of predators (including top predators) in the web', is correlated with the total number of species. Accordingly it is not surprising to find significant correlations with latitude. What is of interest is that there is a strong univariate regression for this dependent variable with rainfall (accounting for 52% of the variance in the dependent variable) (Figure 8.2*c*)

Table 8.5. *Regression analyses among selected food-web statistics and latitude, altitude and rainfall from Australian sites*

Dependent variable	Independent variable(s)	Intercept	Slope (1)	Slope (2)	Slope (3)	Coefficient of determination	F-value	p/F
Number of species in web	Latitude	24.02	-0.45	–	–	0.64	15.8	0.003
	Altitude	16.53	-0.005	–	–	0.168	1.82	0.175
	Rainfall	6.47	-0.003	–	–	0.3	3.88	0.082
	Latitude, altitude	24.32	-0.43	-0.001	–	0.65	7.33	0.016
	Latitude, rainfall	20.22	-0.39	0.0011	–	0.67	8.01	0.012
	Altitude, rainfall	9.55	-0.003	0.003	–	0.35	2.2	0.174
	Latitude, altitude, rainfall	20.71	-0.38	-0.0011	0.001	0.67	4.76	0.041
Number of predators in web	Latitude	6.35	-0.14	–	–	0.56	11.54	0.008
	Rainfall	-0.2	0.0014	–	–	0.53	9.56	0.013
	Latitude, rainfall	3.41	-0.99	0.008	–	0.72	10.21	0.006
	Latitude, altitude, rainfall	2.86	-0.113	0.001	0.001	0.77	7.67	0.013
Number of prey per predator	Latitude	12.87	-0.28	–	–	0.6	13.32	0.005
	Latitude, rainfall	10.31	-0.025	0.0007	–	0.6	5.96	0.026
	Latitude, altitude, rainfall	9.99	-0.26	0.001	0.001	0.63	4	0.06
Number of trophic levels	Latitude	3.9	-0.05	–	–	0.46	7.63	0.022
	Rainfall	1.24	0.0006	–	–	0.52	9.83	0.012
	Latitude, rainfall	2.47	-0.04	0.004	–	0.66	7.6	0.014
	Altitude, rainfall	0.8	0.0005	0.001	–	0.57	5.38	0.033
	Latitude, altitude, rainfall	2	-0.044	0.001	0.0005	0.77	7.87	0.012
Average food-chain length	Latitude	2.95	-0.03	–	–	0.46	7.75	0.021
	Rainfall	1.25	0.0003	–	–	0.38	5.44	0.045
	Latitude, altitude	2.86	-0.046	0.0003	–	0.54	4.75	0.044
	Latitude, rainfall	2.25	-0.028	0.0002	–	0.56	5.14	0.037
	Latitude, altitude, rainfall	1.98	-0.035	0.0004	0.0002	0.68	5.05	0.035
Connectance	Rainfall	0.42	0.00004	–	–	0.3	3.88	0.082

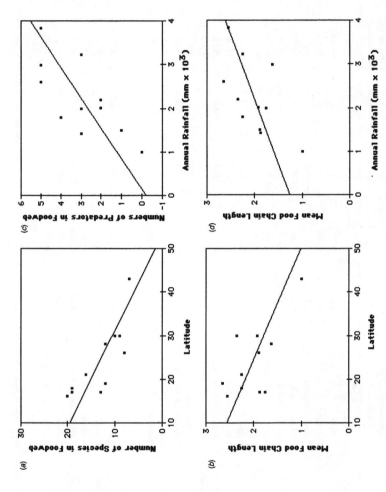

Fig. 8.2. Patterns in tree-hole food webs within Australia: (a) the number of species in tree-hole food webs at different latitudes, (b) average food-chain length at different latitudes, (c) the numbers of predators at sites of differing annual rainfall, and (d) average food-chain length at sites of differing annual rainfall.

where none existed for the total number of species. The other independent variables, added in stepwise fashion, increase the power of the regression, ultimately to account for 75% of the variance in the dependent variable.

Significant regressions do exist for the variable 'number of feeding links', but this is so highly correlated with the values for the total numbers of species in the food webs that they add no additional insight into the analysis.

There are no significant regressions among the values of the predator:prey ratio and latitude, altitude or total rainfall.

The number of prey per predator, used as a dependent variable, generates a highly significant regression with latitude (accounting for 60% of the variance in the dependent variable). Again this accords with the high correlation that exists with the total species in the webs: I return to this point in the discussion below. No other univariate regressions are strong and multiple regressions involving pairs or the set of three available independent variables do not increase the power of the analysis.

The number of trophic levels is relatively weakly correlated with the total number of species but is more strongly related to the number of predators in the food webs. Accordingly it is not surprising to find that the 'best' univariate regression for this dependent variable is with rainfall (as was the case for number of predators). This regression accounts for 52% of the variance in the dependent variable and, again as was the case for the number of predators, this increases to 77% when the three independent variables are employed together.

The average food-chain length is a central variable in some of the theoretical arguments made in Chapter 6 and will be revisited in Chapter 13. Its correlation with a range of other food-web variables (see Table 8.3) makes independent assessment of the results of the regression analyses difficult. There are significant ($p|F$ <0.05) regressions with both latitude and rainfall (Figure 8.2*a* and *d*), but not with altitude. The power of the regressions is increased in the bivariate regression containing both these independent variables (accounting in total for 56% of the variance of the dependent variable) and further increased (to 68%) when altitude is added to the string of independent variables.

When connectance is used as a dependent variable, no strong relationships emerge, as was the case with the global analysis presented in Chapter 7. I have included in Table 8.5 the result of the regression of connectance on rainfall which shows a negative relationship which is marginally significant and of some interest as a trend.

Macrovariables, hypotheses and likely explanations

As with other chapters in which patterns in data are identified and described, I revisit in Chapter 13 these results in light of the sets of general hypotheses presented in Chapter 6. Some less summary considerations, however, are in order with respect to this data set and the analyses just described.

On the basis of Table 8.3, it is reasonable, but neverthless subjective, to redefine the eight food-web variables into three 'macro'-variables:

- 'Species richness': a composite of the closely related variables 'number of species', 'number of feeding links' and 'number of prey per predator' (intercorrelations all >0.83)
- 'Vertical structure': a composite of 'number of predators', 'predator:prey ratio' and 'number of trophic levels' (intercorrelations >0.65)
- 'Connectance' (no intercorrelation >0.52).

I discuss each of these in turn.

Species richness

That the number of species in the food web, or any other faunal assemblage, gets richer with decreasing latitude is no great surprise. Tropical forests are richer in species than temperate ones. The increase in diversity in more tropical tree-hole food webs occurs on a number of levels. Taxa present in the simplest, most 'temperate' of food webs also occur in the tropical ones. But within these taxa there is more diversity. The one species of aedine mosquito in Tasmanian tree holes is substituted by three species of *Aedes* and three of *Tripteroides* in tree holes from the lowland tropical forest of Cape Tribulation. Other higher taxa join the tree-hole food web only in more northerly sites: planarians, odonates and dytiscid beetles. I return below to the point that many of these 'additional' higher taxa are predatory groups.

The reasons underpinning this well-known and 'obvious' pattern are less clear. It seems likely (as has been found for tree holes from western Europe, northern New South Wales and subtropical south-east Queensland) that the actual input of litter per unit area of tree hole differs little from site to site. Undoubtedly the more tropical the site the more equable the environment and the more rapid the rates of turnover as temperature-dependent decomposition processes speed up. Taken at its simplest, this would mean no more than that the organisms dependent upon this resource base would complete more generations in a year and, themselves, turn over more quickly. The actual resource base would not change so why do more species depend upon that base? Why are there eight or ten fine detritus feeders in tropical webs and only one or two in cool temperate communities?

A number of interconnected hypotheses present themselves, separately or in various combinations, as plausible explanations for the observed patterns:

• that the tropical environments present more resource for less effort as micro-organismic activity releases nutrients more quickly than in cooler environments where the saprophagous species must be more directly involved in physical comminution of the resource – in other words that these environments *are* indeed more productive per unit time, in spite of receiving more or less the same input of detritus from external sources;

• that, although each habitat unit represents no more of a resource in temperate as opposed to tropical webs, there may be simply more habitat units available in tropical sites with their greater diversity of tree species, growth forms and range of tree ages;

• that the tropical webs are more predictable, reducing the evolutionary 'risk' of becoming more specialised and, accordingly, leading to a finer division of the resource base by participating species;

• that the much greater representation of predators in more tropical webs (itself a likely product of, on the one hand, their predictability and, on the other, their productivity) reduces the likelihood of competitive inter-actions among the saprophages, allowing more of them to co-exist at lower population levels than would otherwise have been the case;

• that spatial heterogeneity in the tree-hole assemblages allows more species to co-exist within a particular patch of forest without the actual number of species present within any one tree hole increasing greatly.

These hypotheses remain for the most part as just that, awaiting appropriate testing by manipulative experiment. I have analysed further, however, three of the 'best' data sets from the latitudinal sequence: those from Lamington (28° S), Eungella (21° S) and Cape Tribulation (16° S). For Eungella and Cape Tribulation I have selected eleven tree holes at random from the greater number of samples in each case so that the samples become comparable to those from Lamington at which only eleven holes were sampled. In addition I have used just a single month's data from the twelve-month data set from Lamington – that of February – so that that data set in turn becomes comparable with those from the other two sites which were sampled just once (in January or February – the height of the tropical/subtropical wet season in Australia). These analyses shed the weak light of circumstantial evidence on some of these hypotheses. Appropriate results are summarised in Table 8.6.

First it is clear that the tropical tree holes do not necessarily support higher

Table 8.6. *Some analyses of food webs from tree holes at Lamington National Park, Eungella National Park and Cape Tribulation, Queensland*

Site	Lamington	Eungella	Cape Tribulation
Prionocyphon niger			
Mean density per litre	1133.1	142.6	58.00
Standard error	142.6	70.32	37.20
	One-way ANOVAR, ln numbers on 'site':	$F=11.95, p<0.001$	
	Pairwise *t*-tests on ln numbers:		
	Lamington vs Eungella	$t=3.14, p=0.005$	
	Lamington vs Cape Tribulation	$t=5.15, p<0.001$	
Culicoides sp.			
Mean density per litre	44.7	184.1	12.2
Standard error	29.61	166.92	7.47
	One-way ANOVAR, ln numbers on 'site':	$F=0.55, p=0.58$	
Psychodid species			
Mean density per litre	2.45	72.62	33.1
Standard error	1.42	33.88	16.6
	One-way ANOVAR, ln numbers on 'site':	$F=3.30, p=0.051$	
	Pairwise *t*-tests on ln numbers:		
	Lamington vs Eungella	$t=-2.73, p=0.013$	
Number of saprophagous species per tree-hole			
Mean	3.6	3.8	2.27
Standard error	0.43	0.44	0.47
	One-way ANOVAR, numbers on 'site':	$F=3.37, p=0.048$	
	Pairwise *t*-tests on numbers:		
	Lamington vs Cape Tribulation	$t=1.99, p=0.06$	
	Eungella vs Cape Tribulation	$t=-2.39, p=-0.027$	

$n=11$ in each case.
Only significant (or near significant) pairwise tests are shown.

densities of animals than do more temperate ones. Densities of the saprophagous larvae of the scirtid beetle *Prionocypon niger* actually decline from Lamington to Eungella to Cape Tribulation, although the Eungella to Cape Tribulation decline is not statistically significant. For two other selected saprophagous species, *Culicoides* sp. and the larvae of psychodids, densities were higher in the tree holes from Eungella than at either of the two other sites. Densities of *Culicoides* larvae were lower at the tropical Cape Tribulation site than at the subtropical Lamington site. These patterns give little support to the idea that individual tropical habitat units represent resource-richer situations than do temperate ones.

Our studies shed little light on the overall abundance of tree holes across these three sites principally because our overall sampling efforts and strategies have been different at each site. I certainly have the impression that water-filled tree holes are more abundant in some forests than others – and our Cape Tribulation site is one of the richest I have encountered anywhere in the world. This of course may owe more to the state of maturity and preservation of the forest than it does to latitude *per se*.

The number of predators that occur in tree holes in general at the three sites analysed further were almost the same (4, 4 and 5 respectively). Within the more restricted subsets of data analysed further the numbers were also similar (3, 2 and 3 respectively) so few conclusions can be drawn about the role of predators in these three sets of tree holes.

Lastly it is of considerable interest that the numbers of species of saprophages *per tree hole* was more or less identical for Lamington and Eungella ($s=3.6\pm0.43$, 3.82 ± 0.44 repectively, $t=-0.44$, $p=0.665$) and was actually lower for the tropical sites at Cape Tribulation ($s=2.27\pm0.47$; comparison with Eungella $t=2.39$, $p=0.027$; with Lamington $t=1.99$, $p=0.06$). This lends weight to the last of the hypotheses posed, that greater overall richness in tree-hole food webs may owe more to spatial heterogeneity in the composition of the community at each site, rather than to any processes operating within a single habitat unit.

Lastly the strong relationship between latitude and variable number of prey species per predator species is probably merely a reflection of the increasing number of 'prey' species that occur in more tropical webs, given the more or less generalist nature of the predatory species involved. It seems likely that the number of different prey species that occur in the diets of such generalist predators reflects merely the number they may encounter within the same habitat space rather than any learned or evolved choice on the part of the predators.

Vertical structure

This variable essentially reflects the number of predators in the food webs as is seen in derived variables such as the average food-chain length, the number of trophic levels and the predator:prey ratio. So any explanation for why these variables show strong relationships with both latitude and, in most cases, rainfall becomes a search for answers to the question: Why do more predators exist in the more tropical, wetter sites than in others?

A number of hypotheses could account for this pattern, some of which have been presaged in the previous section of this discussion:

- The larger size and, in consequence, more extended life history of most predatory species is better served by the warmer temperatures of the tropical sites, allowing quicker maturation.
- The tropical sites are more productive and hence allow higher trophic levels to exist within these webs as opposed to the less productive more temperate ones.
- The wetter sites are, presumably, more predictable as habitats for aquatic organisms, with a reduced risk of the tree holes drying out before the longer lived predators can complete the aquatic stages of their life histories.
- The wider availability of other tree holes and even other container habitats (even if only as occasional refuges) will reduce the risk of local extinction for ecologically more demanding predatory species and hence increase the likelihood of persistence of a population at any particular site.

The first of these hypotheses is in part a truism given the poikilothermic nature of all tree-hole inhabitants. What needs to be established is that, on average, the predatory species have longer life cycles than the non-predatory species. Anecdotal evidence exists in support of this hypothesis from attempts we have made from time to time to breed out adults from larvae collected in tree holes. Mosquitoes, midges and other detritivorous species emerge rapidly from field-collected samples: tanypodine midges, *Toxorhynchites* mosquitoes and odonates in general take longer. Quantitative information collected under controlled conditions is needed to confirm (or deny) this anecdotal evidence.

The second and third of these hypotheses, that it is the productivity or predictability of the more tropical environments that allows organisms to exist at higher trophic levels, has wider ramifications. As noted in Chapter 6, there has been considerable debate on the mechanisms which may or may not be involved in generating longer food chains within communities and I revisit this issue in Chapter 13. The analyses of the numbers of predators within the Australian continental data set are of special interest inasmuch as they show a likely strong relationship between this variable and the total rainfall of the sites. 'Total rainfall' in these generally wet forests reflects as much the continuity or lack of seasonality in the wetter sites rather than the magnitude of precipitation at any one time (although this may become dramatically high at some of the lowland tropical sites). In turn I take this to be an indicator of the degree to which any particular tree hole will remain wet over extended periods: the principal variable that determines 'predictability' of these habitats as far as their aquatic inhabitants are concerned.

Accordingly this gives circumstantial weight to the 'predictability' hypothesis, although, as Stuart Pimm and I have pointed out elsewhere (Kitching & Pimm 1985, Pimm & Kitching 1987), it does not allow the 'productivity' hypothesis to be wholly discounted.

The final hypothesis, that there are more container habitats available in more tropical environments, allowing at least some of the inhabitants to 'spread the risk' (*sensu* Reddingius & den Boer 1970) of continued existence in an area, has considerable merit. In the tropical sites at Cape Tribulation, for example, there is an abundance of axil waters – of *Alocasia*, of *Zingiber* spp., of the strange climbing pandanaceous *Freycinetia* sp. and of the extraordinary 'backscratcher' ginger *Tapeinocheilus* sp.; there are fallen but water-holding coconuts and the basal spathes of palm fronds; and there is a great variety of tree holes – buttress pans, rot holes, pipes. In contrast the cool-temperate rainforests contain few phytotelmata other than tree holes, and even these are scarce. Between these two extremes the subtropical forests of the Lamington Plateau present a few *Alocasia* axils and fallen palm fronds, but in general few of these, and fewer tree holes than in the tropical situation – even though the forest is mature and undisturbed. The interaction and similarity between faunal assemblages across co-occurring but different classes of phytotelms in Australian forests deserves further attention.

Connectance

Connectance remains an enigmatic food-web characteristic in these analyses as in the global ones presented earlier. It may well be, as Polis (1994) asserts, that measures of connectance made for food webs (or as he would say, *partial* food webs) dominated by arthropods, which are in general generalist predators, are unlikely to be illuminating in comparative analyses. In the present analyses the marginally significant relationship between connectance and total rainfall may simply reflect the fact that although not well correlated with other food-web variables, connectance is more highly correlated with the numbers of predators ($r=0.47$) than with anything else. Hence the connectance/ rainfall relationship may simply be a 'ghost' of the much stronger relationship between the numbers of predators and rainfall, discussed earlier.

The simple connectance ratio (actual/potential feeding links) which I have used throughout as a food-web statistic has complex properties which mitigate against ready interpretation. Consider a simple web with three detritus components and three detritivores. If these detritivores are all generalists then the connectance ratio is one; if they are semi-specialised (i.e. each feeds on two of the three basal components then the value is 0.67, and if each specialises on but one basal component the value is 0.33. These values do not change if the number of detritivores considered is changed. Then add a notional generalist predator to each of these webs: again a set of general values emerges (1, 0.75 and 0.5) which is independent of the number of detritivores in the web. However, if the predator itself is a specialist (feeding on only two or a single prey species) the situation gets much more complex. For a semi-specialist predator (defined as feeding on two species of prey), values for the range of generalist/semi-specialised/specialised detritivores are (1, 0.75, 0.5) for webs with one or two species of detritivore but (0.92, 0.67, 0.42) for webs with three species of detritivore. Lastly, for a totally specialised predator, the sequences are (1, 0.75, 0.50), (0.87, 0.64, 0.38) and (0.82, 0.58, 0.35) for webs containing one, two or three detritivores, respectively. My point here is that values of connectance are very sensitive to *structural* differences in food webs and the interpretation of pattern in any but the simplest webs will always be equivocal. As a general, broad-brush measure of average degree of specialisation across nodal species, they probably have value but more detailed analysis is almost bound to be confounded.

Connections

Again these results will be revisited in Chapter 13. Spatial variation in food webs occurs, of course, at more local scales as well as at the scale of geographical regions and continents. Two different aspects of these more local variations are dissected in Chapters 9 and 10 using both tree-hole and *Nepenthes* pitcher-plant communities as cases in point.

9

Food-web variation at smaller spatial scales: regional and local variation in tree-hole and *Nepenthes* webs
Spatial pattern: smaller scales

Entomologists are well aware of the place-to-place variation in occurrence of the insects they choose to study. As has been discussed in the two preceding chapters, sometimes this variation is on a macro scale: it is unlikely to be profitable to search for *Toxorhynchites* mosquitoes in tree holes in western Europe, or members of the chironomid genus *Metriocnemus* in Australia for example – these genera and the groups to which they belong just do not occur in these geographical regions as far as we know. Even within their ranges, however, particular species are not uniformly distributed. Seeking a particular butterfly, for instance, the experienced collector will generally seek out known sites – often very specific patches of habitat subsumed within the wide geographical range of the species concerned. But even at these 'known' sites the quarry may not be found on or around the same bush that it was found on last season: it may even be absent from the site altogether for more or less extended periods. These familiar reflections on the frustrations of the collector are a clear indication that there is population variation at spatial scales far smaller than the continental or global scales addressed in the two preceding chapters.

Just as with the highly collectible butterflies, so phytotelm organisms show comparable, more or less unpredictable patterns of occurrence within their larger biological ranges and, taken together, these produce emergent properties in the food webs of which these species are integral parts. This chapter, the third of those which examine spatial scale *per se* as an organising variable for food-web patterns, presents data from both water-filled tree holes and a species of *Nepenthes* pitcher plant, collected and analysed at spatial scales ranging from a few hundreds of kilometres down to merely metres.

Metapopulations, structured demes and heterogeneous environments

As we seek understanding of underlying processes at smaller scales in food webs, so the dynamics of the populations of the constituent species become more important. Ideas related to the direct interaction between a species and its food resources within a habitat unit, between different species within the same unit, and the dispersal of individuals among habitat units become critical. Like most naturally occurring arthropods, populations of phytotelm organisms are organised into semi-independent groups which interact, through dispersal, only occasionally. The aquatic milieu of phytotelms and their dispersion as small units across the wider terrestrial landscape demands that the associated arthropods must invade or reinvade habitat units on a regular basis. Inevitably, then, this means that they are organised into what have become known as metapopulations (Levins 1970, Hastings & Harrison 1994, Gilpin & Hanski 1991). An earlier analysis of such structured populations, focusing on the likely relatedness of organisms within and between habitat units and its evolutionary consequences, led Wilson (1980) to coin the, albeit less euphonious, term 'structured deme'. 'Metapopulation' or 'structured deme' – the underlying idea remains the same and leads to an approach which focuses on both the within-habitat unit processes (i.e. at the level of the 'population' or 'deme') and the between-habitat unit processes ('interpopulational' or 'interdemal') rather than the more traditional population approach based on extended life tables which deal exclusively with processes of mortality and natality. The metapopulation approach acknowledges that ecosystems are intrinsically heterogeneous and that the long-term health of a population will be as much (if not more) a measure of the species' success or otherwise in moving between places in which it is established.

This unifying idea (actually more a terminology given that the appreciation of subdivided populations connected by dispersal predates both Wilson and Hanski) is well illustrated by tree-hole species such as *Metriocnemus cavicola* in the United Kingdom (Kitching 1972a). This species of chironomid midge (formerly known as *M. martinii*) is the commonest metazoan inhabitant of water-filled tree holes in southern England. It occurs in most but not all more permanent holes and exhibits a number of different but repeated patterns of larval dynamics within each habitat unit. In some cases a clear summer peak of abundance within tree holes is seen, in others a flat low level of occurrence is seen throughout the year, and in others there is sporadic, unpredictable occurrence in several periods within a year. These contrasting dynamics will be a reflection of the quality of the resource within a particular site, the presence and abundance of potential competitors (larvae of the ceratopogonid *Dasyhelea dufouri*), and the time course of repeated invasion or rein-vasion of the site through the actions of ovipositing females. Once a site inva-sion has occurred, processes of larval dynamics take over and we focus on the resource base and competition (and, in other more complex food webs elsewhere, on predation). Outcomes may be favourable, poor or disastrous for the species depending on other properties and events within the particu-lar habitat unit. But it is the far chancier processes of inter-hole dispersal, extra-hole survivorship, and the seeking of oviposition sites that are crucial to the overall health of the species' metapopulation. These processs are quin-tessentially stochastic and will drive the within-site dynamics to a substan-tial degree. Without new invasion a larval population must eventually decline, however favourable the within-site conditions.

I have commented elsewhere on the spatial structure of such populations in general and on some algebraic approaches to their modelling (Matthews & Kitching 1984, Kitching 1992). We return to ideas of structured popula-tions as we seek the explanation of patterns described in this chapter (revis-ited further in Chapter 13).

Food webs at the regional scale
Tree holes in the Daintree, Queensland

The availability of a particularly rich data set from the rainforest around Cape Tribulation in the Daintree region of far northern Queensland has allowed an analysis of food-web variation on the scale of a few kilometres to a few tens of kilometres. At this scale, variation reflecting topography, drainage patterns and the impact of the littoral zone is apparent in the vegetation and structure of the forest. I examine here the question of whether or not this ecosystem variation is also apparent in the tree-hole communities scattered throughout the district.

The Daintree rainforests

At the regional level it became apparent to us that the sections of rainforest that we had studied could be classified into one of three formations which we called 'maritime', 'riparian' and 'lowland'. These differed in vegetation, topography, distance to flowing freshwater and distance to the coastline.

Maritime forest, as we designated it, was non-riparian forest located on the sandy floodplain within 300 metres of the sea. In separate studies a number of 10 × 10 m plots were surveyed for details of the vegetation, the results from three of which have been included in the analyses of McIntyre *et al.* (1994). Canopy species included *Dysoxylon mollissimum, D. oppositifolium,*

Table 9.1. *Physical dimensions and some results from tree holes from different forest types at Cape Tribulation, Queensland*

	Maritime forest	Lowland forest	Riparian forest
Length (mm)	143.6±21.52	210.6±18.68	231.5±34.10
Breadth (mm)	73.10±9.78	95.5±8.75	91.3±12.57
Depth (mm)	115.2±21.56	118.0±10.88	86.5±7.23
Height from ground (mm)	858.3±123.22	690.9±164.84	1207±152.80
pH of liquid contents	6.1±0.24	5.4±0.17	3.4±0.47
Number of holes sampled	20	31	21
Species (per hole)	3.3±0.38	3.4±0.30	3.4±0.43
Predatory species (per hole)	0.4±0.11	0.4±0.12	0.7±0.15
Species present (total)	10	12	16
Predators present (total)	1	2	5
Number of feeding links	18	27	44
Number of trophic levels	2	3	3
Predator:prey ratio	0.11	0.2	0.45
Prey species per predator[a]	7	7.5	6.5
Average food-chain length	1.73	2.2	2.18
Connectance	0.51	0.53	0.48

[a] Based on the actual number of links recorded in webs.

Sloanea macbridei, Planchonella obovoidea and *Litsea bindoneana*. The area was rich in woody plants with over fifty species recorded within but three 10 × 10 m plots. We sampled a total of 20 tree holes from this forest type, the physical dimensions of which are summarised in Table 9.1.

'Lowland' forest we designated as non-riparian forest at least a kilometre from the coastline, but still below 100 m elevation, located on the coastal plain at the foot of the coastal ranges due west of Cape Tribulation. Again a series of plots were surveyed for their vegetation and McIntyre *et al.* (1994) present detailed results from four of these. Canopy species included *Dysoxylon micranthum, Cardwellia sublimis, Endiandra monothyra* and *Synima cordierorum*. The woody plant assemblages within these sites were different from those at the maritime sites with only a 30% overlap in species according to McIntyre *et al.* (1994). About 60 species of woody plant were encountered in four 10 × 10 m quadrats. Thirty-one tree holes were found and sampled within such lowland forest.

'Riparian' or 'creekside' forest we designated as rainforest located within the other two types but located within a few metres of a permanent freshwater creek. No separate vegetation surveys were done on this vegetation. A small subset of tree species thrived along the creek banks and this included several which showed a marked propensity to form tree holes. This subset included *Xanthostemon chrysanthus*, individuals of which often contained multiple tree holes. A total of 21 creekside tree holes were found and sampled (Table 9.1).

Treehole webs from different parts of the forest

Table 9.1 contains summaries of the food web found within tree holes in each forest type. I have included in this analysis two taxa which I excluded from the analysis in Chapter 8. These were individuals of a predatory veliid bug and of a (presumed) detritus-feeding tetrigid hopper. Both occurred only in creekside holes and form a unique semi-aquatic set that live either on the water surface (as in the case of the veliids) or about the water/air/bark interface (as for the tetrigids). The tetrigids moved freely into and out of the water of the tree holes about which they occurred. In this respect they parallel the behaviour of the *Camponotus (Colobopsis)* ants that occur in mutualistic relationship with the pitcher plants *Nepenthes bicalcarata* which have been discussed in Chapter 7. The results presented in the Table have also required that all the culicid species be combined into a single node.

The food webs derived from these data are illustrated in Figure 9.1. The creekside holes are substantially more complex than those from other areas in the forest: they contain predatory planarians, dytiscid beetles and the surface-living veliids already discussed, in addition to the predatory larvae of the tanypodine chironomid *Anatopynnia pennipes*. This last species of predator also occurred in holes in the lowland and maritime forests. Both creekside and lowland forest contained an odonate top predator. Within the limits of our data set, the larvae of the zygopteran *Podopteryx selysi* appear to be limited to lowland forest sites, those of an anisopteran libellulid to creekside holes. In both instances these top predators occurred only rarely (as would be expected for organisms that sit at the top of food chains) but the lowland/creekside discrimination was absolute within the data set.

A total of 14 different saprophagous taxa occurred across all three webs, nine in the maritime holes, ten in the lowland sites and eleven in the creekside sites. The interaction of these sets of saprophages with the different numbers of generalist predators present within the three sets of holes produced the notable differences in the statistics 'number of feeding links' and 'predator:prey ratio' apparent in Table 9.1.

The question of whether or not the three subsets we have identified are in some real biological sense different 'subcommunities' is deceptively simple. A definitive answer to the question would require experimental manipulations involving transfer of individuals across forest types and even then the

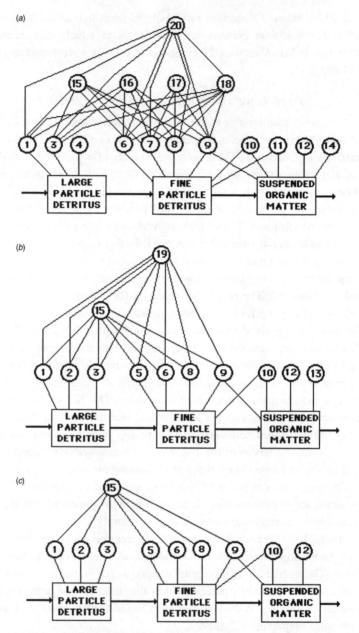

Fig. 9.1. Food webs from water-filled tree holes in lowland tropical rainforest around Cape Tribulation, north Queensland: (*a*) the composite web based on samples from 21 habitat units in creekside situations, (*b*) the composite web based on samples from 31 habitat units in non-creekside, lowland forest, and (*c*) the composite web based on samples from 20 habitat units in maritime ('littoral') forest.

Table 9.2. *Values for Sørensen's Index of Faunal Similarity among species assemblages in water-filled tree holes in three forest types at Cape Tribulation, North Queensland*

	Maritime forest	Lowland forest	Riparian forest
Maritime forest	1		
Lowland forest	0.91	1	
Riparian forest	0.62	0.57	1

Compare with the similarity value of 0.92 obtained using random subsets of tree holes from the overall data base.

credibility of any results would be open to challenge in that any functional differences or differential survival might as well be the result of the intervention process as it might reflect real ecological difference between the three situations. Table 9.2 presents a further analysis which strengthens the idea that the observed differences between the subsets are not merely due to chance. In this table I present the values of Sørensen's Index of Faunal Similarity (see Chapter 5) calculated for all pairwise combinations of the food webs found in each of the three forest types. These values showed a high degree of overlap between the assemblages from the maritime and lowland forest (at 90% similarity). Values for riparian/lowland and riparian/maritime comparisons were much lower (at 57% and 62% respectively). It is impossible to attach a level of statistical significance to these numbers. The second part of the table, however, shows the values obtained when comparisons are made between sets of tree holes selected at random from the full set of 74, with each set containing the same number of observations as in the three 'real' subsets. The similarity values for these artificially selected sets was around 92%. Conservatively, I interpret this to mean that the actual similarity value obtained for the maritime/lowland comparison is hardly less than that obtained between the random sets. The similarities observed between the riparian assemblages and the other two, however, are sufficently low to suggest that there may indeed be an ecological difference marking out the creekside tree holes.

The pattern and explanation

I consider these results, together with the thematic hypotheses identified in Chapter 6, in the general discussion which is Chapter 13. Some consideration is warranted at this stage, simply addressing the question: What biological and/or environmental factors may allow some species to exist in the richer riparian communities while excluding them from tree holes elsewhere? In particular why should the additional predatory species which occur in creek-side sites (and which account for most of the identified statistical differences) occur there and not elsewhere?

The most naive hypothesis to account for observed differences is simply that the sites present different physical and chemical environments to poten-tial inhabitants. The first part of Table 9.1 presents the dimensions of the holes in each of the three subsets of the forest, together with the acidity val-ues obtained for their liquid contents. There is little to distinguish the phys-ical dimensions of the three subsets: the holes in the maritime forest are some-what shorter in their principal surface dimension than in either lowland or creekside holes, otherwise it is hard to detect differences. In particular, there appears no reason to differentiate the riparian holes from the others on any physical basis.

Acidity values however show a clear pattern. The values obtained from the maritime and lowland forests are similar to each other (with values around 5.75) and are in contrast to those obtained from creekside holes which are considerably more acid (with an average pH value of 3.4). If particular species of organism seek out more rather than less acidic situations then this could account for at least some of the observed differences. Acidophilic insects and other organisms are well known in the limnological literature.

In addition to these analyses based on hole characteristics, we need also to consider the biology of the particular groups which produce the food web to food web differences. Some of the issues canvassed earlier in the general discussion of metapopulations are important here. The 'extra' organisms found in riparian holes - the dytiscids, planarians, veliids and tetrigids - are all organisms which might well have ecological needs which demand adja-cent streams for their persistence. Inter-hole dispersal of flatworms, for exam-ple, is hard to contemplate over any distance without occasional inundation. Veliids and dytiscids I suspect are not restricted to tree holes but seek out any pools of static water in which they may find the oligochaetes and insect larvae which make up their prey. Such pools, in the forest we studied, are likely to be more or less restricted to riparian zones. The tertrigids we encoun-tered are also unlikely to be specialist tree-hole dwellers and I suspect are more likely to be found in the more open, exposed situations characteristic of creek banks where there is a rich epiphytic mat on the branches and but-tress roots around the tree holes.

Lastly, and considering only the Odonata, it is of some interest that the zygopteran and the anisopteran appear to have partitioned the tree-hole habi-tat in which they both choose to breed. The riparian/non-riparian distinction observed may well reflect adult behaviour with very different foraging and agonistic behaviours in the two very different odonates. The zygopteran *Podopteryx selysi* is known to be a tree-hole specialist (Watson & Dyce 1978) but we know little of the habitat specificity of the anisopteran which, to this point, has not even been reared successfully to adulthood. Fincke (1994) has demonstrated the likely importance of differential oviposition behaviour in a set of damselflies found in Panamanian tree holes. It is possible that, as with some of the predatory species mentioned earlier, the Daintree libellulids are not tree-hole specialists, merely using tree holes as one of a number of small static water bodies in which to breed. This is the situation for the Bornean tree-hole dragonfly, *Indaeschna grubaveri* (A. E. Orr, unpublished results).

The fauna of Nepenthes albomarginata

Food-web variation on a slightly larger 'regional' scale is observed in data we have available on the fauna occurring in pitchers of *Nepenthes albomarginata* in South-east Asia. *N. albomarginata* occurs throughout Sumatra, West Malaysia and parts of Borneo (Kurata 1976, Clarke 1992). It is common in some parts of its range but is rare in western Borneo. Studies of the fauna occurring within its pitchers have been made from five sites, four in West Malaysia and one in Brunei. The basic information from the Malaysian sites was collected by Roger Beaver (Beaver 1979a, 1980, Kitching & Beaver 1990). The additional information from the Brunei site was collected by Charles Clarke (1992). The Brunei information is also used in the analyses on the role of the host plant presented as Chapter 10.

Pitcher plant habitats in Malaysia and Borneo

Nepenthes albomarginata is a scrambling herbaceous plant characteristic of recently disturbed forest edge situations. In Malaysia Beaver (1979a) described its occurrence in the Penang District from sea level to 1300 m elevation. The island of Penang is more or less equatorial with a natural vegetation which comprises lowland evergreen rainforest. It is floristically extremely rich with a number of extremely local endemic trees characterising the island forests (Whitmore 1975). The study site in Brunei was located in 'kerangas' forest – the heath forests on siliceous soils characteristic of many low nutrient regions of Borneo (Whitmore 1975). Clarke (1992) describes the site at Labi in Brunei in kerangas forest on the margin of a power-line easement. This is one of the easternmost sites for this species. Details of the locations of the pitchers sampled are summarised in Table 9.3. For each site the

Table 9.3. *Locations from which samples of the fauna of pitchers of* Nepenthes albomarginata *have been made*

Location	Position	Elevation (m)	Number of pitchers sampled
Malaysia			
Behang Bay, Penang Island	5° 25′ N 100° 20′ E (1)	*c.* 20	58
Kedah Peak, Kedah	25 mls NNE of (1)	1300	46
Ayer Itam, Penang Island	5 mls WSW of (1)	200	57
Penang Hill, Penang	4 mls W of (1)	950	38
Brunei			
Labi Road	4° 50′ N 110° 52′ E	*c.* 45	20

complete contents of pitchers were sampled, with sample sizes ranging from 20 to 57 pitchers depending on the location.

Food webs from Nepenthes albomarginata

The webs recorded from these sites are illustrated in Figure 9.2. Statistics reflecting the food webs found in each are given in Table 9.4. As in some previous analyses I have excluded the largely terrrestrial predatory spider *Misumenops* sp. from these webs (cf. Kitching & Beaver, 1990). This species of spider appears to be more or less ubiquitous across *Nepenthes* pitchers wherever they occur, so its inclusion in or exclusion from the food-web analyses makes no difference in practice (but see the recent comments of Clarke, 1998, on this topic). The most complex web in this set is that from the pitchers on Penang Hill in West Malaysia at 950 m altitude. This contains three predatory species (one of which is a parasitoid of phorid larvae). In contrast the web from Bahang Bay (which I assume to be at sea level) is lacking in predators entirely and yet is in close proximity to the Penang Hill site and that at Ayer Itam which shows an intermediate level of complexity, containing but two predators. The webs from *Nepenthes albomarginata* on the adjacent mainland at Kedah Peak (1300 m) are also relatively simple, having but one predator present. All of these Malaysian sites are located within about 30 kilometres of each other. In terms of faunal similarity the Penang Hill, Ayer Itam and Kedah Peak webs are all very similar, with Sørensen coefficients varying from 86% to 95% (Table 9.5). The Bahang Bay site stands out due to its relative simplicity, exhibiting but 70% similarity with the Penang Hill site.

   The four Malaysian sites show only minor topographic, altitudinal and vegetational differences among themselves. In contrast the Labi site is 1600 kilo-

(a) **Penang Hill, Penang, Malaysia**

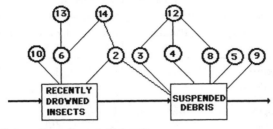

(b) **Ayer Itam, Penang, Malaysia**

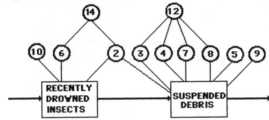

(c) **Kedah Peak, Malaysia**

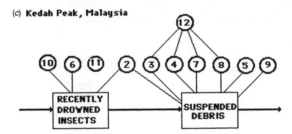

(d) **Bahang Bay, Penang, Malaysia**

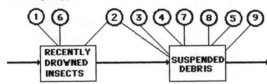

(e) **Labi, Brunei**

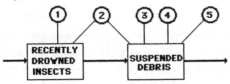

Fig. 9.2. Food webs from pitchers of *Nepenthes albomarginata* at four locations in peninsular Malaysia and one in the Sultanate of Brunei. The Malaysian webs are based on Beaver (1979a, 1980) and Kitching & Beaver (1990); the Bruneian web is after Clarke (1992).

Table 9.4. *Food-web statistics from pitchers of* Nepenthes albomarginata *from five South-east Asian locations*

	Penang Hill	Ayer Itam	Kedah Peak	Behang Bay	Labi
Number of species	11	11	10	9	5
Number of predators	3	2	1	0	0
Feeding links	15	16	15	10	6
Number of trophic levels	2	2	2	1	1
Predator:prey ratio	0.38	0.5	0.1	–	–
Prey species per predator[a]	2	3	4	–	–
Average food-chain length	1.7	1.7	1.4	1	1
Connectance	0.38	0.44	0.5	0.55	0.6

[a] Based on the actual number of links recorded in webs.

Table 9.5. *Faunal similarities as measured by Sørensen's Index among food webs from pitchers of* Nepenthes albomarginata

	Penang Hill	Ayer Itam	Kedah Peak	Behang Bay	Labi
Penang Hill	1				
Ayer Itam	0.91	1			
Kedah Peak	0.86	0.95	1		
Behang Bay	0.7	0.9	0.95	1	
Labi	0.5	0.73	0.53	0.72	1

metres distant from these sites and contains no predators and but five saprophages. The Labi web contains two species of *Tripteroides* mosquito and an anoetid mite which I have assumed to be the same as the taxa so designated in the Malaysian webs. This simplifying (and strictly speculative) assumption leads to relatively high similarity values between the Labi web and the other four (ranging from 50% to 73%; Table 9.5).

The pattern and explanation

Beaver (1979a) notes that *Nepenthes albomarginata* is commonly thought of as a high-altitude species but, as he points out, the species does have a much wider altitudinal distribution. It is of some interest that the higher altitude sites are the most complex in terms of the contained food web and this could be interpreted as reflecting the preference and enhanced survival of this species of pitcher plant for higher altitude sites. This remains, however, mere speculation: an hypothesis for further study. Indeed there are a number of additional species of *Nepenthes* which have distributions that stretch over a wide altitudinal range (see for example Figure 7 in Kurata, 1976) which would repay study from the point of view of how this altitudinal change affects the community within the pitchers.

The essentially fortuitous collection of sites and range of altitudes covered within the data set I have presented can do no more than indicate the particular interest which would attach to such a study. Indeed it is fair to say that the whole question of the role of altitude in determining food-web structure within phytotelms awaits proper attention. Like so many other questions in the study of food webs in general, I suggest that the phytotelms, particularly the pitcher plant, make ideal subject systems for such a definitive studies. In the meantime the results of the analyses presented here and in Chapter 8 can only tantalise!

The strangely simple and 'different' food web that we observed within *Nepenthes albomarginata* in Brunei has been accounted for by Clarke (1992, 1997) on the grounds that this population of *albomarginata* is on the extreme edge of the species' range and also has local morphological features which could restrict access to a number of predatory species. In particular the pitcher form of these outlying populations is elongate and narrowly cylindrical, potentially limiting access to the surface of the contained liquid by larger ovipositing insects such as females of the predatory *Toxorhynchites* species. This, for me, is more convincing than the 'edge of range' argument, given that there is a wide range of other pitcher plant species in the vicinity containing a suite of predators, some of which co-occur in several of the *Nepenthes* species. This theme has been discussed in Chapter 7 and is taken up further in Chapter 10.

The variation within the few tens of kilometres which separate the sites on and adjacent to Penang Island presents a different problem. There is no doubt that all the food webs identified are closely similar. There may be altitudinal differences but we must also consider the possibility that the observed differences have a chance component. It seems unlikely that the differences between the food webs from Penang Hill and Ayer Itam, for instance, represent any more than the presence or absence of a chance encounter with a rare species – in this case the encyrtid parasitoid of *Endonepenthia* and one of four species of mosquito. The differences between the Bahang Bay sites and the others may be more significant. The sea-level sites do not contain *Toxorhynchites*, which may well reflect the overall distribution of this culicid and its possible absence from maritime forests. This too remains a speculation requiring checking.

The role of chance figures even more in the determination of food-web pattern from tree hole to tree hole, pitcher to pitcher, bamboo to bamboo, and it is this theme that I take up in the final parts of this chapter.

Variation at the local scale

The smallest spatial scale I intend to present in this book is that which shows up when the food webs from individual habitat units are compared. I have examined within-unit spatial variation from time to time and have identified physical gradients in pH and conductivity across the width of, and at various depths in, the water bodies within single tree holes. If I had devoted more attention to micro-organisms in our phytotelm studies then it seems likely that these intra-unit differences might prove to be of importance. As it is, at the level of the metazoan food web, I am satisfied that the information I analyse at the level of the unit (presented below) and the sampling methods used are not confounded by such within-unit variation in distribution or relative abundance of the organisms within the food webs.

One further methodological problem arises when the goal is to present the food web here and now in a single habitat unit. Unless the entire contents of the phytotelm are sampled (as is the case usually with, say, pitcher plants or bamboo cups – see Kitching & Beaver 1990) then it is highly unlikely that a simple subsample (as we have usually taken from water-filled tree holes) will contain representatives of the entire fauna of that hole. For analyses at larger levels of spatial resolution this is circumvented by combining the results from several samples, each from a different site. But when the goal is to examine the habitat unit to habitat unit variation in either space or time this is not a permissable option. I circumvented this in Kitching (1987a) by combining the results of three months of repeat sampling from each habitat unit. This does not entirely overcome the problem of missing species but it goes a long way towards doing so.

In order to preserve a greater information content in these data I have used a simple ranking of relative abundance for each organism encountered in a particular habitat unit. Using the approach outlined in Chapter 5, I employed a different-sized symbol to represent levels of abundance falling within each of the four quartiles of the full range of observed abundances (after conversion to logarithms).

Tree holes in south-east Queensland

One of the Australian data sets included in these analyses in Chapter 8 was collected in Lamington National Park in south-east Queensland and was based on the repeated sampling of eleven marked holes in the subtropical rainforest. In Chapter 10 we shall encounter some of the time series data generated by this study. In addition this study permitted me to construct separate food webs for each of the eleven sites included in the study and it is this analysis of food-web variation at the very local level that I present here.

When first presented, in 1987, these results were one of the first cases in which food webs were demonstrated to vary in both space and time. The following account is based on that paper.

The forest and the tree holes

Details of the vegetation, location and climate of Lamington National Park, within which this study was carried out, have been presented in Chapter 7. The eleven tree holes that we sampled were visited monthly over the period from September 1982 to August 1983. They represented a variety of sizes as reflected in their surface area and in their localities within a few square kilometres of forest. On each occasion samples were removed in the fashion described in Chapter 5. The volume of the samples varied within the range 80 ml to 550 ml depending on the size of the hole and the amount of liquid it contained at any one time. From each of those monthly samples for each of the eleven holes, organisms were sorted and counted. The counts obtained were converted to numbers per litre.

Constructing food webs from individual tree holes

For each tree hole included in the twelve-month sampling programme I constructed four webs, each representing a summary of the presence or absence and the level of abundance of those species present in each three-month period of the study. Mean abundances for each species present in each quarter were calculated and these were incorporated into a rank representation as has already been explained. Foodwebs were then drawn using the full web (Figure 7.2*d*) as a template but drawing in only those links defined by the set of species present at the place and time.

A number of summarising statistics were derived from the webs constructed as I have described. For each site and time the number of species present, the number of trophic levels and the number of predatory and top predatory species were calculated. In addition, an abundance index for all the species in the web and the saprophagous and predatory components was computed. These indices were arrived at simply by summing the quartile measures described (0 to 4 in each case). The presence of the top predatory tadpoles of *Lechriodus fletcheri* was incorporated into these indices by recording its presence as one in the sum. A measure of connectance was calculated in the normal manner. These measures have been used to describe the spatial and temporal variation in the web structure within this set of tree holes.

In addition two subsidiary analyses were carried out, each one predicated upon hypotheses implied about the determination of food-web structure within particular tree holes. In the first of these tests the hypothesis was that the pattern of variation among the eleven sites is related to the distance between them: that is, that the pattern in any hole will be most similar to that in adjacent holes and the degree of similarity will fall off with distance. We tested this hypothesis by constructing a correlation matrix for the 44 data sets represented by the webs (eleven sites by four quarters.) Correlation coefficients were drawn from this matrix for selected pairs of holes and the mean values across the four quarters calculated. These correlation coefficients were used as data points to test the null hypothesis that there is no relationship between the degree of similarity (as measured by these correlation coefficients) and inter-site distance.

Secondly I examined the possible role of surface area of the tree holes, as a measure of their size, in determining pattern or structure in the observed webs. Essentially, I looked for significant relationships between the various abundance indices and structural measures with the surface areas of the holes sampled.

The webs

Figure 9.3 (from Kitching 1987a) shows a selection of the 44 webs constructed as described. Each has been overlain on the full potential webs for the region (shown as dashed lines in the figure). This full web is slightly simpler than that presented in Chapter 7 for the simple reason that after these repeated studies of eleven sites had been made we discovered an additional species of midge as an occasional inhabitant of holes in the early stage of assembly – see Chapter 10. Tables 9.6 to 9.9 present summarising statistics on the food webs themselves and the trophic levels within them.

The number of species present in the webs varied from three to seven (Table 9.6). No hole contained all nine species either at any one time or through time. Only two of the eleven sites achieved eight species through time although eight of the remaining sites did have seven of the nine species through time. Site 10 was the poorest overall, having but six species through time. There was much less variation in the number of trophic levels present over sites and times (as would be expected for a variable having such a restricted range of possible values). Eight of the eleven holes had but two trophic levels present on each occasion whereas the remaining three had three levels on some occasions and never less than two such levels (see Table 9.6).

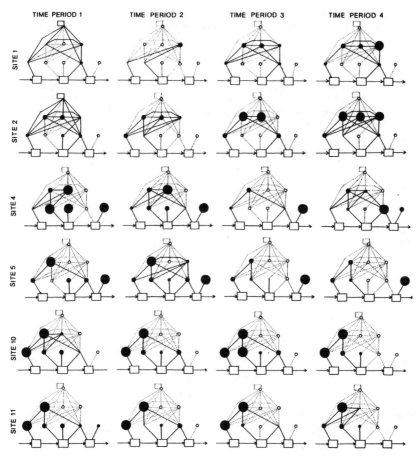

Fig. 9.3. Food webs observed in individual tree holes in Lamington National Park, south-east Queensland. Each of the four webs shown for each hole represents an aggregation of three separate consecutive monthly samples from that site. The different-sized node symbols represent the relative abundance of species observed over the time period at each site (see text for further explanation). The symbol for the top predator, tadpoles of *Lechriodus fletcheri*, represents presence or absence only. The figure is from Kitching (1987a).

The number of species of predators and top predators present varied from one to four, with all sites having at least two present through time. The maximum number of predators present at one time was four, the maximum possible value (Site 2, Quarter 1).

The abundance indices calculated for each of the species can be used as a guide to the numbers as well as the presence or absence of species (Table 9.7). These statistics have the following ranges: for all species, 0 to 33;

Table 9.6. *Summary statistics on the number of species from quarterly food webs from eleven water-filled tree holes from subtropical rainforest in Lamington National Park, south-east Queensland*

Site	Number of species		Number of trophic levels		Number of predators or top predators	
	Maximum	Minimum	Maximum	Minimum	Maximum	Minimum
1	5	3	4	3	4	1
2	7	4	4	3	4	2
3	6	4	3	3	3	2
4	7	4	3	3	2	1
5	7	4	3	3	2	1
6	7	5	3	3	3	3
7	6	5	3	3	3	2
8	6	5	4	3	3	2
9	7	5	3	3	3	2
10	6	4	3	3	2	1
11	6	4	3	3	2	1
Overall	7	3	4	3	4	1

After Kitching (1987a).

saprophages only, 0 to 20; predators and top predators, 0 to 13). The overall abundance levels calculated in this fashion ranged from 5 to 20, whereas the means per hole varied from 8.0 to 16.5. The values of these indices for the saprophages alone ranged in value from 4 to 15. Holes 1 and 5 showed the smallest (3.25) and largest (11.75) mean values, respectively. Lastly, predators and top predators varied in abundance, using this index, from 3 to 12. Overall the saprophage index is more closely linked to the total abundance index with a rank difference across all sites of only 2.1 (Spearmans $R = 0.72$, $0.05 < p < 0.01$), whereas a comparison between the rank order of means across sites for predators and top predators with total species show a mean difference of 4.3 ($R_s = -0.16$, ns). This contrast is not simply a reflection of the number of species involved within the two trophic designations, as there is but one more saprophagous species than there are predators and top predators. In addition there is a weak but significant inverse relationship based on these ranks, between the abundance index of the predators and top predators and that for the saprophages ($R_s = 0.71$, $0.05 > p > 0.01$).

Values of connectance are presented in Table 9.8. Mean values through time range from 0.31 at Hole 5 to 0.59 at Hole 6, with an overall mean value of 0.42. Within-site values range from a grand minimum of 0.15 to a grand maximum of 0.77. The range of values of connectance within any one site

Table 9.7. *Summary statistics based on abundance indices from quarterly food webs from eleven water-filled tree holes from subtropical rainforest in Lamington National Park, south-east Queensland*

Site	All species (33[a])			Saprophages (20[a])			Predators and top predators (13[a])		
	Maximum	Mean	Minimum	Maximum	Mean	Minimum	Maximum	Mean	Minimum
1	14	8.00	5	4	3.25	2	10	6.25	3
2	18	14.50	12	6	5.50	4	12	9.00	8
3	15	12.50	10	7	6.00	4	8	6.00	8
4	0	16.25	10	13	1.75	7	7	5.50	3
5	17	15.75	12	15	11.75	9	6	4.00	3
6	19	15.75	12	11	7.25	4	10	8.50	7
7	17	13.25	12	9	8.25	7	8	5.00	3
8	18	16.50	15	13	10.50	9	7	6.00	5
9	17	13.50	11	12	8.50	6	6	5.00	4
10	17	15.75	13	13	11.25	9	6	4.50	4
11	18	14.50	13	14	9.75	7	5	4.25	4

[a] Maximum value possible in each category.
After Kitching (1987a).

Table 9.8. *Connectance in quarterly food webs from eleven water-filled tree holes from subtropical rainforest in Lamington National Park, south-east Queensland*

Site	Period 1	Period 2	Period 3	Period 4	Mean	Range
1	0.50	0.15	0.46	0.38	0.37	0.35
2	0.77	0.38	0.31	0.62	0.52	0.46
3	0.35	0.62	0.31	0.31	0.40	0.31
4	0.35	0.50	0.23	0.46	0.39	0.27
5	0.27	0.54	0.23	0.19	0.31	0.35
6	0.42	0.73	0.73	0.46	0.59	0.31
7	0.46	0.62	0.35	0.42	0.46	0.27
8	0.50	0.46	0.35	0.46	0.44	0.15
9	0.50	0.69	0.31	0.35	0.46	0.36
10	0.46	0.31	0.31	0.19	0.32	0.27
11	0.35	0.31	0.27	0.31	0.31	0.08
Overall	0.49	0.48	0.35	0.38	0.42	0.29

After Kitching (1987a)

may be taken to be a measure of the temporal change in site-to-site quality which occurred throughout the course of the twelve months of observation. These within-site ranges varied in value from 0.08 (Hole 11) to 0.46 (Hole 2) with an overall mean of 0.29. There was no obvious seasonality in the connectance scores across sites.

In a comparison of the connectance scores with ranked values for the abundance indices calculated in the fashion already described, there was no relationship with the indices either for total species or saprophages alone. A significant positive relationship did exist, however, between connectance and the abundance indices of predators and top predators ($r = 0.71$, $0.05 > p > 0.01$), due to the fact that the presence of predators can contribute up to 75% of the possible value for connectance (see Chapter 5).

There was an inverse relationship between the inter-hole distance and the average degree of similarities shown across webs. There is a significant negative linear regression relationship evident in these data ($y = 0.423 - 0.0037x$, $r = -0.89$, $p < 0.01$). I conclude from this that in general holes that are close together may be expected to have more similar food webs both in terms of the presence or absence of species and the levels of abundance of those species through time.

The pattern and explanation

The results and analyses presented here demonstrate clearly that there can be substantial spatial and temporal variation in the structure of the food webs within a single habitat unit of a particular ecosystem - in this case tree holes in the subtropical rainforest.

The available theoretical explanations from this variation have been discussed in Chapter 6 and in the opening section of this chapter. The question of temporal variation will be revisited at length in Chapter 10 and I shall not dwell upon it further here. Similarly, issues concerning the processes that determine local variation in food webs on a spatial scale will be discussed further in the summarising chapter that is Chapter 13. However, it is worth considering some of the issues in greater detail here.

As has been discussed at length earlier two principal explanatory frameworks are available for explaining food-chain length and food-web complexity in natural ecosystems. The first of these relates to the amount of energy entering a particular system and the second the predictability of that system and the more or less complex food webs that the environmental predictability allows to persist. It is tempting in looking for explanations for variation in the very localised webs that occur in particular habitat units to turn to these explanations as possible structuring processes at this very local level.

From the analyses already presented it would seem there is little support for the idea that the amount of basal energy is determining the complexity of food webs of this scale. If this were the case then I would expect to find relationships between one or more properties of the food webs and the surface area of the tree holes which contain them, as it is through the surface interface that most energy enters tree holes.

The so-called dynamic constraints hypothesis relating to the predictability of the environment provides an apparently plausible explanation for the variation observed in the webs in space and time. If web differences reflect environmental differences due to micro-climate or structural heterogeneity across holes we would expect more complex webs containing longer food chains in those holes showing more equable conditions. As an argument based on the ideas of Pimm and Lawton (1977), however, this notion is fatally flawed. Pimm and Lawton's arguments rest on the premise that on average more complex webs will have longer characteristic return times following major environmental perturbations, before a full equilibrium structure is regained. Accordingly, they argue, *over evolutionary time*, the regular components of the web will reflect the frequency of perturbations at the site. Rephrased, a given degree of environmental predictability will have an associated equilibrial degree of web complexity. The key phrase in their argument, however, is 'over evolutionary time'. The environment presented on a tree-hole to tree-hole basis within a particular patch of forest will vary with season and with a degree of maturity of the trees as the surrounding vegetation changes. These changes will be superimposed on the shorter term variation due to rainfall pattern, incident radiation and litter input into the holes. Even if we adopt a time scale related to the average life time of the trees, this does not allow for an evolutionary explanation of the sort outlined above for local variations described here.

It is possible to recast these ideas of the role of predictability on to the local scale, as has been done by Jenkins (1991). In essence it is possible to think of the 'predictability' (however defined) of a particular habitat unit as permitting, not the long-term persistence of the animal species in an area (as envisaged by Pimm and Lawton), but the much shorter term persistence of the animal within that habitat unit. In other words this is an interaction with predictability which makes itself apparent through the dynamics of the metapopulation, rather than through evolutionary mechanisms.

The behavioural mechanisms have yet to be studied in any particular case but many possibilities exist. For example, the combination of the inevitable chance element affecting the success of search by gravid female insects for available oviposition sites will mean that only some of the suitable sites will be 'found' by individuals of any one species. If this chance process is combined with a specific mechanism in which the oviposition response is triggered by various qualities associated with the water quality and surface (as has been demonstrated, for instance, by Frank *et al.*, 1976, for pitcher-plant mosquitoes in Florida and by Wilton, 1967, for *Aedes triseriatus* in New York) then the period of time that such sites contain water – a measure of their 'predictability' – will contribute to the likelihood of any one site containing that species of animal.

The chance nature of this discovery is further enhanced in species such as mites and microcrustaceans which rely on phoresy as a means of transferring themselves from one tree hole to another. Not only do they need to attach themselves to an emerging adult mosquito, either as a free-riding propagule hanging on to the limbs of the emergent insect or as a temporary parasite within the body of that insect, but they then must take their chances of that carrier insect surviving the vicissitudes of the forest environment and successfully locating another tree hole in which that insect can lay its eggs and into which its passengers can be released.

These processes related to the invasion process will be complemented by others related to post-invasion survival. Aquatic larvae will survive poorly in habitat units which become dry or nearly dry after the initial oviposition has occurred. In habitat units prone to frequent drying out this will reduce the chances of any particular species being found within the aquatic milieu even when it is in place. Again this will be due to the interaction between poor survival in less predictable habitat units (i.e. those that dry out frequently) and the probabilistic nature of any one animal species finding that habitat unit in the first place.

These species-specific mechanisms will produce, conjointly, the different food-web patterns we observe from tree hole to tree hole. Of course the scenarios presented, though plausible, remain conjectural. Experimentation is needed.

In summary then, there is a combination of factors involved in determining local food-web structure within tree holes at least. The chance nature of propagules of a species of organism actually arriving within the tree hole unit is coupled with the greater or lesser likelihood of survival once those propagules are established in the tree-hole habitat. The outcome of these processes will owe a great deal to the dispersion of habitat units in space, the evolved search patterns of the tree-hole organisms in the adult stage and, I suggest, to the predictability and equability of the tree hole as a habitat for the aquatic stages of the organism concerned. I shall revisit these ideas in summary form in Chapters 13 and 14.

Connections

This chapter has illustrated and discussed at length variation in food-web structure at regional and local levels. A variety of factors are implicated at the regional level as being potentially important in producing site-to-site differences in food-web structure. These relate to the vegetation and topography as well as the location itself. In addition, the general biology of the organisms concerned is also an important consideration. At the local level these more deterministic thoughts become less central in any explanatory theory. Chance seems to play an increasingly greater role. Indeed, as a sweeping generalisation as we move from the global to the continental to the regional to the local spatial scale, the role of chance in determining the appearance of the food webs that we see seems to be more and more important as the scale gets smaller and smaller.

One property of phytotelm food webs, of course, is that they occur within plants. In the case of tree holes, for example, there is little apparent difference between the faunas of holes in different tree species. This is not the case, however, with webs in *Nepenthes* pitcher plants and (probably) bromeliads. Chapter 10 presents a short detailed study of webs in co-occurring pitcher plants in Borneo in order to focus on this source of variation in food-web structure.

Some of the results presented in this chapter have indicated the ways in which food webs may change through time but I have not discussed this at length nor have I placed this temporal variation in any more general context with respect to specific data sets. Chapter 11 takes up the theme of the various phenomena which can produce change in community structure and food-web appearance through time.

10

The role of the host plant
Spatial pattern: host plant driven

There is little evidence that phytotelm habitats such as tree holes or bamboo cups held in different species of plant are different from each other in any way that reflects the species of the host plant. For other classes of phytotelm habitat, however, this is not the case. For bromeliads in the Neotropics there is a wide range of plant form and container morphology which reflects through into the animal food web that occurs within them (Picado 1913, Laessle 1961, Fish & Beaver 1978). The other class of phytotelm where the species of host plant is likely to be important is the pitchers of species of *Nepenthes*. There has been little study of this topic, which offers interesting opportunities for investigation of the impact of plant structure, chemistry, longevity, and so forth, upon the animal community. We took the opportunity to make such a study using *Nepenthes* as our subject and, in this short chapter, I present these results.

Food webs from six species of pitcher plant in Borneo

In 1989 Charles Clarke and I examined six co-occurring species of this genus of pitcher plants in Brunei in northern Borneo. As described in earlier chapters, there are more than 70 species of pitcher plants of the genus *Nepenthes* spread throughout the Old-World tropics. In the island of Borneo alone, Kurata (1976) records 30 species. Many of these species occur over only very restricted ranges in inaccessible and poorly known parts of this vast tropical island. Other species are restricted to high elevations, but on the coastal plains of North Borneo several species may be found co-occurring in various combinations among the dry river beds, swamp forests and peat forests. It was for this reason that Clarke and I selected the Sultanate of Brunei for our study of the animal food webs contained within this range of species of pitcher plant.

Six species of *Nepenthes* occur commonly in these lowland forests: *Nepenthes bicalcarata* (the food webs from which have been described in Chapters 5 and 7), *N. ampullaria*, *N. mirabilis*, *N. rafflesiana*, *N. albomarginata* and *N. gracilis*. The highly characteristic pitchers of each of these six species have been illustrated in Figures 2.3 to 2.8. Of these six species all except *N. bicalcarata* are found in West Malaysia and Sumatra as well as on the island of Borneo. Our study site was located along a series of adjacent roadside tracks in the Belait district of south-western Brunei. The climate of this region is more or less constant throughout the year with a temperature range between 23° C and 30° C. The range of *Nepenthes* species we encountered reflects the fact that a variety of vegetation types intersect at this geographical location and all six species can be found growing in close proximity to one another. *N. albomarginata* and *N. ampullaria* were found in clearings along old logging tracks, in heath forest ('kerangas'). *Nepenthes mirabilis* and *N. gracilis* occurred in open dry sandy creek beds surrounded by heath forest and prone to regular flooding. Lastly *Nepenthes rafflesiana* was found along the side of the main road in cleared areas that had been subject to frequent burning. Behind these roadside areas was mixed peat swamp forest, along the edge of which *Nepenthes bicalcarata* occurred commonly. Detailed descriptions of these forest types can be found in Whitmore (1975) and other classical accounts of Bornean vegetation.

Our field work in Borneo applied techniques which I had developed during earlier studies of *Nepenthes maxima* in Sulawesi (Kitching 1987b). Twenty pitchers of each of the six species of *Nepenthes* were sampled. Pitchers were removed intact from the plant, bagged and returned to the laboratory for examination. Most species of *Nepenthes* produced two morphologically different pitcher types termed conventionally upper and lower. Accordingly, we deliberately selected 10 upper and 10 lower pitchers in the sample for each of the six species. As has been our practice in studies of other phytotelms the contents of our sample pitchers were sorted and organisms identified and counted. Subsequently, larvae were incubated until they emerged as adults so that they could be preserved for later more detailed identification. In subsequent work, Charles Clarke carried out feeding trials in glass dishes in the laboratory in order to determine the range of food preferences for each species.

Following these procedures we were able to construct food webs using the basic techniques outlined in Chapter 5 for each of the species of *Nepenthes*. These webs included only the aquatic or amphibious organisms that we found within them. This approach was taken, and is followed here, so that the food webs generated would be comparable to previous studies (Beaver 1983, 1985;

Kitching 1987b; Kitching & Pimm 1985). Several terrestrial or very rare species (such as the spider *Misumenops nepenthicola* and the mycetophilid fly *Xenoplatyura* sp., which entrap emergent adults) were observed in Bornean pitchers but were excluded from the analyses as being non-aquatic. In contrast, we have included the inquiline ants (*Camponotus (Colobopsis) schmitzi*) in the food web of *N. bicalcarata* as they are amphibious and feed direct on the in-fauna at times, although they usually retrieve and consume recently drowned prey from the pitchers (Clarke 1992, Clarke & Kitching 1995). Since this work was done and published in 1993, Clarke has observed the spider *M. nepenthicola* to prey upon aquatic larvae within pitchers from time to time (Clarke 1998). This feeding interaction deserves further study but has not been included further in my discussions here.

We collected information on the size and shape of the pitchers of each species using the same sample of twenty from each species. We measured the height of each pitcher from the base to its lip, the diameter of the opening of each pitcher, both along the mid-line of the pitcher and perpendicular to this. In addition we noted the volume of the contents of the pitchers that we examined and the total volume of each pitcher obtained by filling each with liquid after the contents had been removed for analysis. The pH of the original liquid in each was also measured.

Pitcher patterns

The pitchers of each of the six species of *Nepenthes* that we examined differed markedly in shape, structure, size, fluid, volume and water acidity. These characteristics have been summarised already in Chapter 4 (Table 4.4). Pitchers and fluid volumes of *N. bicalcarata* were substantially larger than that of the other species of *Nepenthes*, those of *N. gracilis* somewhat smaller. The fluid contained in both *N. rafflesiana* and *N. gracilis* was considerably more acidic than that found in the other species of pitchers and one individual of *N. rafflesiana* exhibited the astonishing pH value of 1.6 in one of its upper pitchers. Even this extremely acidic liquid contained active and apparently healthy larvae of *Uranotaenia* mosquitoes.

The organisms found in the pitchers and the trophic roles to which we assign them are listed in Table 10.1. The genera found in these Bornean species of *Nepenthes* were remarkably similar to those that had been encountered by Beaver in his studies of three species in Penang on the mainland of Malaysia (Beaver 1979a, b, 1983). Twenty species of arthropod were recorded from across the set of six species of *Nepenthes*: four macrosaprophages, two general saprophages, eight microsaprophages and six predators. However,

Table 10.1. *The arthropod fauna of pitchers from six species of Nepenthes from the lowland of Brunei, north Borneo, organised by their trophic role in the food webs*

Species	N. ampullaria	N. albomarginata	N. bicalcarata	N. gracilis	N. mirabilis	N. rafflesiana
Macrosaprophages						
DIPTERA						
Calliphoridae						
Wilhelmina nepenthicola	–	–	0.053(0.050)	–	–	–
Syrphidae						
Nephenthosyrphus sp.	–	0.06(0.030)	0.001(0.001)	–	0.11(0.570)	–
Phoridae						
Megaselia campylonympha	0.006(0.006)	–	–	–	–	–
General saprophages						
DIPTERA						
Chironomidae						
Polypedium convexum	0.11(0.052)	–	0.16(0.057)	–	–	–
Ceratopogonidae						
Dasyhelea sp.	0.61(0.089)	0.22(0.073)	0.54(0.091)	0.08(0.051)	0.17(0.069)	0.18(0.073)
Microsaprophages						
DIPTERA						
Culicidae						
Uranotaenia sp.	0.008(0.005)	–	0.03(0.013)	–	–	0.03(0.024)
Tripteroides nepenthis	0.02(0.013)	–	–	–	0.09(0.060)	0.003(0.003)

Tripteroides sp. 2	0.14(0.051)	0.04(0.028)	0.03(0.021)	0.42(0.085)	0.55(0.088)	0.56(0.085)
Tripteroides sp. 3	–	0.65(0.084)	–	–	–	–
Armigeres sp. 1	0.01(0.006)	–	0.02(0.006)	–	–	–
Armigeres sp. 2	–	–	–	–	–	0.002(0.002)
Culex spp.	0.04(0.017)	–	0.15(0.048)	0.11(0.042)	–	0.05(0.031)
ACARI						
Anoetidae						
Zwickia sp.[a]	+	+	+	+	+	+
Predators						
DIPTERA						
Culicidae						
Toxorhychites spp.	0.001(0.001)	–	0.01(0.004)	–	–	–
Chaoboridae						
Corethrella sp.	0.06(0.31)	–	0.001(0.001)	–	–	–
Cecidomyiidae						
Lestodiplosis sp.	–	–	–	0.20(0.062)	0.03(0.028)	–
HYMENOPTERA						
Formicidae						
Campanotus schmitzi[a]	+	–	–	–	–	–

[a] Presence or absence recorded only.

Figures in parentheses are the standard errors of the mean abundance figures (per cc of contents), $n = 20$ in each case.

we encountered substantial variability from one species of *Nepenthes* to another.

Numbers of species encountered ranged from but five in *Nepenthes albomarginata* to sixteen in *Nepenthes bicalcarata*. This difference in number of species is, not surprisingly, reflected in the variability in the other food-web statistics shown in the table. The numbers of predators present in different species of *Nepenthes* varied from zero to five, the maximum food-chain length from one to four links and the average food-chain length from 1.0 (i.e. no predators recorded) to 2.57. The food webs for *Nepenthes albomarginata, ampullaria, bicalcarata, gracilis, mirabilis* and *rafflesiana* that generate these statistics are reproduced from Clarke & Kitching (1993) as are Figures 10.1 to 10.6. The web for *Nepenthes bicalcarata* has already been discussed in Chapter 7 but is re-presented here to allow ready comparision with the other pitcher webs.

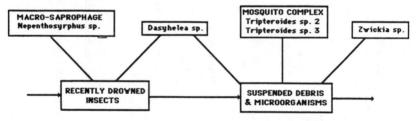

Fig. 10.1. Food web based on the contents of twenty pitchers of *Nepenthes albomarginata* from Brunei (from Clarke & Kitching 1993).

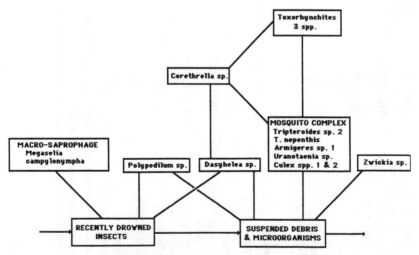

Fig. 10.2. Food web based on the contents of twenty pitchers of *Nepenthes ampullaria* from Brunei (from Clarke & Kitching 1993).

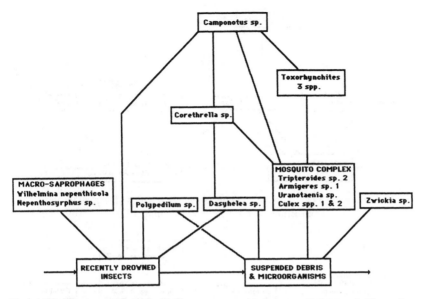

Fig. 10.3. Food web based on the contents of twenty pitchers of *Nepenthes bicalcarata* from Brunei (from Clarke & Kitching 1993).

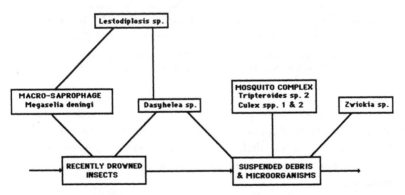

Fig. 10.4. Food web based on the contents of twenty pitchers of *Nepenthes gracilis* from Brunei (from Clarke & Kitching 1993).

The taxonomic composition of the fauna occurring within each species of pitcher varied, but in only a very few cases was any species of inhabitant restricted to a single species of *Nepenthes* pitcher. That having been said, it is also noteworthy that only two species of animals, the ceratopogonid midge *Dasyhelea* sp., and one species of *Tripteroides* mosquito, occurred in all six species of pitcher. The case-dwelling chironomid midge *Polypedilum convexum* was found only in pitchers of *N. ampullaria* and *N. bicalcarata*.

The role of the host plant

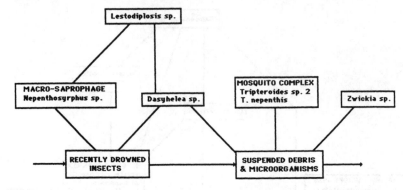

Fig. 10.5. Food web based on the contents of twenty pitchers of *Nepenthes mirabilis* from Brunei (from Clarke & Kitching 1993).

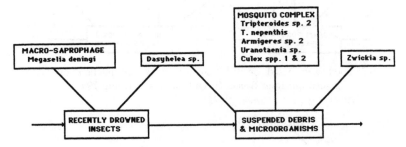

Fig. 10.6. Food web based on the contents of twenty pitchers of *Nepenthes rafflesiana* from Brunei (from Clarke & Kitching 1993).

Filter-feeding mosquito species were common in all the pitchers examined, although the number of species found in each type of pitcher varied considerably. Predatory larvae belonging to the genera *Toxorhynchites* and *Corethrella* were most abundant in *Nepenthes bicalcarata* and *Nepenthes ampullaria*.

Figure 10.7, taken from Clarke & Kitching (1993), is a dendrogram showing the relative similarity based on calculation of Sørenson's coefficients for the fauna of the six pitcher species we studied in Brunei. The webs constructed for *Nepenthes bicalcarata* and *Nepenthes ampullaria* are most similar, sharing over 80% of their fauna. *Nepenthes rafflesiana* and *Nepenthes gracilis* also show high levels of faunal similarity with a Sørenson value of 70%. Low values of similarity between *Nepenthes albomarginata* and *Nepenthes mirabilis* and between each of these species and all the others reflect the fact that the webs in these species contained only few animals, none of which was a predator or top predator.

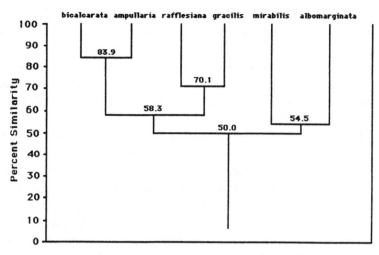

Fig. 10.7. Dendrogram showing the degree of similarity among the arthropod species occurring within food webs in six co-occurring species of *Nepenthes* pitchers in Brunei, north Borneo (from Clarke & Kitching 1993).

I have carried out a series of regression analyses using the range of the food-web variables calculated from these data as the dependent variables and one or more of the variables derived from our measurements of the size and capacity of the *Nepenthes* pitchers as independent variables. It is perhaps surprising that the only one of these to give remotely significant results was that in which the total number of species found in the pitcher was compared with the average diameter of the pitcher mouth for these species ($y = -0.42 + 2.89x$, $r^2 = 0.63$, $F = 6.71$, $p = 0.061$). It seems apparent that at any but the most superficial of levels the explanation for differences that we observed in the food webs across pitcher species does not lie in a simple comparison of pitcher morphology.

Food-web patterns and natural history

In attempting to postulate sensible hypotheses concerning the patterns of food webs that we observed in the six species of *Nepenthes* under study, we must bear in mind a number of key points:

- Each of the species of *Nepenthes* that we studied has a geographical range which extends beyond the immediate area of study; our area of study on occasion lay on the edge of one of these geographical ranges, on other occasions comfortably within it.

- Pitchers on different species of *Nepenthes* represent habitats of different predictability in the sense that in some species such as *Nepenthes rafflesiana* pitchers are short lived, existing on the plant for a matter of weeks at the most; in other cases, such as *N. bicalcarata*, pitchers exist for months. Accordingly we may suppose that the pitchers on each species of *Nepenthes* offer different prospects to potential inhabitants depending on the host plant.
- The morphology of *Nepenthes* pitchers is complex. Each pitcher has differing expanses of glandular tissue around its inner surface and each has an inturned lip or peristome offering different opportunities to potential inhabitants for pupation and shelter. These nuances of morphology are not captured in the simple measures of pitcher dimensions that we made.
- There is a fauna of associated species for *Nepenthes* plants in addition to the organisms which live an aquatic or amphibious life style within the pitcher fluid. We noted, in particular, larvae of several species of Lepidoptera which browse on the tissue of the pitchers themselves, often below the surface of the pitcher fluid. In *Nepenthes mirabilis*, larvae of species of the noctuid moth *Eublemma* occurred frequently, while in the pitchers of *Nepenthes bicalcarata* we found the larvae of the lycaenid butterfly *Cebrella matanga*. In both these instances it seemed as if the larvae preferred a subaqueous existence!

The only differences that we observed between the two most complex food webs, those in *Nepenthes bicalcarata* and *N. ampullaria*, are due to the suite of species of macrosaprophages and the ants (*Camponotus schmitzi*) that are specifically associated with the former species of pitcher plant (Clarke & Kitching 1993, 1995). These ants occur within what seems to be a co-evolved mutualistic interaction in which the plant provides a domicile within the swollen pitcher tendrils. The ants, for their part, act to remove over-large prey items and comminute them, so reducing the likelihood of putrescence and extending the life of individual pitchers accordingly. It is clear from this and previous studies, such as those of Beaver in peninsular Malaysia, that the pitchers of *N. ampullaria* are structured in such a way that some species of flies, notably the larger species of calliphorids and syrphids, have difficulty in pupating beneath the peristomes. Beaver (1983) suggested that this was because there are no sites within the pitcher for pupation if the pupa is to avoid being totally submerged and drowned when the pitcher fills with rainwater. This is not the case with *N. bicalcarata* and it is likely that this is why a rather more complex food web can be maintained in that species and why the additional species observed are larger macrosaprophagous ones.

With three species of the predatory mosquito larvae *Toxorhynchites*, and one of the smaller but equally voracious larvae of *Corethrella*, the communities in *N. bicalcarata* and *N. ampullaria* in Brunei have more aquatic predators than have been observed in any species of pitcher prior to our observations. Furthermore, the species of *Toxorhynchites* may also from time to time prey upon larvae of *Corethrella*, adding an extra link in the food chain by so doing. Of the other four species of *Nepenthes* that we investigated, two of these lacked predators completely and the other two contained the more specialised predatory larvae of the cecidomyiid genus *Lestodiplosis* which feeds upon larvae of the phorid *Megaselia* sp., the syrphid *Nepenthosyrphus* sp. and midge larvae of the species *Dasyhelea* sp. This predator was not found in the webs of *ampullaria* or *bicalcarata* during our study.

Although we encountered a relatively large number of species of predator in *N. bicalcarata* there were in fact fewer individual predators per pitcher than in any of the other three species with predators that we encountered. This is almost certainly because in most cases where a *bicalcarata* pitcher contained a predator, that predator was a larva of a species of *Toxorhynchites* renowned for their cannabilistic attributes and ability to consume any potential competition (Beaver 1983). In *Nepenthes ampullaria* larvae of *Toxorhynchites* are less common and maybe it is this that allows larvae of *Corethrella* to persist in greater numbers in this species of pitcher plant.

The food webs we observed in *Nepenthes rafflesiana* shared several common components with both *N. bicalcarata* and *N. ampullaria*, perhaps because this species of pitcher was collected from a site adjacent to that in which we studied *N. bicalcarata. N. rafflesiana* has several forms and is found in a variety of different vegetation types (Clarke 1997). Its pitchers are short lived, and all but one of the organisms collected from it in Brunei, a species of mosquito belonging to the genus *Armigeres*, appear to be merely opportunistic colonisers reflecting more the adjacent congeneric pitchers than any intrinsic property of the pitchers of *N. rafflesiana* themselves. The lack of predators in our samples from *N. rafflesiana* may well reflect the ephemeral nature of the pitchers rather than any shortage of prey. However, it is possible that predators such as larvae of *Lestodiplosis*, widely distributed in other species of *Nepenthes*, may occur in pitchers of *N. rafflesiana*, especially when this last species is growing together with, for example, *N. gracilis* or *N. mirabilis*.

We have already noted that it may well be physical constraints in terms of elements of pitcher structure that restrict the food-web complexity observed in *N. ampullaria*. Pitchers of *N. gracilis* may similarly impose physical constraints on the type and number of colonising organisms but this time by virtue of their small size and short-livedness. In almost all pitchers of *N. gra-cilis* there is simply not space for larger animals such as mature larvae of *Toxorhynchites*. Although the food web of *N. gracilis* contains seven species, only a very small subset of these is encountered in any one pitcher, perhaps due to the physical space problems to which I have just alluded. None of the species encountered in *N. gracilis* pitchers is restricted to that species of plant and this may well be because *N. gracilis* is widespread – an opportunistic species occurring at all sites examined and hence co-occurring with all other species that we studied.

As discussed in Chapter 9, the food webs we encountered in *Nepenthes albomarginata* are considerably simpler than those encountered in this self-same species of pitcher by Beaver (1985) in peninsular Malaysia (see Figure 5.9). In fact *N. albomarginata* occurs only in disjunct and isolated stands on the island of Borneo and is particularly rare in Brunei. In our study we encountered it only at a single site at the edge of an old survey track in kerangas forest. The nearest known stand of this species is over 50 km to the north in Sabah. Although the pitchers are not short lived, the fluid surface is a relatively long way from the narrow pitcher mouth and ovipositing females of larger species such as *Toxorhynchites* are likely to be constrained from ovipositing at the water surface. In our study site, pitchers of this species trap few insects other than termites, which often completely fill the pitchers, putrefy and accordingly kill most of the species of insect larvae that have chosen to live in these situations.

The sixth species of pitcher plant that we studied, *Nepenthes mirabilis*, is noteworthy in having the widest range of any species within the genus. It is widespread throughout South-east Asia. In Borneo, where it is represented by the anatomically curious *echinostoma* variety, it occurs only in very swampy, usually disturbed areas which are frequently submerged by short-term flooding. The chances of a major disturbance at these sites either by humans or bad weather is always high. The pitchers are short lived, frequently catch large quantities of prey and, in consequence, often become putrid. As already alluded to, many pitchers of this species are damaged by larvae of the noctuid moth *Eublemma* sp., which we encountered in 7 of 20 pitchers of this species examined for the purpose. The absence of the longer lived predatory species of *Toxorhynchites* may well reflect this high level of disturbance. In contrast, predatory larvae of the cecidomyiid *Lestodiplosis* sp. seem able to tolerate more unpredictable habitats, as does the macrosaprophagous species *Nepenthosyrphus* on which this species of predator feeds.

In conclusion, it is difficult to identify a single factor which may be held responsible or largely responsible for causing the variation in food-web structure across these six congeneric species of pitcher plant. Instead we need a complex inter-related set of hypotheses which relate to the structure of the plant, its geography, the surrounding vegetation and number of congeneric species and, lastly, need to take cognisance of non-aquatic insects associated with particular species – in other words, a detailed appreciation of the natural history of both the plants and the fauna.

Connections

Understanding and appreciation of the species-to-species differences in food webs in co-occurring species of *Nepenthes* necessarily draw on an appreciation of variation at other scales. This chapter completes the *Nepenthes* 'story' begun in Chapter 7 with a consideration of food-web variation across the Indo-Pacific region and continued in Chapter 9 with a discussion of variation within *N. albomarginata* across different regions of South-east Asia. The underlying mechanisms suggested at each level of spatial resolution obviously will interact with each other: the strong influence of the number of co-occurring *Nepenthes* (Chapter 7) and the impacts of where in a plant's geographical range a study of the contained food webs is made (Chapter 9) are cases in point. It is at the local level, however, that detailed appreciation of the life cycles of both the plants and animals involved become critical – and it is these factors that I have discussed here. These converging strands are brought together in Chapters 12 and 13. Before this synthesis can occur, however, we need to consider briefly the ways in which food-web patterns change not through space but through time.

11

Variation through time: seasonality, invasion and reassembly, succession
Temporal patterns

This is the last of five chapters in which I describe patterns observed in phytotelm food webs. In this chapter I move away from the theme of variation at various spatial scales and examine variation that is primarily a response to the passage of time.

There are a number of time scales of importance in considering food-web structure . The expanses of geological time and the processes of plate tectonics and resulting biogeographic change determine the basic set of organisms 'available' for invasion of a habitat unit. In addition, this geological time scale is the scale of evolution and speciation in response to environmental change. I revisit these very extended time scales again in one of my concluding chapters (Chapter 13). For the moment though, it is time scales spanning months and years that are of more concern. Not surprisingly it is over these more modest time scales that we have data to illustrate ecological processes and their impact.

I describe in this chapter three processes.

- First is the complex of factors usually referred to as seasonality: in other words the changes which we may observe in community structure within a habitat unit which persists for one or many years. To illustrate this class of change I re-present data from Kitching (1987a) on variation through time in food webs in water-filled tree holes in the subtropical rainforests of south-east Queensland and compare these with earlier results from British tree-hole communities. Any patterns of change will reflect the variation that occurs in communities as organisms go through annual cycles of development.
- There is a second set of processes which operate when a new habitat unit becomes available. These are the processes of invasion and community assembly. To illustrate this set of processes I describe informa-

tion we collected as the community re-established itself over many months following the complete emptying of a set of water-filled tree holes (Jenkins & Kitching 1990). Like the processes that reflect seasonality these processes of invasion and re-assembly assume habitat units that are relatively long lived: units such as water-filled tree holes which exist for periods of decades.

- In contrast many phytotelm habitats are much more short lived. Habitats such as those occurring in pitcher plants and leaf axils occupy plants or parts of plants which have limited life spans. In these instances we observe a succession with invasion to a short period of maturity before a period of community decline as the habitat unit itself ages and deteriorates. To illustrate the sorts of patterns that are produced in this situation I present data collected from *Nepenthes ampullaria* by Roger Beaver (presented previously in Kitching & Beaver 1990) and a hereto unpublished set of data I collected while working with Dr Matthew Jebb in northern New Guinea. This last study examined the very simple food web which occurred in the water-filled bracts of the inflorescences of the zingiberaceous plant *Curcuma australasica*.

Seasonality

Relatively few classes of phytotelmata are good candidates for the study of seasonality simply because the container plants themselves are subject to seasonal changes in growth or may not even be sufficiently long lived for seasonal phenomena to be apparent. Some of the longer lived bromeliads, bamboo internodes and, in particular, water-filled tree holes do present appropriate candidates for the study of seasonality. Although a number of studies present data on population phenology of selected species over one or more annual cycles (e.g. Kitching 1972a,b, Istock *et al.* 1975, Sinsko & Craig 1979, Frank *et al.* 1984, Bradshaw & Holzapfel 1984), relatively few studies have examined the way in which the whole community and the food web that it comprises vary through even a single twelve-month period.

In examining patterns of seasonality within phytotelm communities we must draw a distinction between features of the community as a whole, and population phenology of individual species. I present here examples of both types of data.

In the early eighties I examined the variation in the entire fauna of a set of eleven tree holes in subtropical rainforest in south-east Queensland. Results of these studies have been published in Kitching (1983, 1987a) and Kitching & Beaver (1990).

The two later papers referred to presented results in which the monthly samples taken from each of nine holes were combined to present quarterly results for particular habitat units in order to identify within-unit variability of food webs (see Chapter 9). For my present purposes I have re-analysed these data, retaining the individual monthly data points in order to identify any annual cycles which may occur in food-web structure. This analysis is summarised in Figure 11.1.

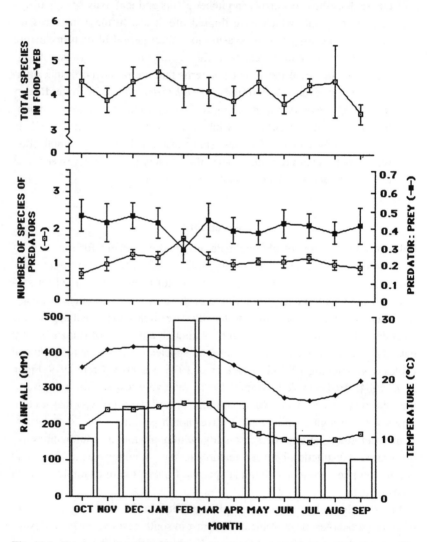

Fig. 11.1. Food-web changes and climate over a twelve-month period based on studies of eleven habitat units in subtropical rainforest, Lamington National Park, Queensland.

Figure 11.1 presents the means of the number of species, the number of predators and the predator:prey ratios present in the tree-hole food webs varied over an annual cycle. In addition I include in the figure the climate data from the study site (redrawn from Kitching 1983). There is distinct seasonality at this site with very marked dry and wet seasons and a correlated temperature regime. There is, however, hardly any pattern in the food-web statistics. Only in the late wet season (late summer) is there any marked change. Only in this month does the average number of predatory species present rise sufficiently to have an impact on the predator:prey ratio. This lack of seasonal pattern in community attributes reflects observations in tree holes elsewhere. Even in the dramatically seasonal temperate zone of southern England the number of species present in a set of tree holes did not change over the annual cycle (Kitching 1971). In both the subtropical rainforests of southeast Queensland and the moist deciduous woodlands of the United Kingdom all but the smallest tree holes remain moist throughout the year. The lack of seasonality in species richness may be due in no small part to the slow rates of development imposed upon the substantial number of saprophagous species as a result of the low nutritional value of the detritus resource. I anticipate that only in locations where tree holes are predictably and almost invariably dry over a segment of the year will this force any variation in this pattern (or, rather, lack of pattern!). I am unaware of community-level studies of these very seasonal holes, but those in which the American mosquito *Aedes sierrensis* occurs in California would no doubt fall into this category (Hawley 1985a,b).

The lack of any clear seasonal pattern in these food-web statistics from tree-hole studies does not mean, of course, that seasonal patterns are not present – simply that they are not translated into community-level metrics. Clear patterns of seasonality occur in population phenologies of many species of tree-hole organism. These are reflected in changing levels of abundance through time, and in associated changes in age structure. Examples of such patterns are the annual variation in abundance levels of the midges *Metriocnemus cavicola* (Chironomidae) and *Dasyhelea dufouri* (Ceratopogonidae) which I examined in water-filled tree holes in England (Kitching 1972a,b; Figures 11.2, 11.3). Larvae of the ceratopogonid midge are clearly more abundant during the winter months in those holes in which they occurred. In contrast the commoner chironomid midge is present throughout the year but in some holes reaches very high densities in the summer months. Where both species co-occur at high densities their phenologies may be complementary (Kitching 1969). It remains a moot point whether or not this complementarity owes anything to co-evolved competitive exclusion. This general point is

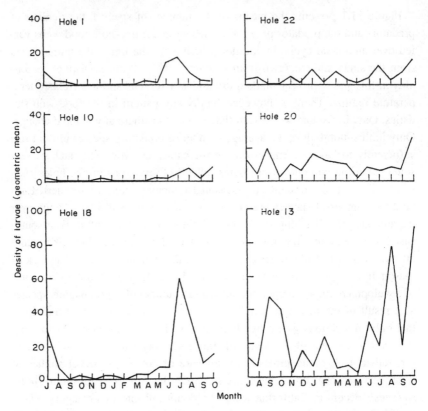

Fig. 11.2. Changes in abundance (numbers per 100 ml of sample) over a sixteen-month period of larvae of the chironomid midge *Metriocnemus cavicola*, in six selected tree holes in *Fagus sylvaticus* in Wytham Woods, Berkshire (from Kitching 1972a).

taken up again in Chapter 12, in which I also present selected population-level results from the subtropical Queensland studies alluded to earlier in this chapter.

Invasion and reassembly

In order to examine the processes of invasion and reassembly in water-filled tree holes in subtropical rainforest, Bert Jenkins and I removed all the solid and liquid contents from ten holes located in buttress roots of a range of species of trees (Jenkins & Kitching 1990). We washed the holes out with deionised water several times. Then we filled all ten holes to their lips with deionised water. No further manipulation of the sites was made and all rein-vasion and establishment processes that followed were the result of natural

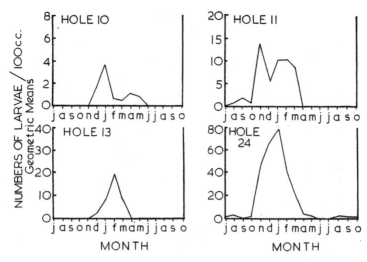

Fig. 11.3. Changes in abundance (numbers per 100 ml of sample) over a sixteen month period of larvae of the ceratopogonid midge *Dasyhelea dufouri*, in four selected tree holes in *Fagus sylvaticus* in Wytham Woods, Berkshire (from Kitching 1972b).

processes occurring within the rainforest ecosystem. There were undisturbed tree-hole sites in the vicinity of the experimental sites and we did nothing to either enhance or impair productivity of organisms from these sites.

We removed monthly samples of 100 ml of material from each of the ten tree holes over a period of 12 months from June 1984 until May 1985. We sieved and sorted these samples in the laboratory and counted the numbers of each organism present.

Food webs were constructed for each site for each month of observations. We followed the way in which the recovering food webs converged upon the food-web structure that we had observed in undisturbed holes within the same patch of forest. Data were collected on the number of species in each site, the number of predators, the number of trophic levels, the number of trophic links and predator:prey ratios. Statistical tests were carried out comparing the values for each of these variables at one month after the establishment of the experiment, at seven months and twelve months. Significant differences were observed between the values obtained for all five food-web statistics at the five per cent level when Months 1 and 7 were compared, and at the one per cent level when Month 1 and Month 12 were compared. Jenkins and I took these differences to indicate that the food web had developed in both structure and composition during the twelve months of observation following our perturbation.

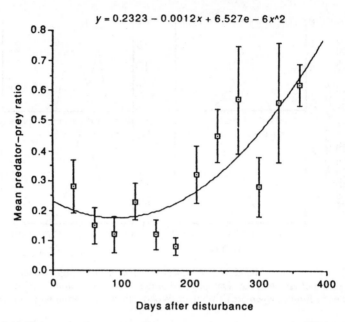

$y = 0.2323 - 0.0012x + 6.527e - 6x^2$

Fig. 11.4. Changes in the mean predator:prey ratio across ten water-filled tree holes in the subtropical rainforest of Lamington National Park, Queensland, following complete clearing of the hole contents (from Jenkins & Kitching 1990).

Figure 11.4 shows the changes that we observed in the predator:prey ratio. There was a significant non-linear regression relationship between the predator:prey ratios and the number of days following disturbance. In contrast there was no significant regression between the mean number of species in the web and the number of days after disturbance, indicating that the number of species alone is but a poor indicator of community reassembly. The predator:prey ratios did reflect the ways in which the community structure itself was changing through time. The ratio for the ten experimental holes had been calculated over a ten-month period before our experimental manipulation of these sites and produced a figure of 0.76±0.04. A year after we had carried out our disturbance of each of these sites we observed a mean ratio of 0.62±0.07. This value did not differ statistically from the pre-disturbance value, leading us to conclude that in this aspect at least the community had recovered from the disturbance at the end of the twelve-month period.

We followed the development of predator:prey relationships following disturbance in more detail by examining the relationships between the number of predator species and the number of prey species from our data set. These two quantities were positively correlated ($r = 0.76$) and these results are presented in Figure 11.5 as the plot of number of predator species against number of prey species and vice versa. In both instances there is a significant linear regression relationship but there are substantially different standard errors of estimate in each case. We took this to mean that prey species are better predictors of predator diversity during food-web reassembly than prey species are of predators.

To investigate further the recovery of the community we examined the relationship between the number of feeding links in the period following disturbance and other food-web statistics. We observed, not surprisingly, that the number of trophic links increases significantly as the mean number of species in the web increases. This relationship persists when the number of species of predators alone are considered as the independent variable. Lastly we were able to show that the number of trophic levels present in the food webs in each of our ten replicates was very strongly related to the number of species of predator. Again this is not a surprising result but does quantify the way in which the mean number of trophic levels increases as a direct response to increase in the mean number of predator species in the web.

What these results show is that, following a major disturbance – in this case imposed artificially – upon a series of phytotelm habitat units, there is a gradual restoration and re-establishment of community organisation and food-web structure. As the ratio of predators to prey species gradually recovers, so specific predator–prey links observed in undisturbed webs are re-established. Broken trophic links are restored across trophic levels and gradually the disrupted food web is reassembled. The fact that the mean predator:prey ratio we obtained twelve months after the disturbance was statistically not different from the mean predator:prey ratio before disturbance is of special interest. Mithen & Lawton (1986) suggested that any food web will develop through time until a characteristic value of its predator:prey ratio is reached. These authors suggested that this predator:prey ratio will in some sense be a characteristic of certain types of community. The results we obtained from our studies of tree-hole reassembly conformed to the predictions of Mithen & Lawton's (1986) model by approaching the predator:prey ratio of 0.76 characteristic of undisturbed webs in the region. Mithen and Lawton suggested that two interrelated mechanisms will be involved to produce the characteristic and unvarying predatory:prey ratio observed in food webs for particular types of communities. Firstly, the number of kinds of predacious species will be determined to a significant effect by the range of prey species available. Secondly, they suggest there will be competition for enemy-free space among prey species (Jeffries & Lawton 1985), allowing the number of species of predators to determine to some extent the number of species of prey within a community.

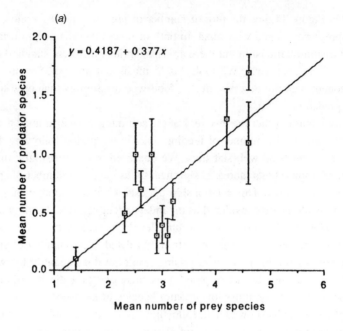

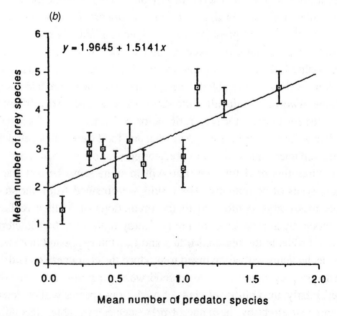

Fig. 11.5. Relationships between the mean numbers of predators in ten water-filled tree holes in the subtropical rainforest of Lamington National Park, Queensland, following complete clearing of the hole contents. (*a*) Numbers of predators vs numbers of prey. (*b*) Numbers of prey vs numbers of predators (from Jenkins & Kitching 1990).

During the process of food-web reassembly successive trophic links are restored and the community recovers. At any point in time the number of trophic links in the recovering food web will reflect the total number of species in the web but, in particular, will reflect the number of species of predators present. Hence the largest proportion of increased links during reassembly can be attributed to interactions between predator species and their prey.

This observation underlines the obvious fact that an understanding of the natural history of particular species, especially when those species are predators, remains one of the keys for understanding of tree-hole communities. The appearance of the first predatory species and, in due course, the first top predatory species in the recovering community causes a quantum leap in the values of a number of food-web statistics.

Succession within ephemeral habitat units

As we have already seen, many phytotelm food webs occur within containers that are themselves relatively short lived. In these instances the development and very existence of the food web depends to some extent on the age of the plant container. The following two studies illustrate these points with great clarity.

Processes of succession in Nepenthes pitchers in Malaysia

Beaver (1985) studied the way in which the food web within pitchers of *Nepenthes ampullaria* in West Malaysia changed over a period of several months from the time at which the pitchers opened. The results he obtained are presented in Figure 11.6 and Table 11.1 (from Kitching & Beaver 1990). The figure shows the pattern of change in food-web connectance (calculated as the number of feeding links present at any one time divided by the number of links present at the time of maximal development of the food web), predator:prey ratio, total number of species, and what Beaver calls the number of trophic types. This last statistic is arrived at by combining species of similar trophic status to obtain a set of guild designations which can be used as a reasonable summary of the trophic structure of the food web at any point in time.

We observe from the figure that in contrast to the tree-hole situation described above there is no 'stable' web towards which the overall structure develops. Rather, the web structure, as indicated by the various statistics measured, increases to a maximum and then goes into a long slow decline as the pitcher ages and the community as a whole senesces. These changes through

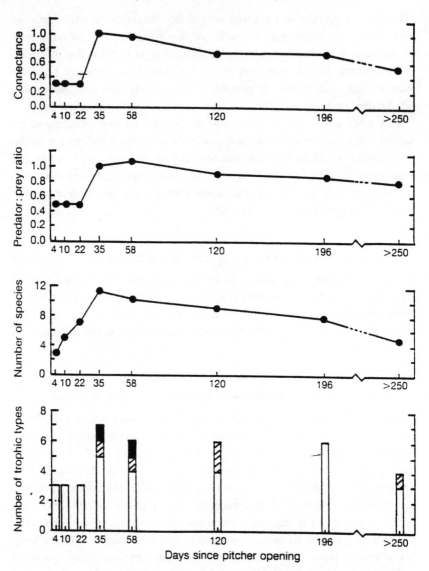

Fig. 11.6. Changes in food-web characteristics within pitchers of *Nepenthes ampullaria* following opening (from Kitching & Beaver 1990 based on data collected by Beaver in peninsular Malaysia). Trophic types are indicated as open bars (saprophages), hatched bars (predators) and solid bars (top predators).

Table 11.1. Guild structure changes through time in pitchers of *Nepenthes ampullaria* in Malaysia

	Days after pitcher opening															
	4		10		22		35		58		120		196		>250	
Guild	Genera	Species	Genera	Species	Genera	Species	Genera	Species	Genera	Species	Genera	Species	Genera	Species	Genera	Species
Filter feeders	2	2	3	4	3	6	3	6	3	6	3	4	3	3	1	1
Carrion feeders	1	1	1	1	1	1	1	1	0	0	0	0	0	0	0	0
Detritus feeders	0	0	0	0	0	0	3	3	3	3	3	3	3	3	3	3
Aquatic predators	0	0	0	0	0	0	1	1	1	1	1	1	1	1	1	1
Terrestrial predators	0	0	0	0	0	0	1	1	1	1	1	1	1	1	0	0
TOTALS	3	3	4	5	4	7	9	12	8	11	8	9	8	8	5	5
Number of guilds occupied	2		2		2		5		4		4		4		3	

After Kitching & Beaver (1990).

Ages is in days since the opening of the pitcher. The number of genera and species of arthropods in each guild is presented separately for each pitcher age.

time probably reflect changes in the available food supply. Certain guilds will be present in quantity as the quality of the energy base represented by the insect detritus within the pitcher changes. The presence and relative abundance of the detritivorous guilds will facilitate the presence of predators and top predators.

As in the case of our tree-hole manipulations the earliest invaders observed in pitcher plants were principally saprophagous. Predators and top predators came only later in the development of the community. In the case of the pitcher plant where the whole habitat unit is itself ageing, it may be that egg-laying adults of predatory species of insect are deliberately seeking out and using pitchers of particular ages – rather than the sequence simply being the result of differential discovery and survival following constant challenge of that environment by visiting female insects.

The sequence of changes we observe in the guild structure within pitcher plants shows a gradual change in the available feeding opportunities within the community. Carrion feeders are able to exploit dying and freshly dead material which enters the pitchers and these are among the very first invaders. The pitchers are suitable environments for such species just as soon as they are open and acting as pitfalls. These early invading saprophagous species tend to 'fade out' of the community at later stages, possibly as the pitchers themselves become less attractive to particular fractions of their insect prey. Filter-feeding species within the pitcher food web build up gradually and persist throughout the life of the pitcher. Aquatic and the associated terrestrial predators appeared in Beaver's observations from 35 days onwards, by which time a substantial saprophagous fauna had been established.

Taxonomic diversity increased most dramatically over the first 35 days, during which time the number of genera present rose steadily to the maximum of nine. Within the guilds of saprophages there were fewer genera present but, in many cases, these were represented by more than one species.

Community succession in the water-filled bracts of Curcuma australasica

Curcuma australasica is a fleshy herb with a widespread distribution in Cape York, Australia, and throughout New Guinea (Smith 1987). It produces highly characteristic erect inflorescences up to about 180 mm in length which are made up of a spike of green and pink bracts, within the axils of which complex zygomorphic yellow flowers occur. The bracts define small watertight cups forming a temporary aquatic habitat.

Like other water-holding floral bracts (Seifert & Seifert 1976a,b; Mogi *et al.* 1985), the water bodies held by *Curcuma* are inhabitated by a range of invertebrates making up a characteristic food web (Kitching 1990). The short-lived inflorescences and the co-occurrence of inflorescences of different ages within a single plant population presented an opportunity to examine succession within the relatively short-lived habitat unit. The changes in *Curcuma* occur over an even shorter time scale than those described above in pitchers of *Nepenthes ampullaria*. The *Nepenthes* pitchers studied by Beaver in Malaysia persisted for up to eight months, in contrast to the estimated two-month life of *Curcuma* inflorescences. The aquatic habitat within the inflorescences may be even more short lived, retaining its integrity for a little as a month.

During a period of research in 1988 at the Christensen Research Institute near Madang, New Guinea, I discovered a mature bamboo grove in an abandoned 'garden' which, in addition, contained many individuals of *C. australasica*. Multiple visits to the site were made between January 20th and February 18th 1988.

I sampled inflorescences by cutting the whole spike below the first bract and inverting it into a plastic bag. At the laboratory the length of the inflorescence could be measured and the total volume of water that it contained could be estimated. I then washed the inflorescence thoroughly through a series of sieves in order to collect all of the living metazoan inhabitants of the bract axils. The sieved washings were searched using a stereo-microscope and all the aquatic animals encountered identified and counted. I took three sets of samples, each comprising eight or nine inflorescences chosen to represent a sequence of ages from newly opened to senescent. These sequences were defined subjectively using characters such as the degree of openness of the bracts, the colour of the inflorescence, the presence or absence of flowers and their water-holding capability.

Correlates of this method of ageing based on the amount of water held by the inflorescence per unit length were obtained and are described below. In the third series of observations, conductivity and pH were measured to act as indicators of water quality.

Table 11.2 lists the aquatic fauna encountered in *Curcuma* habitats with their taxonomic designations and trophic roles. The saprophages were divided into macro-, micro- and suspension feeders and, in general, this reflects the relative size of the larvae concerned.

The analyses I present here describe counts of the four most common saprophagous species: that is, larvae of the mosquito *Uranotaenia diagonalis*, ceratopogonid larvae (a mixture of *Culicoides sumatrae* and *Dasyhelea* sp., dominated by the former species in the ratio of 6:1 based on 73 adults reared),

Table 11.2. *Aquatic insect larvae recorded from the water-filled inflorescences of* Curcuma australasica *from Madang, Papua New Guinea*

Species	Order and family	Trophic role	Authority
Uranotaenia diagonalis	Diptera: Culicidae	Suspension feeder	Brug 1934
Culicoides sumatrae	Diptera: Ceratopogonidae	Microsaprophage	pers. obs.
Dasyhelea sp.	Diptera: Ceratopogonidae	Microsaprophage	pers. obs.
Chironomine sp.	Diptera: Chironomidae	Microsaprophage	pers. obs.
Tipulid sp.	Diptera: Tipulidae	Saprophage	Alexander 1931
Psychodid sp.	Diptera: Psychodidae	Saprophage	Brauns 1954
Eristaline sp.	Diptera: Syrphidae	Surface scraper	Hartley 1963
Graphomya sp.	Diptera: Muscidae	Predator	Skidmore 1985

larvae of the eristaline syrphid, and the predatory larvae of the muscid *Graphomya* sp. As the inflorescences senesced and dried, a variety of terrestrial insects became associated with them, notably stratiomyid larvae, aphids and ants. These were not included in the study.

Physical aspects

In general as the *Curcuma* inflorescences aged there was an increase in the amount of water per unit length as the inflorescence opened and expanded. This peaked when the inflorescence was mature. Subsequently this quantity declined as the bracts gradually rotted and became, accordingly, less watertight. Both pH and conductivity gradually increased as the inflorescence aged and, indeed, these two quantities were significantly correlated one with the other ($r^2 = 0.75$, $0.02 > P > 0.01$). This pattern was not without exception, of course, as individual inflorescences may suffer accidental damage and 'age' prematurely in consequence.

Food webs

Figure 11.7 shows the food webs adduced for 'young', 'mature' and 'old' inflorescences for the aquatic community listed in Table 11.2. The 'young' web is based on the first in each of the three series of samples, the 'mature' web on the central age grade in each case (Sample 5), and the 'old' web on the final samples.

Three features of importance emerge from the food-web diagrams as presented. First, species richness increased from the youngest to the oldest webs ($n = 2$, 6, and 6 respectively); second, the number of trophic levels was greater in the mature web than in either the young or old webs (3 as opposed to 2);

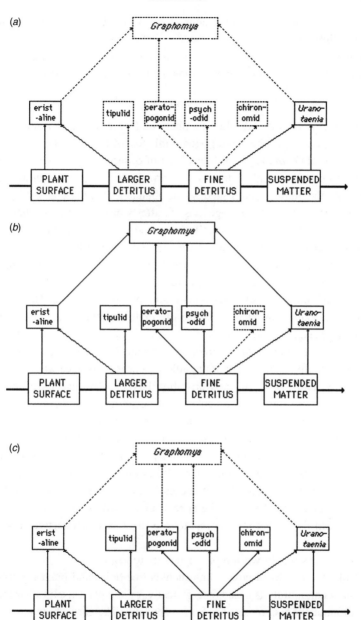

Fig. 11.7. The food webs adduced for (*a*) 'young', (*b*) 'mature' and (*c*) 'old' inflorescences for the aquatic community living in bracts of *Curcuma australasica* near Madang, Papua New Guinea. The full regional web for *Curcuma* inflorescences is indicated in each diagram with sections that are absent from particular age grades indicated by dotted lines and boxes.

and, third, the degree of connectance in the web, measured as the number of trophic links present (x) as a proportion of the total links possible in the regional web ($x/13$), rose from 0.23 in the young web to a maximum of 0.85 in the mature web and declined somewhat to 0.62 in the old web.

Insect abundances

Figure 11.8 shows the absolute numbers of larvae of the mosquito *Urano-taenia diagonalis*, of the ceratopogonid midges, of the eristaline syrphid, and of the muscid *Graphomya* sp. In each set of observations the youngest inflorescences contained only a few larvae, principally of the mosquito *U. diagonalis*. As the inflorescences aged, so the abundances of these larvae increased to a maximum around plant age-class 5, after which numbers declined. Larvae of the eristalines also gradually increased in numbers, although showed a clear decline towards the end of the age sequence in two of the three sets of observations. The peak of abundance for this species was less well marked, probably because, overall, numbers are much less and any one set of results will reflect chance oviposition events on the part of adult flies. The other relatively abundant saprophages, larvae of ceratopogonids, were characteristic of the later stages of the inflorescences' life although low numbers were present throughout. Larvae of the predatory muscid *Graphomya* sp. became most numerous in the mature stage of the inflorescences' existence when their principal prey items, mosquito and eristaline larvae, were also most abundant. Numbers on any one inflorescence were never large (probably because of cannibalism or potential cannibalism) and any one larva can range over all the inflorescence, moving from bract to bract on the moist surface of the plant.

In interpreting the results of community change within *Curcuma* inflorescences it seems likely that facilitation is a key process. As in early, now somewhat discredited, models of plant succession, a sequence of events occurs in which the presence of a particular species in an early seral stage modifies the habitat such that it becomes acceptable to the species characteristic of later stages in the sere.

The process would work more or less as follows.

Unlike food webs in phytotelms such as tree holes and pitcher plants, the *Curcuma* community is based, energetically, on the plant itself rather than on allochthonous inputs of detritus from elsewhere. The earliest invaders of the water body held in the bracts are eristaline larvae and the larvae of *Urano-taenia*. The eristalines probably feed by scraping the living plant surface (Hartley 1963) although in this regard they may differ from most other eristaline larvae which are saprophagous (F. Gilbert *in litt.*). I suggest that the activities of the early invading eristaline larvae release waste materials

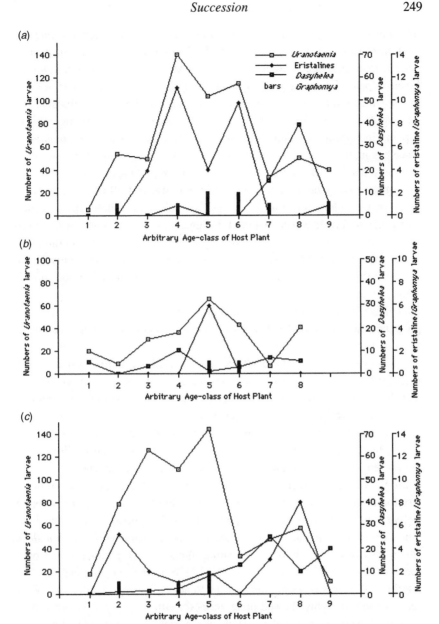

Fig. 11.8. The absolute numbers of larvae of the mosquito *Uranotaenia diagonalis*, of the eristaline syrphids, of the ceratopogonid midges, and of the muscid *Graphomya* sp. in three aged sequences of the water-filled inflorescences of *Curcuma australasica* from Madang, Papua New Guinea.

into the water body both through egestion and simple breakdown during the feeding process. Very rapidly this will lead to an increase in the suspended detrital load. It is such suspended organic matter on which the *Uranotaenia* larvae depend. Presumably the demands of a few very young larvae are met by the small amounts of such material present in the fluid adventitiously. The feeding of the eristalines will generate sufficient material, subsequently, to support increasing populations of the filter-feeding larvae of *Uranotaenia*. Once begun, this process is cumulative: more saprophages produce more detritus, more detritus allows participation by more saprophages. Further, disruption of the plant's epidermis by eristaline activity, as well as general senescence of the inflorescence through time, will allow microbial activity which will generate further detrital material. I suggest this may account for the gradual increase of the number of saprophagous species seen in the community as the inflorescence ages. The presence of the predatory species only in the mature inflorescences is also consistent with a biological facilitation model: simply stated, the predator enters the system at a time which coincides with the availability of an adequate food supply and the likelihood of such a supply persisting at least in the short term. That such a large and voracious predator is involved is perhaps surprising but the mobility of the larva, and its ability to move from bract to bract in the water film usually present in the perhumid climate of *Curcuma* in New Guinea, will allow it to maximise its prey intake and, presumably, minimise the actual larval period – a necessary feature for a predator in such an ephemeral habitat.

This sequence of events remains an hypothesis at this stage and the system certainly deserves more attention. Its geographical inaccessibility makes the present observations interesting but tantalisingly incomplete.

Mogi, with others (Mogi *et al.* 1985) examined the comparable community found in water-filled axils of *Alocasia* sp. in the Philippines. They found twelve species of dipteran larvae plus mites and copepods in the contained fluid. The seven culicids included the predatory *Toxorhynchites splendens* which they showed to have a substantial impact on the populations of prey species (cf. the work of Lounibos *et al.* 1987b, in their experiments involving the axil waters of *Heliconia* spp. in Venezuela – see Chapter 12). Most species showed highly skewed preferences for either 'younger' or older axils although the predatory *T. splendens* showed a peak of occurrence in the young to middle-aged class of axils, not unlike that shown by *Graphomya* larvae in our results.

The other habitat which bears comparison with *Curcuma* inflorescences is the inflorescences of Neotropical species of *Heliconia* (Musaceae) which have been much studied by Seifert and his co-workers (e.g. Seifert 1975, 1982;

Seifert & Seifert 1976a, 1979a,b; Seifert & Barrera 1981). These situations contain a food web also based on the plant parts within the inflorescence. Seifert (1982), in reviewing the whole body of work, notes that 'facilitation is more important than competition' among the saprophages, with hispine beetles being the most important facilitators, much as eristalines appear to be in *Curcuma* communities. Seifert & Seifert (1979b) show that there is an increase in complexity of the food web from the younger to the older bracts (as observed in *Curcuma*) although I am unaware of any results dealing with changes in the overall community in whole inflorescences of different ages.

Connections

The processes of temporal change in the structure of phytotelm food webs reflect both bottom-up and top-down structuring processes. In Chapter 12, I consider key works on the bottom-up processes of competition and predation in phytotelm organisms. These processes, together with the stochastic processes of foraging by adult insects for suitable habitats in which to oviposit, drive temporal change in community structure. Such change, of course, occurs within the top-down constraints of habitat availability and predictabilty, together with factors associated with the species pool itself, from which community members are drawn to produce the food webs we observe. These processes and their interactions with bottom-up population processes are described in Chapter 13, in which the theories and predictions of Chapter 6 are revisited.

Part IV:

Processes structuring food webs

Sitting quietly in the bird blind overlooking Gannet Pond remains one of my lasting images of northern Florida. Fox squirrels – slightly portly, bigger versions of the more familiar grey squirrels – were perched on a huge tupelo stump sharing seed with brilliantly coloured indigo buntings and purple finches against the backdrop of the lagoon, itself a riot of water lilies and water hyacinth. But this was no pastoral idyll and the opaque surfaces of the lake were frequently disturbed by the slow and confident passage of large alligators. Indeed I sat staring at the water for over an hour once before a minute movement caught my eye – the mess of 'twigs' snagged on the end of a fallen branch on the water's edge was actually a pile of basking baby alligators. Tall Timbers Research Station contains a mix of hardwood hammocks and pine plantations gradually descending to a network of causeways, tiny islets and shallow lagoons. Initially a hunting preserve for the raising of turkey and quail, it is now an idyllic field station for wildlife and ecological research. As I walked the causeways between islands looking for water-filled tree holes, the rest of the fauna seemed to make a point of presenting itself to me. A baby armadillo bulldozing its way through the deep leaf litter in search of juicy invertebrates actually walked into my foot; a band of masked raccoons fled at my approach looking back in guilt as if surprised in the middle of some nefarious heist; and, the often-heard rustle and splash of water moccasins were a frequent reminder that this was not the benign landscape of the casual glance.

I spent some weeks at Tall Timbers in 1988 following up on the earlier excellent work of Bill Bradshaw and Christine Holtzapfel, who studied the dynamics of mosquito populations in the abundant water-filled tree holes of the reserve. I was anxious to complete the food web that their work indicated was present and I expected this to be the same straightforward task that it had been in Sulawesi and New Guinea. But 1988 was a drought year and

253

although I could find Bill and Christine's sites without too much difficulty, they were, for the most part, dry. Northern Florida's tree-hole community is the richest in North America, with an array of predators and saprophages of all sizes but, on that particular occasion, finding even a few sites was challenging.

The forests of Dumoga-Bone National Park in northern Sulawesi are even richer in life than those of northern Florida. Troupes of macaques watch, and occasionally pelt, the unwary field worker poking around in tree buttresses looking for water-filled hollows. The creak and 'whoosh' of hornbill wings as they forage from one fruiting tree to another are among lasting sensory impressions. And the wealth of butterflies confirmed the wonder expressed by Alfred Russel Wallace who visited these self-same forests in the mid-1800s. Drought was not a problem in these forests in the wet season of 1985 when I sought tree holes there – but storms were. Howling thunderstorms ripping through the forest several times a week played havoc with the trees – on one occasion, as I was caught within the forest in such a storm, it was unnerving to hear trees falling all around. This is one contributing factor to the generally small stature of this equatorial forest. Few trees were greater than half a metre in diameter and, accordingly, tree holes were again hard to find. Fortunately an enterprising timber poacher had preceded Expedition Wallace in 1985 and removed twenty or thirty palm stems, leaving stumps which rotted out to produce water-filled cavities – a most satisfactory if unplanned outcome for the phytotelm student!

Every forest is different yet underlying themes remain the same. The field ecologist seeks specific information from each forest yet hopes for a common pattern to emerge when these disparate bits of information are juxtaposed.

This section of the book brings together the theory of food-web structure and maintenance set out in Chapter 6, with the patterns observed in food webs in phytotelms that have been described in Part III. Chapter 12 examines ideas and data on the bottom-up approach to understanding food webs – through studies of the interspecific processes of competition and predation. Chapter 13 complements closely Chapter 6 and is structured in the same way. Each of the ideas, hypotheses and predictions set out in the earlier chapter is re-evaluated in light of the results described in the intervening sections of the book.

12

Competition and predation – basic forces structuring the community?
Population-level structuring processes

Until relatively recently virtually all theories which attempted to explain the structuring of communities of organisms within ecosystems invoked one of two processes for explanation: competition between species or predation.

Some theory

Competition is conventionally defined as the dynamic process by which two or more species which share similar resources divide up that resource. Accordingly along a particular resource dimension of the habitat in which such species occur we may expect to find that the resource has been subdivided into a series of overlapping fractions by the species utilising that resource. This division is classically considered to be the result of competition or more accurately the avoidance of competition (see, e.g., Whittaker 1975, Whittaker & Levin 1975, Keddy 1989, Strong *et al.* 1984 and references therein, Putman 1994). Conventional models of community organisation define each species involved in an assemblage of this sort as either being intrinsically better at exploiting a particular segment of a resource dimension or having evolved to do so in order to avoid competition with other species. The resulting segregation and avoidance of competition may be seen in both spatial or temporal terms. In other words visiting a particular habitat at one point in time we may expect to find a set of organisms exploiting different portions of the resource base or we may simply find that a particular resource base is exploited by different organisms at different times during the course of a season or on some other time scale.

There are many implications of this 'naive' theory of the structuring of communities by interspecific competition. Communities may be considered to be more or less 'full' – more or less tightly 'packed' – depending on how many co-occurring species share a particular resource base. Where there is a large number of species sharing such a single resource dimension then the range over which each exploits that resource would be expected to be narrower than in a case involving fewer species. Similarly for a particular set of species we might expect that the presence or absence of any one of those species would affect the range of resources that may be exploited by the others. So if six species exploit the detritus filling a water-filled tree hole, for example, then we might expect that each would make use of a smaller segment of the overall resource base than if that same resource base were being exploited by only three species. The degree to which a particular resource base can be more or less subdivided is often supposed to be determined by other environmental factors. A more predictable or equable environment will 'permit' finer subdivision, and the species involved may have narrower individual resource bases, than a less predictable one. This may be reflected in latitudinal or altitudinal patterns of community 'complexity'.

This basic theory is deceptively simple and deceptively attractive. The problem associated with it is demonstrating whether or not such interspecific competition occurs in nature or, indeed, whether a pattern observed in nature is competition induced or merely a chance outcome of stochastic processes of invasion and survival (the so-called 'null' model of recent authors). This conundrum translates, for my present purposes, into a series of more specific questions of the kinds:

- Does a particular species of tree-hole mosquito do better in a situation where there are no other species of mosquito sharing the same resource base?
- Or will it exploit essentially the same segment of the resource?
- Do we see competition in action now, determining the food choices of a particular species available at a given site 'now', or is the current diet of that species a preprogrammed characteristic?
- In this latter case, are we seeing what is called the 'ghost of competition past' (Connell 1980, Strong 1984), in which particular feeding preferences have evolved as part of the programmed behaviour of a species resulting in avoidance of competition with other organisms which might share the resource?

This last idea assumes that a particular species has co-evolved with other species over a long period of time and even if those other species are no longer present, the behaviour of the first species remains constrained by its history of interaction, or more correctly avoidance of interaction, with these other species.

A further feature of theories which use ideas of competition to account for observed structure in communities is that they assume that the sets of organisms that have been studied, and from which conclusions may be reached concerning competition, are at equilibrium. In other words that populations of sets of species that occur together do so at the maximum level consistent with the overall set of resources available in that habitat. It has become commonplace to suggest that many communities, including those that occur in phytotelmata, are in fact non-equilibrium communities (e.g. Bradshaw & Holzapfel 1983). In these instances we assume that population levels of the inhabitants are maintained by other factors than competition at levels below those at which competitive interactions would begin to be significant.

Non-equilibrium communities may be maintained in this state by a number of factors. The scarcity, unpredictability or unpalatability of the resource base itself may be such that populations never rise to levels at which they will begin to compete with other species. More interestingly it is frequently suggested that the presence of predators within the communities maintain prey populations at levels sufficiently low that competition among prey species is seldom a reality (Paine 1966, Kitching 1986b). As with primary considerations of competition this addition of predation as a structuring force is also modified by the wider environment in which such interactions occur. In early successional stages or less predictable environments, it has been suggested, we may find a reduction in the importance of predation as a competition-modifying factor. In highly predictable 'climax' communities, in contrast, predation will be important, competition dampened and the whole resulting community more diverse and richly structured.

So in summary we have a 'traditional' set of population theories which starts by suggesting that one sort of two-species interaction – competition – is a fundamental force in structuring communities but a second sort of two-species interaction – predation – modifies this interaction so that its full force is seldom observed. Again both competition and predation may be used in explanation of the pattern we see in a habitat unit here, reflecting on-going two-species interactions, and/or they may serve as historical explanations for why organisms behave the way they do at the present time.

This background of classical theory demands the collection of data and the performance of experiments to test its reality. In this regard many authors have seen the small, circumscribed, relatively simple communities contained within phytotelmata as ideal experimental microcosms. Not only are the natural communities relatively simple but the multiple-unit nature of the habitat type allows for ready replication and manipulation. So the numbers of selected species can be artificially increased or decreased, predators can be introduced or removed, and the basic detritus food base can be enriched or impoverished.

This chapter examines some of the issues that have emerged over many years of study of species interactions within selected phytotelms. In particular there is a large body of work dealing with the assemblage of container-breeding mosquitoes in North America and this work is selectively reviewed. In addition population patterns that are interesting and open to interpretation in a variety of ways are also presented from other locations.

It should be noted from the outset that several different forms of evidence from insect populations within container habitats have been used as the bases for hypotheses or to test these hypotheses. The distribution of species across, say, tree holes of different kinds within a region may be used to imply habitat segregation, with the implicit assumption that this has been based in some fashion on historical competition. Further, within particular habitat units population phenology may be used to indicate temporal responses either of avoidance of competition or of direct interaction between competitor and competitor or competitor and predator. Lastly, and all too rare, is evidence derived from manipulative experiments in which very specific hypotheses are erected, testing situations devised, and critical statistical analyses carried out to test the veracity or otherwise of the original hypotheses.

The mosquitoes of North America – the ghost of competition past?

I set out this account in a stepwise fashion. First I describe studies which have attempted to establish that the species contained within co-occurring assemblages of mosquitoes partition the available habitat space in a predictable, and presumably deterministic, manner. Then I describe studies establishing the *possibility* of both intraspecific and interspecific competition before returning to the question of whether we may reasonably suppose such competition occurs in the field.

Habitat partitioning in tree-hole mosquitoes

A set of species of detritus-feeding mosquitoes occur in water-filled tree holes in the Great Lakes region of the United States. Robert Copeland and George Craig made an extensive study of the distribution and occurrence of these five species (Copeland & Craig 1990). The five species comprise *Aedes triseratus, Ae. hendersoni, Anopheles barberi, Orthopodomyia alba* and *O. signifera*. The larvae of all five species feed on suspended detritus or, in some instances, by ingesting larger particles of decaying organic matter within the tree holes in which they occur. Copeland and Craig made an intensive search for water-filled tree holes at a number of rural and urban sites in Indiana. In this fashion they discovered 65 holes. They sampled each by siphoning out the contents, adding more water, and siphoning them a second time. As a result, they could be almost sure of obtaining all of the mosquito larvae contained within each.

They subdivided the 65 holes that they sampled into different classes depending on a number of factors. The basic division they used was the classification into pans and rot holes, reflecting the presence or absence of a bark lining to the hole, and discussed in Chapter 2. Further classification was made into basal and canopy holes depending on whether they were located in the buttress roots of the trees that were studied or in more elevated situations. Elevated rot holes were further subdivided into 'shallow' and 'deep'. Thus they were able to examine samples from 13 basal pans, 10 elevated pans, 23 elevated shallow rot holes and 19 elevated deep rot holes. They encountered no basal rot holes. The holes that they examined were contained almost exclusively within either maple trees (*Acer* spp.) or beech trees (*Fagus grandifolia*). At an early stage in their analyses they determined that the actual species of tree was not important (as other authors have found) and in all further analyses the holes from both species of trees were combined.

Figure 12.1 reproduces the results they obtained, showing the occurrence of each of the five species of mosquito across the four hole types that were examined.

The most common of the species, *Aedes triseriatus*, occurred across all four types of tree holes but was the only species found in basal pans. *Aedes hendersoni*, in contrast, was a species of elevated habitats, particularly rot holes. The two species of *Orthopodomyia* both favoured elevated deep rot holes over all other situations. *Anopheles barberi* also showed a preference for this sort of hole but this preference was less pronounced than for the *Orthopodomyia* species. Copeland & Craig (1990) carried out a series of other analyses on this data looking at levels of association between all pairs within the set of five species and defining what they called subcommunities in this regard. The details of their analyses can be found in the published work and need not concern us here. They concluded that their data demonstrate a segregation of habitat by these five species which in some instances at least is very clear. So, for instance, of the two species of *Aedes*, *triseriatus* may be taken to be a species of low elevation within the canopy, *hendersoni* one that prefers canopy sites. This observation is backed up by other work on the oviposition behaviour of these two species (Scholl & DeFoliart 1977, Sinsko & Grimstad 1977).

It would seem to be a reasonable hypothesis that the habitat segregation so elegantly demonstrated by Copeland & Craig (1990) is the result of inter-

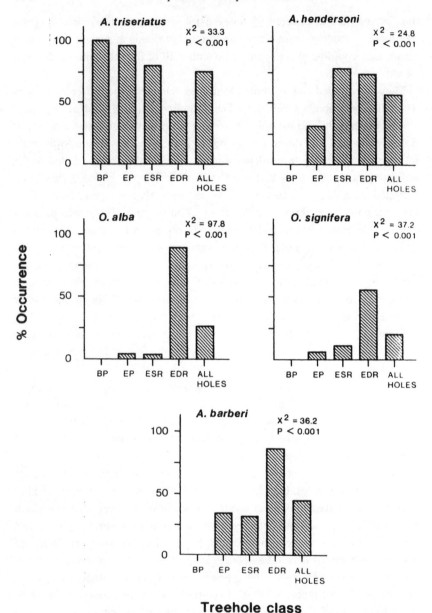

Fig. 12.1. The relative abundance of larvae of *Aedes triseriatus, A. hendersoni, Orthopodomya alba, O. signifera* and *Anopheles barberi* in tree holes of various kinds and positions. Key: BP, buttress pans, EP, elevated pans, ESR, elevated shallow rot holes, EDR, elevated deep rot holes. See text for further discussion. (After Copeland & Craig 1990.)

specific competition. In this part of the United States there are no predatory larvae occurring in tree holes and densities of mosquito larvae reach very high levels.

The existence of competition

There can be little doubt that the potential exists for resource competition when mosquito larvae occur in high densities. This has been demonstrated clearly in a number of studies examining tree-hole species kept in laboratory cultures. The experiments of Todd Livdahl and his co-workers have been exemplary in this regard.

In elegant laboratory manipulations using larvae of *Aedes triseriatus*, Livdahl (1982) set up treatments in which he combined different densities of young larvae with different concentrations of tree-hole water. He argued that water removed from naturally occurring tree holes contained the suspended organic matter on which these larvae depend for food. By diluting this natural tree-hole water, he was able to control levels of resource availability for the experimental treatments that he established. In addition, he repeated these larval density/food availability treatments, including in one set a fixed number of older larvae of the same species of mosquito. There was a clear decline in survivorship of the initial larvae of *Aedes triseriatus* with increasing initial density of young larvae, exacerbated by the decreasing concentration of the liquid medium in which they were reared. The declines observed were more pronounced in the replicates that also contained older larvae. These results are summarised in Figure 12.2 taken from Livdahl's (1982) paper.

Livdahl's findings demonstrate clearly that intraspecific competition can have a significant effect upon the well-being of tree-hole mosquito larvae. In a later set of experiments, Livdahl, this time in collaboration with Michelle Willey (Livdahl & Willey 1991) further demonstrated that competition could occur between two different species of container-breeding mosquito. In this second study they examined populations of *Aedes triseriatus* and *Aedes albopictus*, a container-breeding mosquito that has recently invaded North American ecosystems. This imported species was brought into North America in used car tyres, which frequently contain water residues. Livdahl & Willey were able to estimate the impact each of these species would have on the per capita growth rates of the other. In order to do this they established experiments in which they set up all combinations of 7, 14, 28 and 42 larvae of each species. Each set of combinations was set up in both water-containing tyres and in simulated tree holes (using water imported from naturally occurring tree holes). Survivorship to adulthood was recorded and compared with

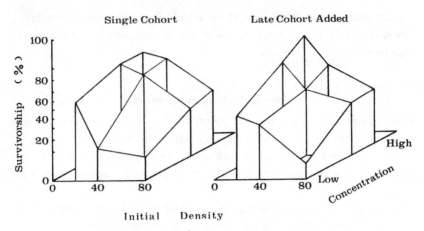

Fig. 12.2. The survivorship (arcsin √ transformed) of larvae established at different initial densities at different concentrations of tree-hole water. See text for further details. (From Livdahl 1982.)

results from similar larval densities maintained in single species cultures. Emerging adults were measured so that estimates of the reproductive rate could be obtained. (The egg-laying potential of female Diptera can usually be estimated with considerable accuracy from the size of the adult female.)

In this fashion they were able to estimate the relationship between the per capita growth rate observed in each of their experimental treatments and the initial numbers of *Aedes triseriatus* on the one hand and *Aedes albopictus* on the other. Figure 12.3 summarises their results, indicating that each species' growth rate will decline in the presence of the other species and that this decline will be related to the density of the competing species as well as the overall density of the combination of two species. Livdahl & Willey's results showed, further, that for the cultures established in tyres, there was no combination of *triseriatus* and *albopictus* numbers that would lead to co-existence of the two species: in all situations in the presence of *Aedes albopictus*, *Ae. triseriatus* would go extinct. In the simulated tree holes, however, it seems that there is an equilibrium point at which these two species could coexist.

The results of Livdahl & Willey (1992) paralleled those found by Copeland & Craig (1992) who set up comparable experiments in simulated tree holes using *Aedes triseriatus* and its naturally occurring congener, *Ae. hendersoni*. They too found that the results obtained showed an asymmetry in the impacts of competition on each of the species. Larvae of *Ae. hendersoni* were able to compete better with larvae of their own species than with larvae of *Ae. tris-*

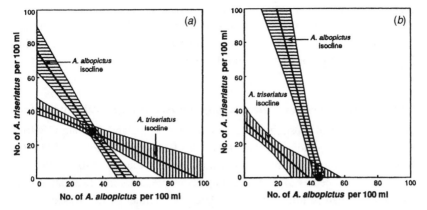

Fig. 12.3. Density isoclines for numbers of *Aedes triseriatus* and *A. albopictus* which allow for coexistence. The hatched areas define the 95% confidence intervals about the equilibrium isoclines for each species. The results presented are for (*a*) tree-hole habitats and (*b*) simulated tyre environments. (From Livdahl & Willey 1991.)

eriatus. Conversely *Ae. triseriatus* faired relatively poorly in competition with conspecific larvae, performing better in contest with larvae of *Ae. hendersoni.* Other related studies on competition among tree-hole mosquito larvae have been made by Ho *et al.* (1989)

The results we have discussed to date demonstrate firstly that there appears to be habitat segregation among co-occurring mosquito species in the northern part of North America. In addition the potential for both intraspecific and interspecific competition resulting in a reduction in performance in populations of the competing species has been clearly demonstrated experimentally. The question remains: Has the habitat segregation that was observed by Copeland & Craig (1992) and others actually been determined by competitive interaction?

No unequivocal answer is available. Bradshaw & Holzapfel (1986b) examined the set of three mosquito species occurring in water-filled tree holes in Europe – *Aedes geniculatus, Anopheles plumbeus,* and *Culex torrentium.* Examining field populations of these species in France and southern England they were able to show that field-collected pupae generally were of smaller size and weighed less than would be expected of that species reared in the laboratory. This observation, they suggested, could well indicate that all three species of mosquito are competing for resources in the field. However, their population results suggest that this competitive limitation in population performance is as likely to be due to intraspecific effects as it is to interspecific effects. Each species of mosquito frequently occurs in near monocultures and that any chance encounter between one mosquito larva and the next is as likely to be intraspecific as it is to be interspecific. They conclude that only field manipulation of well understood populations will resolve this question. To date these experiments have not been done.

Enter *Toxorhynchites* – the dramatic impact of a generalist predator

The assemblage of mosquito species studied by Copeland & Craig (1990) in the Great Lakes region and that studied by Bradshaw & Holzapfel (1986b) in Europe contained no obligate predatory species. Larvae of *Anopheles barberi* and other species of mosquito may act as predators in a facultative fashion and we shall return to this point later, but neither of these assemblages contained larvae of the genus *Toxorhynchites*, the voracious predatory mosquito found in most other parts of the world. Steffan & Evenhuis (1981) present an exhaustive review of the biology of species of *Toxorhynchites*. These mosquitoes are large and impressive as adults, more nearly resembling crane flies to the untutored eye. This has led frequently to alarmed reactions but, in fact, species of *Toxorhynchites* do not feed as adults. The reasons they are able to forego the protein feed so essential for the females of other species of mosquito is that their large larvae are predators. The 69 known species of *Toxorhynchites* are primarily container breeders, occurring in rock holes and a variety of artificial containers, as well as the full range of phytotelmata. They occur in virtually all parts of the world except western Europe and northern and western North America. As predators these larvae will tackle any other arthropods with which they co-occur: indeed they have been recorded to attack small tadpoles! Because the most common and apparent species found in the same habitats as larvae of *Toxorhynchites* are larvae of other species of mosquitoes, they have become known as mosquito predators. In fact their prey include larvae of chironomids and tipulids, the aforementioned tadpoles, dragonfly nymphs and syrphid larvae. Larvae are also known to be cannibalistic, feeding readily on individuals of their own species whenever encounters occur.

The role of larvae of the genus *Toxorhynchites* in structuring communities within water-filled tree holes in particular and the impact that this may or may not have on competition between other species of mosquito was the subject of an influential paper by Bradshaw and Holzapfel published in 1983.

William Bradshaw and Christina Holzapfel studied populations of mosquito larvae occurring in 115 water-filled tree holes scattered across eleven species of hardwood tree within the Tall Timbers Research Station in northern Florida. The species of mosquito that they encountered were *Culex restu-*

ans, Culiseta melanura, Aedes triseriatus, Orthopodomyia signifera and *Anopheles barberi*. These primarily filter-feeding species were joined in the Floridan tree holes by two species of predatory larvae – *Toxorhynchites rutilus* and the chaoborid *Corethrella appendiculata*. During this study they removed mosquito larvae from a subset of the tree holes that they had encountered, returned them to the laboratory where they were counted and estimates made of the age structure of each population. Samples were then returned to their original tree holes. In this fashion the processes of population development and change could be followed in a great range of holes throughout a summer season.

Many fascinating observations and conclusions emerged from this landmark study. Bradshaw and Holzapfel were able to show that within individual tree holes extinctions and reinvasions were commonplace and, important for later discussions within this book, extinctions and immigrations occurred in particular tree holes irrespective of the community composition already present within that tree hole, tree-hole size or stability, or the average number of species present. From the point of view of our present discussion, however, their analyses of the existence of density-dependent effects whether they are intraspecific or interspecific is the point of interest. As in their later European work, Bradshaw & Holzapfel (1983) examined the total productivity of pupae for each of the holes that they studied. They were then able to compare these figures with the expected productivity based on laboratory cultures of the species of mosquito. They argued sensibly that if the pupal productivity was less than that observed in laboratory cultures then this might be the result of competitive interaction, and if so there should be a relationship between the change in pupal weight and the density of larvae of the species involved in the competition.

What they showed was fascinating. In situations which had no larvae of the predatory *Toxorhynchites rutilus* present, there was a *positive* correlation between the numbers of larvae of *Aedes triseriatus*, the commonest species of mosquito present, and the pupal production. When larvae of *Toxorhynchites* were present then pupal production was correlated only with the density of the predatory larvae. They concluded that there was no evidence that larvae of *Aedes triseriatus* experience either intraspecific or interspecific competition at any time.

In contrast they identify predation as the most important factor in determining community structure and the relative success of other species of mosquito larvae within tree holes. They suggest for the relatively complex food web found in these north Florida tree holes that predatory species such as larvae of *Toxorhynchites rutilus* are keeping populations of other mosquito species at a level below that at which competition would become significant. Accordingly, although it is possible to imagine circumstances in which the food resources of tree holes is limiting and where competitive reduction in performance of mosquito species in these situations is possible, the actual densities achieved are simply not sufficient for this to occur. Bradshaw & Holzapfel (1983) go on to suggest that it is the ability of individual species to deal with particular feeding situations and the likelihood of encountering drought situations in particular tree holes that determines their success or otherwise in particular sites. They suggest that the habitat segregation observed in these species results from adaptations and ecological capabilities possessed by the individual species of insect before it invades particular tree-hole situations and that this segregation of habitats owes nothing to interspecific competition. It remains a moot point as to whether or not competition plays an important role in the ecology of the assemblages found in situations such as the Great Lakes region and western Europe where there are no co-occurring predatory species.

This division of mosquito assemblages into a with/without predator situation, however, is shown to be potentially simplistic by a more recent piece of work. Edgerly *et al.* (1999) have examined what they call intra-guild predation among three species of *Aedes* mosquitoes in experimental situations. Although conventionally thought of as filter feeders and detritivores, larvae of *Aedes aegypti, Aedes triseriatus* and *Aedes albopictus* are all on occasion predatory, especially on small individuals of their own and other species. In an elegant study these authors were able to show that this sort of predation occurred in all situations of food availability or environmental conditions but that some of the species were more vulnerable and others more likely to be predacious. These results are important even though based on laboratory studies because they indicate that even in natural situations where assemblages of mosquito larvae occur in the absence of obligate predatory species such as *Toxorhynchites*, it may well be facultative predation of the sort exhibited by these species of *Aedes* larvae that has the ability to structure the community and the populations that occur within it. Edgerly *et al.*'s (1999) paper opens up a whole field of enquiry examining this sort of interaction among species of tree-hole mosquito in those regions considered to be lacking in predatory species.

Lastly in this consideration of the role of predation in modifying and determining the community structure in phytotelmata, two additional studies expand our view from that restricted to mosquitoes in tree holes to the full set of insect species occurring in other habitats. In a fascinating study on Venezuelan habitats, Lounibos *et al.* (1987b) examined water bodies held in the bracts of *Heliconia*, in bamboo internodes and in the axils of two species of *Aechmea* bromeliads in lowland rainforest in eastern Venezuela. They introduced first instar larvae of the native *Toxorhynchites, T. haemorrhoidalis*, into each of these situations and followed the development of the community in comparison with control situations without the predatory larvae. In some instances the introduced predatory mosquitoes faired poorly and did not survive. In general this poor performance, where it occurred, was related to characteristics of the environment into which they were released – its propensity to dry up, size and so forth. Lounibos *et al.* demonstrated, nevertheless, that the presence of predatory larvae reduced the numbers and skewed the age structure of larvae of other mosquitoes, of ceratopogonids, psychodids and other dipterous larvae in the bracts of *Heliconia* and the bromeliads studied. In the bamboo internodes the age structure of other species of mosquitoes was skewed as a result of the presence of the predators, although actual abundance levels of these larvae was not affected. This study is noteworthy as it counteracts any suggestion that studies such as those of Bradshaw & Holzapfel (1983) are relevant only within the narrow confines of culicid assemblages in tree holes. The introduction of a keystone generalist predator such as the larvae of *Toxorhynchites* clearly can impact upon the whole metazoan food web across a range of phytotelm types.

Such general impacts were also observed in studies by Mogi & Yong (1992) who examined the organisms found within the pitchers of *Nepenthes ampullaria* in Malaysia. They examined pitcher food webs at two locations: at Malacca in peninsular Malaysia and at Kuching in Sarawak. At one of the mainland sites the communities were relatively simple and dominated by the larvae of the filter-feeding mosquito, *Tripteroides tenax*. Densities of larvae in these simple communities reached high levels. The food web at the Sarawak sites was much more complex and contained a complex of predators in addition to filter-feeding and other detritivores. Mogi & Yong (1992) note that the *summed* densities of all the filter feeders in these more complex communities remained much below those achieved by *Tripteroides tenax* in the almost predator-free situation at Malacca (and also, incidentally, confirming the trends in food-web complexity in *Nepenthes* pitchers introduced and discussed in Chapter 7).

With this work we have confirmation of the potentially central role of predators in structuring communities in all five principal classes of phytotelms: in tree holes (Bradshaw & Holzapfel 1983), in bromeliads, bamboo internodes and leaf axils (Lounibos *et al.* 1987b) and in pitchers (Mogi & Yong 1992). This work adds further weight to the notion that the role of predators, where they occur, is critical in structuring phytotelm communities.

Table 12.1. *Summary statistics for the abundance levels of larvae of Prionocyphon niger and Anatopynnia pennipes in 11 tree holes in subtropical rainforest in south-east Queensland*

Site	Prionocyphon niger				Anatopynnia pennipes			
	Minimum	Maximum	Mean	Standard error	Minimum	Maximum	Mean	Standard error
1	0.00	150.00	66.00	11.20	0.00	50.00	12.40	5.07
2	10.40	593.30	228.20	48.85	5.00	280.00	83.00	28.39
3	50.40	790.00	368.80	63.43	0.00	122.20	21.00	10.56
4	0.00	274.30	103.60	22.69	0.00	62.50	23.80	5.91
5	45.00	1431.30	426.80	114.13	8.90	237.50	54.40	17.57
6	6.10	350.00	91.00	26.49	0.00	137.50	40.10	12.72
7	0.00	5236.40	646.70	419.76	0.00	100.00	16.70	8.70
8	20.00	886.70	380.10	74.12	10.50	52.30	26.80	4.17
9	40.00	1466.40	345.30	110.41	0.00	32.00	13.90	2.51
10	594.60	3120.00	1386.00	192.47	25.60	266.70	91.70	20.00
11	120.00	2166.70	877.10	147.53	20.00	173.30	98.60	13.56
All	0.00	5236.40	447.20	56.35	0.00	280.00	42.94	4.95

Abundances are in larvae per litre of sample.

A cautionary note – *Anatopynnia pennipes* and *Prionocyphon niger* in Australian tree holes

I conclude this chapter with an account of a predator–prey example of a different sort designed to show that not all common predators within tree holes play the keystone role we can assign to *Toxorhynchites* species.

In our studies of the tree-hole communities in the subtropical rainforest of Lamington National Park (see Chapter 7) by far the most common saprophages are larvae of a species of scirtid beetle, *Prionocyphon niger*. These are preyed upon by a variety of predatory midge larvae, mites and tadpoles. In mature communities, however, larvae of the tanypodine chironomid *Anatopynnia pennipes* outnumber the other predators both in numbers and frequency of occurrence.

Elsewhere in this book I have discussed our monthly sampling regime in a set of eleven water-filled tree holes in subtropical rainforest in south-east Queensland. As part of these studies, samples of detritus and water were taken from each site by hand dipping. These were sorted to species and the counts, in each instance, were standardised to densities per litre.

Some of the results of these studies are presented in Table 12.1 and Figure 12.4. The mean abundances over all sites for each of the twelve months of study are shown in Figure 12.4(*f*) with the associated standard deviations. The other graphs in the figure show the results for five selected sites from among the eleven sampled. Sites 5, 8, 9, 10 and 11 were selected because all maintained relatively high numbers of both species throughout the study period.

Estimates of the abundance levels of *P. niger* larvae varied enormously, from zero to over 5000 l^{-1} within a single site and at least by an order of magnitude within every site. Mean levels for the twelve months of sampling varied from site to site, with a range of 66 to 419 l^{-1}. Variability across sites is summarised by the coefficient of variation values given in Table 12.1. Values of this statistic (expressed as a simple proportion) ranged from 0.59 to 2.25, indicating that there is variation in the predictability of sites as habitats for the species: that is to say, some sites represent a more equable environment for beetle larvae than do others, as measured by their population responses.

For *A. pennipes*, levels of abundance were less dramatic, ranging overall from 0 to 280 l^{-1}, again with changes of an order of magnitude or more observed within each site (Table 12.1). Values for the coefficients of variation in the eleven sites ranged from 0.48 to 1.66. Again this is an index of spatial variation in suitability of the environment for this species. It is noteworthy that the average value for the coefficient of variation was higher (1.07 ± 0.491)

Competition and predation

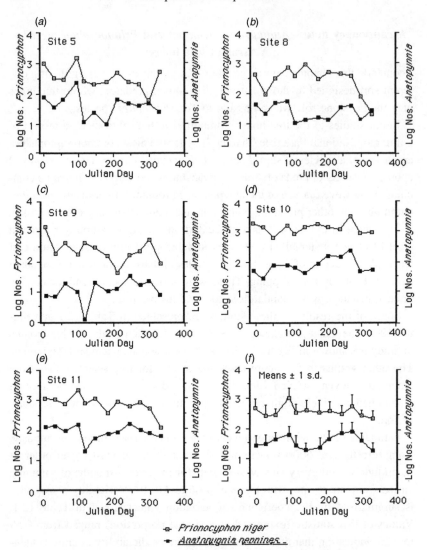

Fig. 12.4. The numbers of larvae of the predatory tanypodine chironomid *Anatopy-nnia pennipes*, and those of its principal prey, larvae of the scirtid beetles *Priono-cyphon niger*, through a twelve-month period in selected water-filled tree holes in subtropical rainforest in south-east Queensland.

Table 12.2. *Linear regression analyses on the numbers of larvae of* Prionocyphon niger *and* Anatopynnia pennipes *in 11 tree holes in subtropical rainforest in south-east Queensland*

Site	Intercept	Slope	F-value	Probability
1	−10.56 (6.93)	0.35 (0.092)	14.45	0.003
2	−7.49 (36.82)	0.40 (0.133)	9.12	0.013
3	33.82 (22.72)	−0.035 (0.055)	0.41	n.s.
4	19.16 (10.33)	0.045 (0.080)	0.31	n.s.
5	−1.87 (14.34)	0.13 (0.025)	27.48	$p<0.001$
6	4.95 (11.36)	0.39 (0.090)	18.55	0.00
7	4.90 (4.89)	0.018 (0.0032)	32.39	$p<0.001$
8	23.17 (7.94)	0.009 (0.018)	0.30	n.s.
9	15.07 (3.58)	−0.0034 (0.0071)	0.23	n.s.
10	−7.01 (36.50)	0.071 (0.024)	8.87	0.14
11	57.63 (25.15)	0.047 (0.025)	3.47	n.s.
All	25.48 (5.24)	0.041 (0.0067)	37.77	$p<0.001$

The simple model $y=a+bx$ is used where y represents the numbers of *A. pennipes* per litre and x the numbers of *P. niger* per litre. Numbers in parentheses are the standard errors of the estimates of intercept and slope.

for *A. pennipes* than for *P. niger* (0.88 ± 0.493). It must also be observed that samples from six of the eleven holes gave zero readings and, for two of the holes, these zero counts occurred in six of the twelve monthly samples. In general, larvae of *A. pennipes* were present at about a tenth of the abundance levels of those of *P. niger*, their principal prey item.

I used these and other data to examine the statistical relationship between the levels of abundance of the larvae of *A. pennipes* and its principal prey species, *P. niger*. Table 12.2 presents the results of a simple regression analysis of the abundance levels of the two species on a site-by-site basis. In 5 of the 11 sites there is a significant linear regression between abundance levels of *Anatopynia* and those of *Prionocyphon*. The actual level of significance varies but, given the coarseness and acknowledged inaccuracies of the sampling method, the analysis shows overall a strong positive interaction between the two species. Figure 12.5 plots the abundances of one species against the other, for four of the sites. In three of the sites, the regression analysis yielded significant results (Sites 1, 5 and 10; see Table 12.2) and the fourth (presented in the interests of evenhandedness) indicated no relationship of value between the two variables.

There is little doubt that in a statistical sense there is a direct numerical response between the predator *Anatopynia pennipes* and its principal prey,

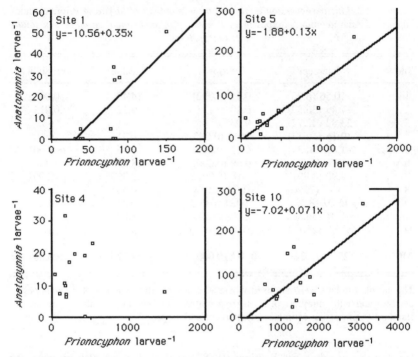

Fig. 12.5. The numbers of larvae of the predatory tanypodine chironomid *Anatopynnia pennipes*, plotted against those of its principal prey, larvae of the scirtid beetles *Prionocyphon niger*, in selected water-filled tree holes in subtropical rainforest in south-east Queensland.

Prionocyphon niger. The significant correlations obtained between the abundance levels of these two species are remarkable results for two reasons. First, the methods used to estimate abundances are inherently likely to produce estimates with large variances. Secondly, the five sites in which non-significant results were obtained were among those with very low levels of abundance of *A. pennipes*, with frequent zero counts in the monthly samples from those locations. In other words, levels of one of the species were so low that no significant relationship might be expected for computational reasons from those sites.

These regressions may be compared further by examining the standard errors of the estimates of slopes and intercepts in each case. Of the six sites which showed significant relationships the slopes grouped around two values. Those for Sites 1, 2 and 6 have an average value about 0.38 and are clearly not significantly different from each other using a ± 1 s.e. criterion. Similarly, the slopes for Sites 4, 7 and 10 group around a value of −0.59 and, again, are clearly not statistically different. If we use a 95% confidence interval (± 2 s.e.) then Site 5 groups with Sites 1 and 2. The values of the intercepts are all statistically similar at the 95% level with an overall value around 12.

So, what do these relationships mean, and how might they have come about? First, they do *not* mean that there is a direct density-dependent relationship between the two species – a concurrent relationship in the sense of Solomon (1949). Were this the case one would expect the relationship to be non-linear with some key process rising to a maximum value. In other words some functional response (*sensu* Holling 1964) would come into play and the curve relating the two species would level off.

It is also hard to imagine what sensory mechanisms could produce such a relationship, if indeed there were a causal relationship between the two species. For such a relationship to exist an ovipositing adult *A. pennipes* would need to be able to detect the presence of larvae of *P. niger* in a site before oviposition. This is unlikely and, even if it were to happen one would expect a lag in the relationship equivalent to the egg development time of the midge. No such lag is evident in the data.

The most likely explanation is that both species are responding, as ovipositing adults, to indicators of the 'quality' of a site based on evolved preferences and linked to physico-chemical cues available to the ovipositing adult insects. Such cues have been demonstrated in other tree-hole insects (e.g. Wilton 1967 for the mosquito *Aedes triseriatus*). So long as the egg periods of the two species are not greatly different this would produce an instantaneous relationship of the sort observed. Of course the evolutionary feedback which may have fixed these preferences for the predator could well have included the enhanced survival of larvae due to the ready availability of prey species. That no avoidance response in the prey species has evolved suggests that the approximately 1:10 predator:prey ratio does not produce a drastic effect on the prey species. There is little evidence that the tree-hole community in general is energy limited (see Chapter 13) and this provides a further line of circumstantial support for this explanation.

I conclude from these results, and the very clear indications from other work, that *Toxorhynchites* larvae play keystone roles in phytotelm communities; such a role cannot be assumed for other predators even when these reach high densities (as does *Anatopynnia pennipes* on occasion). There are significant biological differences to be taken into account – the voracity of the species involved, their propensity to be cannibalistic, their generation time relative to that of their prey, and so on. The relative importance of predation as a community-structuring process must be judged on a one-off basis. This having been said, if the community of interest contains *Toxorhynchites*, then it is a not unreasonable assumption that this species is having major structuring impacts.

Connections

The bottom-up processes considered in this chapter are key building blocks in understanding how phytotelm communities are put together. But it is the interaction of these population-level processes with the larger-scale top-down processes at the habitat and faunistic level that produce the food-web structures that we observe in nature. Chapter 13 revisits the hypotheses proposed in Chapter 6 drawing on some of the processes described in this chapter. Then the two levels of process are synthesised in the concluding Chapter 14.

13

Stochasticism and determinism: processes structuring food webs in phytotelmata
Chapter 6 revisited

So what do all these patterns tell us about the underlying dynamics that form food webs within phytotelmata? Chapter 6 is a personal attempt to bring together those ideas which seem to me to be relevant to this 'big' question. The four boxes within Chapter 6 summarise these ideas as a series of emergent hypotheses and associated predictions. These can now be put to the question in light of the patterns described in the preceding chapters.

It must immediately be said that some of the hypotheses identified can be tested only anecdotally from available information; others are but weakly tested; a few we can be a little more confident about. By bringing together the theoretical base (from Chapter 6) with available information (all other chapters) I hope to be able to define, at least by implication, the state of the art as far as the ecological understanding of phytotelm communities is concerned. If our understanding is found wanting (as it certainly is on some issues) then such a treatment should identify future areas for further investigation.

I shall use the numbered hypotheses of Boxes 6.1 to 6.4 as the structuring device for this chapter.

The global scale

Four separate processes are of interest when considering the determination of food-web pattern along gradients defined at the global scale: continental drift, host plant biogeography, adaptive radiation, and synoptic environmental conditions.

Continental drift

H1. Hypothesis: *The fauna of phytotelmata in a particular geographic location will reflect the tectonic history of the location at least in some instances.*

275

P1. Prediction: *Some faunal elements will be unique to particular locations at levels higher than the species.*

H2. Hypothesis: *The fauna of phytotelmata may include faunal elements from adjacent biogeographic areas due to dispersal.*

P2. Prediction: *Examples of particular phytotelm types close to the boundary between two biogeographic regions will contain elements not found further from the boundary.*

To present tectonic history (and, in particular, continental drift) as a hypothesis for explaining observed patterns in food webs is somewhat meretricious as the 'hypothesis' comes close to being a truism so long as we accept the reality of the geologic processes involved.

At the anecdotal level, we can pick faunal elements from the bestiary presented in Chapter 3 and the Annexe and place them in the context of what is or is not known about the particular taxon at the global level. Where this taxon, in general, is restricted to a particular geographic region this will obviously constrain its presence or absence within phytotelms in that region or another.

So, for example, the unique occurrence of the podonomine *Parochlus bassianus* in *Richea* axils in Tasmania reflects the general southern distribution of this particular subfamily of the Chironomidae. It seems likely that a similar explanation can be adduced for the occurrence of the Neurochaetidae in axil waters in a number of widely distributed Gondwanan localities. Indeed the occurrence of a single Malaysian species recorded within the family is evidence for the second prediction in Box 6.1, that phytotelmata close to major biogeographic boundaries may well contain elements originating from the adjacent region. The Malaysian neurochaetids could very well have a southern origin, perhaps sharing Australasian ancestors with species currently occurring in New Guinea. Last, the distribution of container-breeding species among the frog families reflects the continental distributions of these families in general, which in turn is a consequence of the tectonic history of these land masses. So the Myobatrachidae in Australia, the Leptodactylidae and Dendrobatidae in the Neotropics, the Rhacophoridae in the Oriental region, the Hyperoliidae in the Ethiopian region, and the Mantellidae in Malagasy subregion, all contain species with a variety of breeding habits, and each contributes to the container fauna in their regions.

It is clear we can identify endemic taxa within container faunas at least at the continental level. One of the striking features of phytotelm faunas, however, is that there are many taxa at the level of the genus and above which are cosmopolitan across massive sections of the globe. Ceratopogonid midges of the genera *Culicoides* and *Dasyhelea* are found in virtually every type of container habitat wherever in the world they occur. Mosquito subgenera such as *Aedes (Finlaya)* and *Culex (Culex)* are similarly cosmopolitan. Other container-breeding culicid genera such as *Wyeomyia* and *Tripteroides,* and chironomid genera such as *Metriocnemus*, enjoy immense distributions without being cosmopolitan. In other instances the association with container habitats may occur at a higher level. A great range of container habitats contains species of eristaline syrphids or scirtid beetles. These close associations suggest a long-standing existence of the container habit in these groups and, indeed, a co-radiation between the animal taxa and the container plants themselves. I return to this point below. As Laird (1988) suggests, for the culicids at least, the container-breeding habit may date as far back as the Carboniferous. It appears also to establish a basal role for rainforest ecosystems in the evolution of container breeding – although this may be yet another truism, relevant to most ancestral ecological strategies within the Insecta, not merely container-breeding.

Host plant biogeography and metahabitats

H3. Hypothesis: *The food webs found within particular plant groups will reflect in part the biogeographic history of the container plants.*

P3. Prediction: *Where a particular plant taxon has a well-defined range, the contained food web may vary from the middle to the edge of the range of the taxon.*

H4. Hypothesis: *The richness and complexity of a food web within any one phytotelm type will be increased if a range of other phytotelm habitats is also present.*

P4a. Prediction (1): *Foodwebs within a particular phytotelm type will be more speciose and complex when a range of other phytotelm types is also present.*

P4b. Prediction (2): *Where a set of phytotelm types occur in a particular place there must be at least a partial overlap in species composition across phytotelm types.*

This set of hypotheses is well addressed by some of the data sets presented earlier. In general there seems little doubt that, all else being equal, then the history and distribution of the host plant has a major impact on the richness and complexity of the food webs that occur within phytotelmata. Primary evidence for this is provided in our studies of *Nepenthes* pitcher plants, with more anecdotal support from studies of tree holes and bamboo internodes.

The global data on *Nepenthes* food webs presented in Chapter 7 shows that food webs in species of *Nepenthes* in the peripheral part of the range of the plant genus – in Madagascar and northern Australia for example – are less complex both horizontally and vertically than those occurring in the centre of the distribution of *Nepenthes* – in Borneo and Malaysia. Prediction P3 is accordingly well borne out.

Prediction P4a seeks a mechanism for the pattern just described. The wide range of food-web statistics from species of *Nepenthes*, which are well predicted by the independent variable 'number of co-occurring' species, strongly suggests that it is the 'metahabitat' provided by the set of species of *Nepenthes* which contributes to the high levels of richness and complexity observed in at least some of the pitcher species. Where the 'metahabitat' is in fact composed of different types of phytotelmata rather than merely containers in different species of the same general phytotelm type (as in, say, pitcher plants, or bromeliads) evidence can also be adduced to support the general hypothesis, albeit more anecdotally.

It is in this sort of comparison that we examine the degree of overlap between the faunas of different phytotelm types. In the case of the six species of *Nepenthes* discussed in Chapter 10 there was considerable overlap among the faunas of particular species – although even within the different pitcher plants, the degree of overlap varied. Moving to comparisions across phytotelm types a similar result is observed. In work carried out in Sulawesi and New Guinea respectively, a range of phytotelm types was examined. Considerable overlap existed between the faunas of tree holes with those of bamboo internodes. Less commonality existed between the faunas of these habitat types and, say, co-occurring pitcher plants or axil waters (Kitching 1987b, 1990).

There seems little doubt that the co-occurrence of a range of phytotelm habitats enhances the community of any one of these habitats. Only where particular phytotelm habitats present very special ecological challenges to potential inhabitants will this be other than marked.

Adaptive radiation

H5. Hypothesis: *Regardless of other patterns, phytotelm communities in general will have a 'local' flavour, reflecting the occurrence of some taxa particularly associated with the land mass or region on which they occur.*

P5. Prediction: *The most speciose clades within the higher taxa occurring in phytotelmata will generally be those best represented in phytotelm communties, and these will differ from biogeographical region to region.*

Again there is a real risk that this sort of hypothesis and its consequent predictions revert to mere truisms. The prediction (P5) as such is not borne out simply because the most speciose clades within virtually all phytotelm communities examined are the culicids. Most of the genera involved have extensive distributions, transcending traditional biogeographic regions. Some support for the idea that the phytotelm fauna will reflect the wider evolutionary radiations that have occurred around them is detected when the 'cosmopolitan' taxa are set aside. The most obvious support for this proposition is found within the Anura. The mantellid frogs in particular are a Malagasy radiation and are also a major component of axil and tree-hole communities in Madagascar.

Latitude, synoptic climate and productivity

H6. Hypothesis: *Foodwebs respond in richness and complexity to the environmental gradients correlated with latitude.*

P6. Prediction: *There will be clear gradients in food-web statistics with latitude, higher latitudes having simpler webs.*

H7. Hypothesis: *Production is correlated with latitude, and higher levels of productivity allow for the existence of more complex webs, containing longer food chains.*

P7. Prediction: *Foodwebs in more productive areas should be more complex with greater average food-chain lengths.*

H8. Hypothesis: *More predictable environments allow more complex food webs to occur, containing longer food chains.*

P8. Prediction: *Environments with climate or other environmental variables that are less variable should contain more complex webs and longer food chains.*

This suite of hypotheses and associated predictions have received much attention by Stuart Pimm, John Lawton, Bert Jenkins and others (see Chapter 6). Of the three predictions, the first relates simply to the patterns observed in the data. The second and third have often, perhaps erroneously, been regarded as competing alternatives.

The clear gradients in species richness, number of feeding links and the average food-chain length that occur within the tree-hole data presented in Chapter 7 clearly support the first prediction. Even at the continental level these patterns are clearly present (Chapter 8). Such gradients are not present in, say, the *Nepenthes* data of Chapter 7 simply because these pitcher plants, and most other types of phytotelmata other than tree holes, are essentially of the moist tropics. Our studies of tree holes and the results presented in Chapters 7 and 8 clearly demonstrate the utility of the habitat for such a comparative study of food-web structure.

The relationship between productivity and food-web characteristics in phytotelm systems remains 'not proven'. There is an argument which suggests that because all interactions in food webs (as I draw them at least) are trophic, then basal productivity within food webs must determine the nature of the contained food web. This would be compelling if, and only if, it was merely the availability of food energy which determined food-web characteristics. If this book demonstrates nothing else it does show that this is not the case.

If we assume that the volume of detritus entering a container habitat is a measure of the available basal energy for the food web within, then the comparisons presented between, say, Australian and English tree holes undermine any notion that the availabiity of energy is important. The two food webs observed in these locations differ greatly in complexity in whatever way that metric is defined, yet the annual inputs of leaf litter and other detritus are closely similar. I have argued elsewhere that the pattern of influx and the annual patterns of temperature and rainfall are such that the energy available within the tree holes in Wytham Woods is in fact much less than in the subtropical ones of Lamington National Park; mere gravimetric measures are not a reflection of energy availability. The levels of activity of poikilothermic organisms must be a reflection of ambient temperature. For the tree-hole habitat this means that in warmer situations there will be faster decomposition of the detritus (micro-organisms of all kinds are ectothermic!), the feeding activities of metazoan organisms upon the products of this decomposition will be faster and more efficient, as will associated rates of development, inter-site movement and reproduction. All of these factors will make the subtropical tree hole, in evolutionary terms, a more likely candidate as the basis for a more complex community. However, this modification of the 'energy-availability' hypothesis is achieved by turning from the mass of energy entering the tree holes to a consideration of the temporal patterns and behavioural and physiological availability of that energy to potential inhabitants – which is a reflection of the predictability of the location as much as it is of gross productivity. This unavoidable concatenation of ideas of productivity and predictability is the reason why I see the two predictions (P7 and P8) as complementary rather than mutually exclusive.

In fact there is considerable evidence in support of the idea that environmental predictability is important in determining food-web complexity and food-chain length in phytotelms, even when the mechanism imputed is one of food availability as outlined above. This evidence is contained within the data sets on tree-hole food webs at both the global and continental levels. The results presented in Chapter 8, showing that a significant proportion of the variance in most food-web statistics can be accounted for by using over-all rainfall figures, is circumstantial evidence for this: high overall rainfall may be interpreted as an index of the 'availability' of tree holes as aquatic habitats for their fauna.

There are, however, two general problems with testing hypotheses concerning the role of predictability in determining food-web structure. These relate, on the one hand, to interactions between variables and, on the other, to problems in determining appropriate temporal scales for determining environmental variability.

A simple consideration of rainfall alone, for example, is unlikely to provide a wholly satisfactory test for the predictability hypothesis. Areas with similar patterns of rainfall in, say, the south-eastern USA and north-western Europe may produce a comparable physical availability of tree-hole habitat for potential inhabitants, but the much higher ambient temperatures of the Floridan location will open a much longer effective window for this exploitation to occur.

Further, in trying to put the predictability hypothesis to the test I have collated monthly rainfall averages for a variety of locations. The inherent variability in these figures is not a good predictor of food-web variables. The flaw in this approach though is the assumption that it is the intra-annual variation in the environment which is the key determinant. In fact much longer time scales are inherent in the Pimm/Lawton hypothesis, which raises the issue of environmental predictability as a community determinant. It is more likely that such short-term variability could produce impacts at the short-term, local level and I return to this point briefly below.

In summary, although it seems likely that environmental predictability is important in affecting food-web characteristics within phytotelmata, this predictability is likely to be a reflection of a composite of physical variables and may reflect a time scale of decades or centuries. If such longer time scales are indeed the key then experimental manipulations, much vaunted as the way forward in food-web studies, will be inherently limited.

The continental scale

Some of the issues, patterns and conclusions discussed with respect to the global spatial scale will also apply at the continental scale. I shall discuss these here only when they add new insights to the overall discussion.

The finer grain of continental patterns

H9. Hypothesis: *Long-distance latitudinal patterns will be overlain and complicated by other environmental gradients such as rainfall patterns, altitude, distance to coastline.*

P9. Prediction: *Significant proportions of the variance associated with the values of food-web statistics along continental gradients will be accounted for by variables other than latitude.*

The existence of clear latitudinal patterns within continental confines is clearly demonstrated in Chapter 8. The analysis presented in Table 8.5 partitions the observed variance in food-web variables as responses to rainfall and altitude, as well as latitude. In several instances these additional variables do add power to the analysis: the variation in the numbers of predators present in the webs, for example, is 'explained' to the 56% level by 'latitude', but the addition of rainfall and altitude to the analysis takes this explanatory level up to 77%; the comparable 'improvement' when average food-chain length is the dependent variable is from 46% to 68%. This is evidence in support of the hypothesis that the large-scale patterns seen across communities spread over many degrees of latitude are overlain at the finer continental scale by other factors, reflecting aspects of the environment which are masked within global analyses.

Each continent represents a different environmental templet within which communities have evolved (see Chapter 14). At the coarsest level, the latitudinal range represented by the particular continent, the location of major mountain ranges and gross variation in altitude are likely to be the crucial factors which affect local synoptic climate. In turn this synoptic climate will dictate the general environmental equability represented by sets of tree holes, pitcher plants or other phytotelm units. In the analyses of tree-hole webs within continental Australia presented in Chapter 8 the location of the Great Dividing Range close to the eastern coastline of the continent restricts mesic environments to the coastal strip and ranges. Superimposed on this is a north-to-south temperature gradient which shortens seasons and is associated with lower diversity in the cooler south. These two interacting phenomena almost certainly contribute more than anything else to the variation in food-web properties observed from Tasmania to the Daintree.

Adding altitude to these considerations highlights one of the continent's peculiarities. Australia overall offers the range from sea level to 2229 m only. Altitudinally driven differences in vegetation and fauna do exist and there is a true alpine zone in the Snowy Mountains of the south-east, and in Tasmania. However, our tree-hole studies covered a relatively limited range of altitudes (up to 1320 m only) with some of the higher altitudinal values coinciding with lower latitudes (see Table 8.1). Accordingly, the role of altitude remains to be investigated critically. The question of its importance within phytotelmata remains to be settled. I suggest that the much greater altitudinal range seen in other continents will provide a better testing ground for any hypothesis relating altitude to food-web structure. The occurrence of *Nepenthes alata* from sea level to 8000 m on Mt Kinabalu in Sabah (Kurata 1976, Clarke 1997), or of tree holes from low elevation to montane levels in parts of Africa, New Guinea or South America, provide exciting prospects for such investigations.

Climate change and species pumps

H10. Hypothesis: *Phytotelm communities in ecosystems which have been subjected to pulsed climate change, which nevertheless allowed patches of mesic forest to persist, will be richer than those in areas where climate change has been either less periodic, or more severe.*

P10. Prediction: *Tropical rainforests should contain the richest and most complex of food webs within particular types of phytotelmata.*

Simply restated, a first test of this hypothesis would comprise demonstration that areas of extremely high biodiversity would show comparable exceptional diversity in their phytotelm communities. The difficulty arises in assembling data sets in which such pattern is not confounded by differences in latitude or synoptic climate. Accordingly we seek data sets from comparable habitat types (such as water-filled tree holes) that are widely distributed in comparable latitudes. Then we look to correlate overall diversity within the habitat type with whatever is known about the region overall.

Tree holes in equatorial rainforests present one of the more obvious opportunities for this sort of comparative analysis. Such forests occur in South America, West Africa, South-east Asia and on the island of New Guinea. Community data are available from New Guinea, northern Borneo and the island of Sulawesi (see Chapter 7). The key analysis, though, demands comparable data from South America and West Africa where community data is at best patchy and, more often, non-existent (although community data from Panama tree holes are about to be presented, O. Fincke pers.comm). So, again, the proper analysis must await further data. The data that are available, from New Guinea in particular, give some support to the 'species pump' idea. The tree-hole communities in both New Guinea and northern Borneo are very speciose and rich and occur in areas which, by virtue of their topography and location, are supposed to have experienced the Tertiary and Quarternary advance and retreat of tropical rainforest that is central to the hypothesis.

Habitat segregation and intraspecific interactions

H11. Hypothesis: *Parallel evolution of sets of ecologically similar species within a single land mass will lead to clear patterns of habitat preference and consequent segregation.*

P11. Prediction: *Where a particular, ecologically uniform taxon is diverse within a phytotelm type we expect to see marked differences in behavioural patterns in adult insects leading to (at least partial) segregation in space within particular ecosystems.*

H12. Hypothesis: *The presence or absence of 'keystone' predators within phytotelmata will have major qualitative and quantitative impacts on the food webs they contain.*

H12. Prediction: *Dramatic differences in food-web structure will occur across the boundary of the geographical ranges of keystone predators when these are not uniformly dispersed over an entire continent.*

There is persuasive evidence in favour of both these hypotheses from the patterns described in Chapter 12.

The differentiation in preferred habitats displayed by mid-western tree-hole mosquitoes in North America described by Copeland & Craig (1990) suggests, almost unequivocally, that the five species of mosquito partition the available tree-hole habitat so that each species has a preferred type and location of breeding sites. To jump from these field observations to the conclusion that interspecific competition is the underlying process requires, in addition, a leap of faith. However, the clear and elegant demonstrations of such competition in controlled, laboratory situations by Livdahl (1982) and others, and the circumstantial evidence of such competition (for European tree-hole mosquitoes) is persuasive.

I suggest the most parsimonious conclusion we can draw currently is that, in the absence of keystone predatory species, closely related species partition the available habitat, usually through the non-random oviposition behaviour of ovipositing adults. This partitioning has the effect of reducing the potential negative effects of interspecific competition on the fitness of the animal species concerned. Like all such conclusions, a few caveats, need to be added to this one. First, the persuasive evidence derives from only a very few studies, restricted geographically and taxonomically. Second, it may well be that situations in which predators are unimportant are in fact very few, and that the conclusion, although of value, may be accordingly limited in applicability. Lastly, habitat partitioning could be achieved by non-random larval survivorship rather than (or in addition to) non-random oviposition behaviour. All these caveats represent fruitful avenues for further investigation.

The role of predators as major agencies conferring structure on food webs in phytotelms is clearly demonstrated. The work of Lounibos *et al.* (1987b) on axil waters in Venezuela, described in Chapter 12, clearly demonstrates the dramatic role predatory larvae of *Toxorhynchites* can have on overall community structure. Mogi & Yong (1992) confirmed this in their work on *Nepenthes ampullaria* in Malaysia. I have no difficulty with the notion that voracious predators and top predators such as larvae of *Toxorhynchites,* odonate larvae and, possibly, tadpoles of some frogs, can exert major impacts within phytotelm communities. Two cautionary points need to be made.

First, not all predators, even when they reach high levels of abundance in units of particular phytotelmata, play such a 'keystone' role. Larvae of the tanypodine chironomid *Anatopynnia pennipes* in Australian tree holes, for example, have an impact on their prey numbers (inevitably) but do not appear to change overall food-web structure (Chapter 12). It may be that more generalist, larger, more mobile predators are pre-adapted to fill these keystone roles. More investigation is warranted.

Second, dramatic impacts of predators have been observed in smaller rather than larger phytotelm units: in pitcher plants and axil waters rather than water-filled tree holes. The role of habitat size in limiting the effectiveness of predators as modifiers of food webs has not been fully investigated. Some preliminary work on this question described in the last part of Chapter 9 failed to find clear relationships between surface area of tree holes and food-web statistics exhibited within that tree hole. Experimental manipulation introducing predators into carefully controlled situations would shed further light on this question which, for the moment, I regard as far from resolved.

The regional scale

Metapopulation dynamics

H13. Hypothesis: Populations of phytotelm organisms and the food webs within which they occur will vary at the scale of the landscape, reflecting population-level processes of extinction and establishment.

P13. Prediction: Populations of selected species will occur in some, spatially defined, subsets of phytotelm units and not in others: local extinctions may occur and these patterns may be more marked in rarer species such as predators.

Appropriate data for testing this hypotheses has been presented in Chapter 9. It remains largely unsatisfactory. So although it seems logically inescapable that such metapopulation dynamics will be important in determining phytotelm community structure, the evidence is not convincing at this point in time.

The differentiation of the tree-hole community that we studied in the lowland rainforest of north Queensland into three overlapping subsets, each correlated (*post hoc*) with features of the physical environment (Chapter 9) is open to a number of interpretations. The three subsets could represent chance assemblages reflecting metapopulation variation for each of the participating species at the scale of the local landscape. It seems unlikely that such chance assemblages would correlate so clearly with non-biological features of the environment (riparian/non-riparian, coastal/non-coastal). Alternatively the three emergent 'sub-communities' could represent separately evolved sets of biological interactions suggesting separate histories. The most likely explanation, however, is a combination of the two. Most of the saprophagous species in this set of food webs occurred over the whole landscape. Absences of particular saprophagous species from particular holes no doubt reflect metapopulation processes. It is among the predators and top predators that clear differentiation along habitat lines occurred and, for these species, each subset of tree holes and the surrounding environment no doubt represents a limited, specialised resource. The differentiating dimension in these cases is likely to be features of the environment that are unrelated to the tree holes themselves – which are merely larval habitats for most of these predators. Adult odonates, for example, require particular configurations of open water, riparian vegetation and tree cover to provide them with access to prey during their adult stage.

Metapopulations of phytotelm insects will be defined both by the unit of container and by larger scale dynamics reflecting characteristics of both the animal species and the container plants themselves. In an extreme example, the bract axils of the inflorescences of *Curcuma australasica* (described in Chapter 11) divide the populations of *Uranotaenia diagonalis* into those within each axil. These are grouped into the subpopulation within each inflorescence (comprising 20–30 axils). In turn the inflorescences occur in groups representing the distribution of the plant itself: characteristic, in northern New Guinea at least, of disturbed ground with an intermittent canopy cover. The plant species has a very wide distribution within tropical Australasia. Such nested compartmentalisation of larval populations is not uncommon for specialist phytotelm organisms. In no cases that I am aware of have the implications of this complex population structure been studied from a genetic or ecological viewpoint.

Island biogeographic processes

H14. Hypothesis: *The presence of particular phytotelm animals within a patch of forest (say) will reflect the island biogeographic processes of invasion and extinction.*

P14. Prediction: *Phytotelms of particular types occurring in larger, less isolated ecosystem patches should have richer faunas and consequent food webs than those occurring in smaller and/or more isolated patches of the same ecosystem type.*

There is clear evidence that island biogeographic processes are factors contributing to the structure of phytotelm food webs, although these may often be masked by the outcome of other 'grander' processes. Direct and circumstantial evidence has been described in the global data on webs in tree holes and *Nepenthes* pitcher plants in Chapters 7, 8 and 9.

The most direct analysis described in this connection is that in Chapter 7 in which I examined the relationship between various food-web variables observed in *Nepenthes* pitcher plants and the size of the land mass on which they occurred. In general the relationships, although suggestive, were weak (see Table 7.7). Perhaps of more importance is the more anecdotal observation that in the most isolated location from which data were collected (from pitchers of *N. pervillei* from the Seychelles) the food web was exceptionally simple and 'out of line' with that which might have been expected for a site lying more or less midway between the sites in Sri Lanka and Madagascar. The remoteness of the island group resulting in few immigrant species (the distance effect of the Theory of Island Biogeography) presents the most plausible explanation for these results. The single data point represented by the Seychellian data is masked in the overall regression analyses presented in Chapter 7 where the very strong and significant effects of 'distance to the centre of *Nepenthes* distribution' and 'number of co-occurring host-plant species' predominate.

In a similar fashion the tree-hole data collected in Sulawesi and presented in the earlier part of Chapter 7 are also out of line with the global latitudinal trends identified in the wider analysis. The most plausible explanation as to why the webs encountered in Sulawesi were somewhat simpler than expected for such an equatorial site (cf. New Guinea for instance) is because of the peninsular effect producing faunal attenuation (an island biogeographic effect) at the study sites which were located towards the tip of the Minahasa (northern) peninsula of the island of Sulawesi. The food web elucidated from Sulawesi contrasts with that recently observed in northern Borneo (at about the same latitude) which contains three species of anuran and three odonates as well as a full range of saprophagous and predatory species (Kitching & Orr 1996). The Bornean results came to hand too late to be incorporated fully in this book's analyses but are mentioned here to point up the poverty of the Sulawesi tree-hole community.

I conclude that island biogeographic processes are universal but are often overlain by other evolutionary, ecological and behavioural processes. I predict further that universal phytotelms (such as tree holes) on isolated islands such as Hawaii or other islands of the Pacific in general will have simpler food webs than those occurring elsewhere at the same latitude and climate. Such a pattern, of course, could be complicated by the autochthonous evolution of 'special' inhabitants of such phytotelms which may not occur elsewhere. Data from such sites are tantalisingly incomplete.

The local scale

Succession

H16. Hypothesis: *Successional processes will occur on a habitat unit to habitat unit basis, reflecting the history of that particular unit.*

P16. Prediction: *Distinct changes in community structure should occur through time following either the appearance of a new habitat unit or catastrophic 'clearing' of a pre-existing one.*

H17. Hypothesis: *Given the simple nature of phytotelm food webs, a facilitation model of succession within habitat units seems most likely.*

P17. Prediction: *Predators and other higher trophic levels will appear in developing food webs only after a set of saprophages ('prey') is well established.*

Data on tree holes and *Curcuma* axils show clearly an incremental series of changes following, on the one hand, catastrophic clearance and, on the other, natural ageing of the habitat unit. In each case there is a sequence of species' establishments which lends credence to a facilitation model of succession. It must be pointed out, however, that this notion of facilitation is based simply on the observation that predators do not become established in the phytotelm units until prey species are present within them.

The data from *Curcuma* axils suggest further complications in which food webs are seen to senesce as the habitat unit itself senesces. In such situations (and the *Curcuma* instance may well be a model for other axil, pitcher-plant and, even, some bromeliad systems) the changes observed within the phytotelm food webs almost certainly owe more to changes in the container than to any internal community dynamic.

Seasonality

H18. Hypothesis: *Seasonal patterns within phytotelms are only possible in long-lived containers: even in these habitats seasonality is unlikely to be obvious at the level of the food web, although it may be apparent in the population dynamics of particular species within the communities.*

P18. Prediction: *Values of food-web statistics sampled regularly should show little variation which can be correlated with change of season although within more complex webs the relative abundance of some species may show such patterns.*

Larger tree holes represent persistent aquatic milieux for their inhabitants. As described in Chapter 9, however, their low-grade energetic nature and dependence on allochthonous energy mitigate against clear seasonality in food-web characteristics. This is evident in the Australian-based data presented in that chapter. These data also indicate, however, that in some instances populations of particular participants in the community do show phenologies which are seasonally correlated. So, for example, anurans occur in the communities we studied in the subtropical rainforests of south-east Queensland only in a tight temporal window in the summer months. Similar seasonal patterns are described in Chapter 12 for saprophagous species occurring in complementary seasons within the predator-free tree-hole communities of southern England.

Seasonality in non-woody phytotelmata is 'forced' due to the phenology of the container plants themselves. Annual growth of *Dipsacus* in the northern hemisphere obviously restricts the axil communities within these plants to a short existence owing to the equally short period of maturity of the plants. Similarly phytotelmata and the communities they contain that are associated with particular, non-permanent plant parts will also be temporally restricted. Many examples have been alluded to earlier. Bract axil waters associated with *Curcuma australasica* (Chapter 9), flower axils of the north Queensland ginger plant *Tapeinocheilus ananassae*, and others, have been included in our studies over the years.

Stochastic discovery and survival

H19. Hypothesis: *At a single point in time the presence or absence of particular species and consequent web structures within individual habitat units will reflect the chance processes of discovery and subsequent survival by the dispersing propagules of that species.*

P19a. Prediction: *Individual habitat units will contain random subsets of the species that make up the regional food web.*

P19b. Prediction: *No individual habitat unit will contain all the specialist organisms which make up the regional food web for that phytotelm type.*

Two sets of data have been presented which illustrate the food webs occurring within single units of particular phytotelm types: from water-filled tree holes, discussed in Chapter 9, and from bamboo internodes, illustrated in Chapter 5. Both data sets confirm the predictions set out in P19a and P19b (above). Although the food webs drawn for each habitat unit, in these examples, contain an apparently unpredictable subset of the species that make up the overall fauna, locally, for that habitat type, some constraints are likely to apply which would 'forbid' the persistence of certain subsets. In particular it seems unlikely that predatory species could exist alone in either tree holes or bamboo internodes: indeed none of the individual webs identified presented such situations. I suppose it is feasible that adults of predatory species could oviposit within tree holes or bamboo cups which, by chance, contain no other species, but any resulting larvae are unlikely to persist in that situation. Our observations on succession and reassembly, discussed already in this chapter, support this contention.

Part V:

Synthesis

*Treeholes abound in the primary dipterocarp forests of northern Borneo –
but so does everything else. A hectare of forest we surveyed contained over
260 species of tree. The same patch of forest is home to almost 60 species
of bird, 320 species of butterfly, 50 amphibians and forty-five reptiles. This
superfetation of life is apparent to the field ecologist as nowhere else I have
worked. Staggering up the steep track on the eastern shore of the Belalong,
we were dive-bombed by dragonflies, a range of large and small butterflies,
and by the more sinister large vespids that abound in these forests. We stepped
over a flat-nosed pit viper coiled like a triangular spring beside the path, and
over the very visible products of the resident herd of bearded pigs. We heard
the overhead honking of rhinoceros hornbills, the accelerating crescendo of
the call of the helmeted hornbill, and disturbed a flock of bushy-crested hornbills. Pigmy squirrels the size of mice headed up the tree trunks waggling
their bottoms as they scent marked en route; flying lizards launched themselves and headed downwards in graceful swoops adjacent to the same trunks.
The thing about the Bornean forests is that one can go on and on: the giant*
Campanotus *ants, the fantastic fungi, the dipterocarps themselves, forest
dragons, pangolins, spiderhunters and tree holes.*

*Nowhere have I come across tree holes that contain such an abundance
of life as in these Bornean forests - and a visible abundance at that. Adjacent
to the eastern path about half a kilometre from the Kuala Belalong Field
Studies Centre is a tree hole in the buttress roots of a forest tree. It nestles
into the groove formed by two adjacent roots and, as we first saw it, had a
brilliantly white mass of foam perched above it – the product of the rhacophorid frog whose tadpoles we found within the tree hole; we actually
watched as the tiny wrigglers emerged and fell into the waiting tree hole
beneath. Two dragonflies and a damselfly specialise in these habitats, joining
larvae of* Toxorhynchites *and* Paramerina *to exploit the dozen or more detritus*

feeders. In these forests too I first discovered ladder-like tree trunks with horizontal stress 'planks' between adjacent buttresses, each 'plank' containing a fully functional tree hole – five or six such holes stacked one over the other.

I am assured that the most recent observations on tree holes in the similarly super-rich forests of Panama point to a similarly diverse and trophically complex food web. To understand these webs is to understand the tropical rainforest, and it is to that goal that this book has been directed.

From the deciduous woodlands of southern England to the primary forests of Borneo is a great distance both geographically and ecologically, but it is that journey which has begun to produce the glimmerings of understanding that I think I have about how the community dynamics of these places work. This final part of the book tries to draw together earlier disparate threads and build a structure around these glimmerings.

14

A food-web templet
A summarising statement

All ecological phenomena can be viewed in light of key axes along which selection has occurred and against which various emergent properties of ecological systems can be matched. In their simplest (and perhaps most useful) form these axes are the universal ones of space and time. They can be transformed to represent changes in the magnitude of key processes such as competition, disturbance and stress. These axes have been used to construct so-called templets (to retain Southwood's, 1977, deliberately archaic spelling) as synthesising devices which have been very effective in providing an heuristic framework for understanding complex, multivariate phenomena.

Other templets

The templets of Southwood (1977) and Grime (1979) introduced the ideas of semi-diagrammatic summaries of ecological ideas. In both these cases the authors were making statements about the life-history strategies and processes of selection in animals and plants respectively. Southwood's model defines habitat heterogeneity in terms of time ('durational stability') and space. Against the temporal scale he places the pre-existing notion of r and K selection – basically the ends of a continuum of species-specific strategies ranging from the rapid-breeding, highly dispersive, short-lived, poorly competitive through to the slow-breeding, sedentary, long-lived and highly specialised. Against the spatial axis he erects another selective continuum which he calls 'adversity selection' – a concept devised by Whittaker (1975) to describe the syndrome of selective pressures and adaptational responses observed in plants growing under varying degrees of stress. The idea of 'adversity selection' and its use within the habitat templet concept was further developed by Greenslade (1983). Any species can then be placed within this Cartesian framework (although some regions are less likely to be occupied than others – see Kitching 1986b for further discussion). Southwood (1977) claims this structure forms a sort of 'ecological periodic table' against which both pre-existing and new information can be viewed. Southwood's achievement, in my view, was not only to place the two forms of co-occurring selective pressures into the same framework, but to indicate the emergent community-level properties that were likely to be with each of these points in the template space. Restated, although based on population level appreciation of the life-history strategies of organisms it has produced insights about multi-species assemblages – communities by any other name.

In parallel with Southwood, Grime (1979) developed the idea of a triangular space, the edges of which represented the degree of application of three basic evolutionary forces acting upon plants. He described these as competition, stress and disturbance. At any particular place or time one or other of these selective forces will prevail – in productive undisturbed vegetation, competition will be a key process; in areas of continuous low productivity, development of stress-related adaptations will be the key; and in frequently disturbed environments a weedy so-called 'ruderal' strategy will prevail. Grime's triangle, as it has become known, allows particular life forms (or sets of life-history characteristics) to be associated with particular regions within the three-space. Again, community-level implications can be drawn by associating particular life forms with particular environments.

Both Southwood and Grime, then, take an approach which examines the characteristics of particular species and places them in a general context. In 1997 I published a first attempt to apply such a 'templet' approach to understanding biodiversity (Kitching *et al.* 1997). This synthesis was built around the question: What set of processes are involved in generating a particular set of species which the ecologist might encounter in a sample at a particular moment in time? Setting aside methodological issues (that is: that one sampling method may well produce a different set of organisms from another), this clearly involved time-based processes from the plate tectonic to the very recent (was the forest burned, or thinned, or harvested yesterday?). Trying to define a spatial axis was more difficult so I was reduced to defining a series of columns covering the continental and geological, the climatic, patchiness, and lastly the degree of species–species interactions involved. Under each of these headings, community characteristics could be defined at different time scales.

Trying to capture all of the processes in biodiversity generation and maintenance involved a major extension of the life-history-based approaches of Southwood and Grimes. Here I take a slightly less ambitious path by trying to define a templet for just one aspect of community dynamics – the maintenance of food-web structure.

A food-web templet

I present here (Table 14.1) a templet which provides both a context and summary for the ideas on food webs in phytotelms that have been discussed in
this book. The templet represents a complex synthetic hypothesis about the
ecology and evolution of food webs in these habitats. As such it is the organised combination of the hypotheses first stated in Chapter 6 and revisited in
Chapter 13 and elsewhere in this book. Strictly, being conceived within the
universe of discourse concerned with phytotelmata and their fauna, it is not
applicable to any wider set of habitat types. Undoubtedly, however, many of
the processes and direct or indirect connections that are implied may have
relevance to a wider set of communities.

In trying to connect together sets of phenomena within community ecology we are faced with the problem of coping with the phenomena we observe
(such as food-chain length, high-level taxonomic composition or number of
species in a web) and the biological processes which may have produced
these phenomena (such as competition, environmental predictability, or stochastic discovery events). Both of these classes of objects occur across all
spatial and temporal scales. Sometimes the distinction between the processes
and the end results of those processes cannot clearly be made and a sort of
hybrid synthesis emerges.

In Table 14.1 I have attempted a stricter separation between process and
product by using different type-faces – simple Roman for the process and
bold face for the product. In addition in each occupied cell of the spatio-temporal two-space I have indicated where in this book the back-up information
and further entries into the literature can be found.

I divide the spatial and temporal axes of the templet into the segments –
of space and time respectively – first identified and defined in Chapter 6 and
further elucidated in Chapter 7. The spatial scale ranges, accordingly, from
the global, through continental and regional scale to the strictly local. The
temporal scale subdivides into three categories: the geological, ecological and
behavioural.

Of course the divisions I have made in both the spatial and temporal scales
are not linear. The geological and global end of the scales cover vast times
or distances in marked contrast to the behavioural and local extremes of the
framework which may deal with events that last minutes over distances of
centimetres! Further, the divisions that I have introduced in both the spatial
and temporal scales are merely descriptive conveniences. Neither scale is in
practice clearly differentiated into segments: each represents a continuum.

Table 14.1. *A spatio-temporal templet synthesising pattern and process observed in phytotelm food webs*

		TIME	
	Geological time	**Ecological time**	**Behavioural time**
Global space	**Intercontinental patterns in complexity** Higher taxa faunistics, Biogeographic history *Chapter 7*	**Correlations with host plant range** **Tropical/temperate gradients** Host plant range Environmental predictability, *Chapters 7, 8*	–
Continental space	**Large-scale pattern** Synoptic climate *Chapter 8*	**Latitudinal/altitudinal patterns** Environmental predictability Host plant range *Kitching & Pimm 1985; Chapters 8, 10*	–
Regional space	**Composition of the 'regional' food web** Microclimate Local topography *Chapter 8*	**Topographic patterns** Metapopulation dynamics Interpopulational exchange *Chapter 9*	**Movement among populations** Long distance displacement *Chapter 9*
Local space	**Evolved search behaviours** Chance encounter *Chapter 9*	**The faunal set from which a unit web is drawn** General environmental equability *Chapter 9*	**Inter-habitat unit variation** Stochastic search and survival processes *Kitching 1987a; Chapter 9*

S
P
A
C
E

Within the body of the table boldface entries indicate observed patterns, plain typeface indicates processes, and italicised entries refer to chapters within this work (and some references to earlier works) in which the entries are discussed further (see text).

Two summarising questions

The complex of ecological processes and the rationale for their inclusion in this summarising template is discussed at length in Chapter 6 and revisited, in the light of the patterns we have observed in phytotelm food webs, in Chapter 13. The underlying arguments and discussion are no more than long and convoluted ways of answering two deceptively simple questions which may occur to us as we peer into a particular container habitat:

- Why is this particular animal here, now?
- Why is this particular food web present here, now?

I address these two questions by way of encapsulating the results and thoughts presented through most of this book.

Why is this particular animal here, now?

The presence of a particular species of animal in a phytotelm food web is an outcome of many processes: some obvious, others less so.

First the species must represent a lineage that has persisted and/or evolved in the biogeographical region. It may represent a clade that has radiated in that biogeographical region or may be a more recent 'invader', having 'leaked' from adjacent biogeographical regions.

The particular species may be present in the phytotelm of interest because the ecosystem in which this habitat unit occurs is rich in phytotelm habitats, both of the same and different kinds to that under study. In addition the species may be present because the location presents a more equable and/or predictable physical, chemical and biological environment than elsewhere.

The species will be present in the particular tree hole, pitcher plant, axil water or whatever because the habitat unit we happen to be examining is located at the preferred height, exposure, aspect or whatever, consonant with the habitat-partitioning processes which have evolved among the wider community of phytotelm organisms present in the particular location. In other words, the species in question has evolved to select more efficiently or survive better in this particular subclass of the phytotelm type. In addition the habitat unit obviously does not contain any predatory species inimicable to the presence of our subject. If this is a community in which a keystone predatory process is in operation, our subject species has survived the keystone impact, or perhaps has not yet encountered the keystone predator in action.

The presence of our subject species reflects the fact that a population of this species exists at this particular place and time. Elsewhere the species may be absent as populations wink in and out over the landscape across which the species' metapopulation occurs.

On a temporal scale we encounter the species because its presence is consistent with the particular stage in succession or community reassembly occurring within the habitat unit. In the unlikely event that we have encountered a highly seasonal species, an encounter with it means that we have targeted that particular season.

Lastly, our species is present in this particular habitat unit because, quite by chance, propagules of this species encountered the particular tree hole, pitcher plant, bromeliad tank or bamboo internode in the recent past. Equally, all of the chance processes which can lead to local extinction have not occurred up to the point at which we encounter the species.

Why is this particular food web present here, now?

After sampling, efficiently, a particular class of phytotelmata in a particular place we obtain a construct, the food web, which summarises many of the important features of the community of organisms within those phytotelmata. The set of processes that produce the food web we observe are somewhat more complex than those we may adduce to account for the presence of any particular species.

First, the set of contributing species represents a co-occurrence of all those conditions, discussed above, which may control the presence (and, by implication, the absence) of any one species.

The complexity of the food web will be influenced by the broader distribution of the phytotelms themselves. Where a wide range of phytotelm types is available, more complex and richer food webs will be present within each one of these types than would be the case if the habitat type occurred alone. In addition, where a food web is restricted to particular taxa of container plants then a number of features related to the occurrence of that taxon will be important: the number of congeneric plant species and the distance to the centre of distribution of the plant taxon come to mind.

The richness of the phytotelm community we observe will be enhanced when particular genera (or other taxa) within them have undergone rapid speciation and radiation in the surrounding region. In addition the food web may be enhanced both in terms of richness (total number of species involved) and structure (number of feeding links, predator–prey ratios, number of trophic levels, connectance, etc.) with decreasing latitude and increasing rainfall. This pattern, in all probability, reflects increased predictability of the environment. This correlation between what we suppose to be predictability and the richness and complexity of food webs occurs at global, continental and local scales. The formative processes and history are likely to be different but the emergent properties are comparable.

Where biotic regions have achieved great diversity by the operation of mechanisms such as so-called 'species-pumps', as a result of the climatic and geological history of the area, we expect to find that the food webs we observe are more complex and rich across the board, including within phytotelm habitats.

The food web we observe within a selected phytotelm unit may well reflect the presence or absence of 'keystone' predators within it. In addition the web will owe its richness and structure to the operation of biogeographical processes. Insular or peninsular situations will contain simpler webs than 'mainland' locations. Our selected web may also be simpler than others because it reflects a particular stage in a successional or habitat reassembly process.

Conclusions

The template of processes and phenomena, and consequent summarising statements concerning the existence of species and the food webs of which they are part, that I have described in this short chapter merely summarise the more detailed discussions that have been presented earlier in the book. It is, in my view, important to be able to relate available sets of data and hypotheses about the occurrence of specific processes into a wider framework. This sets each piece of data and associated speculation in context and points to the need for further work or further thought required to sharpen ideas in our understanding of the determination and persistence of particular food-web characteristics.

Annexe: The phytotelm bestiary

In this Annexe I present as near complete a record as I can of the aquatic animals (above the level of the Protozoa) recorded from phytotelmata and the literature which has grown up describing their natural history and ecology. In general I cover publications up to mid-1997 although with such a scattered, polyglot and sometimes obscure literature there will no doubt be many missed records. In many instances, as elsewhere, I have found it convenient to summarise large blocks of literature as tables. Representatives of some the most commonly encountered families, particularly of dipterous larvae, are illustrated wherever possible with species, or at least genera, that have been encountered in phytotelmata.

As appropriate I use the 'biontic' and 'philic' designations described in Chapter 3.

Phylum: Platyhelminthes

The Platyhelminthes include three major classes, two of which are wholly parasitic. The third, the Turbellaria or flatworms, are free-living and widespread in freshwater, intertidal and marine situations. Examples of this last class have been recorded from plant containers. It is likely that some examples of the endoparasitic Platyhelminthes occur within phytotelm frogs but these have never been explicitly studied.

Flatworms are carnivorous, feeding on carrion and living prey. They catch prey items, often much more active and frequently larger than themselves, by wrapping their bodies around those of their prey and extruding copious quantities of mucus in which the prey may become entangled. Bennett (1966), writing of intertidal species, describes the extensible oral frill which may be used to enclose partly the prey, forming a receptacle within which external digestion can begin. Such literature as exists on the topic suggests turbellarians are very catholic in their choice of prey. Mellanby (1963) says they eat 'animals of any kind but particularly insects and crustaceans'. Extensive work by Reynoldson (1964) and Reynoldson & Davies (1970) on an assemblage of flatworms from Welsh lakes confirms this dietary catholicism but suggests, as might be expected, that each species exploits some food items more efficiently than others. These selected food items provide a 'refuge' resource in hard times so that apparently generalist predators have the ability to act as specialists under some circumstances. According to Reynoldson and his colleagues (and most later commentators) it is this flexibility that has permitted the stable coexistence of the set of ecologically similar species of flatworms over extended periods of time.

In phytotelmata tricladid flatworms have been recorded from bromeliads, water-filled tree holes and some moist plant axils. In the early studies of both Picado (1913) and Laessle (1961), flatworms are recorded from bromeliad waters. Picado (1913) in his studies in Costa Rica recorded *Geoplana picadoi* (Geoplanidae), *Geocentrophora metameroides* (Prorhynchidae), *Rhynchodemus bromelicola* and *R. costarricensis* (Rhychodemidae). In Jamaica Laessle (1961) added *Geocentrophora applanata* to this list as well as extending the known range of *G. metameroides*. Laessle commented that the flatworms he encountered appeared to be restricted to the outer parts of the bromeliad water bodies where dissolved oxygen levels were higher (see Chapter 4). Neither author comments on the prey items of the flatworms they collected.

We have encountered flatworms, tentatively identified as belonging to the widespread genus *Dugesia* (Planariidae), in water-filled tree holes in rainforests in tropical Queensland. These are common, sometimes abundant in lowland sites but have also been found at higher altitude sites at Eungella (22° S). In captivity they fed readily upon the living mosquito larvae with which they were confined.

Thienemann (1934) records four species under his cryptic heading 'Tricladida terricola' from the water-filled leaf axils of species of *Musa* in Indonesia: *Bipalium penzigi, Pelmatoplana bogorenses, Desmorhynchus ochroleucus* and *D. nematopsis*.

Tricladid flatworms are readily visible using low power microscopy and highly distinctive. Accordingly lack of records from other phytotelmata or tree holes (for example) elsewhere can be taken to indicate, perhaps surprisingly, a restricted occurrence in these apparently ideal habitats.

Rhabdocoel flatworms are much less easily detected, being smaller, more transparent and altogether less visible than their tricladid relatives. They too are predators feeding, according to Mellanby (1963), on microcrustaceans.

Some species feed on minute algae and yet others are parasitic. They are recorded by Fish (1983) from bromeliad waters and from axils of *Pandanus* in Sumatra and Bali by Thienemann (1934). They may well occur in other phytotelms. They have been seldom sought and absence of records is not significant.

Phylum: Rotifera

The Rotifera are a phylum of minute multicelled organisms which are ubiquitous in freshwater habitats. Like rhabdocoels and gastrotrichs (see below) their minute size means they are often overlooked in general surveys using techniques more adapted to larger organisms. Rotifers are distinguished by possession of a corona of cilia, the actions of which have led to their common name: wheel animalicules. These cilia are also responsible for directing water and suspended food particles towards the mouth, as well as propelling the animal through the water.

Reflecting their widespread occurrence in aquatic situations, rotifers have been recorded from all major classes of phytotelmata.

Picado (1913) notes the lecanid *Monostyla* sp. from Costa Rican bromeliads and Torales *et al.* (1972) record the philodinid *Rotaria rotatoria* from bromeliads in Corrientes in Argentina.

In water-filled tree holes the most extensive comments on rotifers are those of Röhnert (1950). She recorded three species of rotifer from her north German sites: *Habrotrocha thienemanni, H. tripus* and *Metopidia lepadella*. Röhnert regarded only the first of these, which occurred in 'almost every hole' [transl.], as dendrolimnetobionts: the other two she classed merely as accidental inhabitants. In an earlier list pertaining to all European records Thienemann (1934) following Hauer (1923) includes, in addition, *Colurella gastracantha, Monostyla arcuata, Macrotrachela quadricornifera, M. ehrenbergi, Squatinella stylata* and *Lepadella patella*. Beattie & Howland (1929) recorded species of *Philodina and Metopidia* from British tree holes.

Following their extensive studies of water-filled bamboo internodes at Nagasaki, Japan, Mogi & Suzuki (1983) say: 'Rotifera were frequently recognised when fresh material... was examined microscopically, but they were not quantified'.

There are several records of rotifers from axil water communities. Alpatoff (1922) records *Callidina symbiotica* and *Adineta vaga* from water-filled axils of *Angelica sylvestris* in Russia. *Callidina tridens, Diglena mustela, Philodina roseola, Rotifer citrinus* and *R. vulgaris* were found in the axils of *Dipsacus sylvestris* (Varga 1928) and unnamed bdelloid rotifers in axils of *Scir-*

pus sylvaticus by Strenzke (1950). I am confident that other axil waters contain rotifers awaiting discovery!

Lastly, rotifers have been recorded in both *Nepenthes* and *Sarracenia* pitcher plants. Thienemann (1932) records rotifers from species of *Nepenthes* in South-east Asia and, in addition, draws attention to the work of Menzel (1922) on *N. melamphora*. Beaver (1983), in his comprehensive review of the *Nepenthes* community, suggests that these records, along with those of Nematoda and Oligochaeta (see below), may reflect simply organisms to be found in other phytotelmata or, even, water-logged soil. He cites their marked commonness in ground (as opposed to aerial) pitchers as supporting, if circumstantial, evidence for this contention. There is undoubtedly merit in this suggestion and many casual records of phytotelm organisms can be placed in this xenic category. However, Röhnert's (1950) identification of one among three species of rotifers from water-filled tree holes as a specialist inhabitant of these situations, suggests caution against discarding records too casually. In *Sarracenia purpurea*, in Michigan, Addicott (1974) not only recorded both loricate and illoricate rotifers from the pitchers but used features of these organisms as a response variable in his manipulations of larval densities of the mosquito *Wyeomyia smithii*. He showed clearly that relative frequency of occurrence of rotifers, particularly the illoricates, declined with increasing larval mosquito densities.

Phylum: Gastrotricha

The gastrotrichs (or 'hairybacks') are multicelled organisms like large ciliated Protozoa in appearance. They are little known but probably feed on minute particles of organic matter. They are known from a variety of aquatic habitats including tree holes and bromeliads.

Torales *et al.* (1972) recorded the chaetonotid *Chaetonotus similis* from bromeliad water bodies in Argentina and Röhnert (1950) recorded another *Chaetonotus* species from tree holes in Germany. She considered these as casual members of the tree-hole community and not as regular, specialised, inhabitants.

Phylum: Nematoda

The roundworms are one of the most diverse and ubiquitous of phyla on earth and have been put forward as one of the 'mega'-diverse groups (along with the Arthropoda and Fungi) in recent arguments about global biodiversity (May 1990). They occur in every conceivable ecological situation from the guts of insects, to abyssal mud, to all manner of freshwater and terrestrial situations. They may be decomposers, parasites or herbivores and are undoubtedly of great significance in the dynamics of nutrient cycling and decomposition in all ecosystems.

All this having been said, the number of records of named nematodes from phytotelmata is few: where they have been noted it is usually by passing reference to 'nematodes'. At this level at least they have been observed in all classes of plant containers including bromeliads and pitcher plants. As with some other groups the lack of detailed study no doubt reflects the specialised techniques required, and associated taxonomic difficulties.

Species recorded from European tree holes include *Rhabditis* sp. and *Plectus cirratus* (Thienemann in Benick 1924, Thienemann 1934). Varga (1928) recorded *Tripyla setifera* from *Dipsacus* axils in central Europe. Goss *et al.* (1964) recorded large numbers of worms belonging to the genus *Panagrodontus* from the American pitcher plant *Sarracenia alata*. Beyond these few records the most complete list is that presented by Thienemann (1934) for material collected from the leaf axils of bananas (*Musa* spp.) in Indonesia. A sample of 30 individual nematodes contained thirteen species of twelve genera including species congeneric with those from European tree holes (see Thienemann 1934 for details). It remains moot as to whether any of these species are specialists or whether the accumulation of moist detritus or access to plant epidermis presented by these phytotelmata are an attractive general habitat to a wide range of nematodes.

Phylum: Tardigrada

The Tardigrada are a small phylum of somewhat enigmatic microscopic organisms that occur across marine, freshwater and moist terrestrial habitats often associated with moss, algae or other dense mats of vegetation. The great majority feed on the contents of green cells which they pierce and extract with their stylets and suctorial mouthparts. Rarely some species feed on rotifers and nematodes and there are a very few specialised predatory and even parasitic species (Cuenot 1949).

Few tardigrades are recorded from phytotelmata and I suspect the records which exist reflect accidental 'contamination' from adjacent material. Alpatoff (1922) records *Macrobiotus tetradactylus* from the leaf axils of *Angelica* in Russia and Varga (1928) noted the predatory species *Milnesium tardigardum* from axils of *Dipsacus* in central Europe.

Phylum: Annelida

The great phylum of segmented worms, the Annelida, are ubiquitous over the face of the earth, occurring from mountain tops to oceanic abysses, from deserts to rainforests. In general, consistent with their relatively unprotected epidermal layers, annelids are either aquatic or moisture loving in one sense or another. The phylum divides clearly into three Classes: the Polychaeta, the Oligochaeta and the Hirudinea. The first of these, the paddle-worms, are generally speaking marine and predatory. In contrast the oligochaetes are quintessentially decomposers, processing organic material in soils, leaf litter and other substrates in marine, freshwater and terrestrial situations. Lastly the Hirudinea, the leeches, are haematophagous and occur in freshwater and terrestrial ecosystems.

All three classes of annelids have been recorded from phytotelmata, although it seems likely that the records of leeches, summarised by Frank (1983), in bromeliads represent organisms merely using the equable moist environment of bromeliads as temporary shelters (as do many species of frog – see below). Table A.1, based largely on Thienemann (1934), summarises the records known to me.

The predominantly marine Polychaeta contain one subfamily (the Namanereidinae) that are freshwater and one species that appears to be a phytotelm specialist. This species, *Lycastopsis catarractarum*, is recorded from axils of *Colocasia indica* and *Musa* sp. in South-east Asia (Thienemann 1934) and from water-filled tree holes in lowland rainforest in New Guinea (Kitching 1990, Glasby *et al.* 1990). This predatory species presumably feeds on the range of insect larvae and oligochaetes which also live in these habitats.

Oligochaetes occur commonly in all classes of phytotelmata acting, presumably, as generalist decomposers. No species have received separate study by ecologists working on container communities although Thienemann's (1934) work describes the extensive collections made during the Deutsche Limnologische Sunda-Expedition to South-east Asia in the 1920s. As indicated in Table A.1 six families of oligochaetes and more than a dozen species are known from tree holes, bamboos, bromeliads and/or leaf axils. No doubt more assiduous survey would extend the list considerably. Unlike *Lycastopsis*, none of the species recorded is likely to be a container specialist. Oligochaetes must generally reach phytotelmata in run-off and mobile debris of one sort or another and, accordingly, are likely to be associated generally with moist or water-logged detritus, wherever it may lodge.

Phylum: Arthropoda
CRUSTACEA

Second only to the insects, the Crustacea are a highly diverse and successful group of arthropods that have penetrated most of the globe's habitats (although, unlike the insects, they are still in very large part aquatic). Trophically, crustaceans span the whole range of feeding opportunities from free-living predators, through herbivores and decomposers to some of the most highly specialised internal and external parasites in the animal kingdom.

The classification of the subphylum Crustacea is complex, being conventionally divided into six Classes. Members of four of these are recorded from phytotelmata but, of these, only the copepods (Class Maxillopoda) and ostracods (Class Ostracoda) are common. Table A.2 is a compilation of records of Crustacea from phytotelmata. Many of the species involved are minute and, although commonly encountered, they have been largely ignored as members of the container communities. Again it is unclear to what extent any of the species recorded are phytotelm specialists.

Cladocera

The Cladocera are ubiquitous planktonic organisms which are usually characteristic of more eutrophic waters with abundant phytoplankton. They feed by filtering algal and other material out of the water column (Bernardi *et al.* 1987). Unlike copepods and ostracods they seldom occur in phytotelmata, although Gurney (1920) recorded three species from British tree holes.

Copepoda

The copepods are a complex and diverse group of organisms that occur widely in water bodies of all kinds. They have been recorded from all Classes of phytotelmata from most parts of the world. Most of the records in which species are identified are of cyclopoids and harpacticoids.

Cyclopoid copepods in general fall into planktonic and benthic groups. The greater part of the planktonic groups are filter-feeding decomposers but some are carnivores. Dussart (1967) summarises their feeding habits as: 'omnivores predatory upon micro- and nannoplankton as well as benthos' [transl.]. Known prey items for some species include *Daphnia*, ostracods, harpacticoid copepods, diatoms, algae, rotifers and even larval fish.

Eleven species of cyclopoids have been recorded from phytotelmata: from tree holes, bromeliads and axil waters. The feeding habits of most of them are unknown. *Diaptomus bisetosus* from tree holes in Europe feeds on detritus

Table A.1. *The Annelida and their association with container habitats*

Taxon	Site	Authors
POLYCHAETA		
NEREIDAE		
Lycastopsis cattaractarum	*Colocasia, Musa* axils, Indonesia; tree holes, New Guinea	Thienemann 1934, Kitching 1990
OLIGOCHAETA		
NAIDIDAE		
Pristina rosea	*Zingiber, Pandanus, Crinum, Colocasia* axils, bamboos, Indonesia (cosmopolitan, generalist)	Thienemann 1934
Aulophorus furcatus	*Colocasia* axils, tree holes, Indonesia (also water-tank, India)	Thienemann 1934
A. gravelyi	Tree holes, Indonesia (also littoral habitats)	Thienemann 1934
A. superterrenus	Bromeliads, Costa Rica	Picado 1913
'naidid'	Bromeliads, Jamaica, Brazil, Florida	Laessle 1961, Fish 1976, Fish & Soria 1978
AEOLOSOMATIDAE		
Aeolosoma bengalense	*Colocasia* axils, Indonesia	Thienemann 1934
ENCHYTRAEIDAE		
Fridericia bulbosa	*Colocasia* axils, Indonesia	Thienemann 1934
F. striata	Bromeliads, Costa Rica	Picado 1913
Enchytraeus harurani	*Colocasia* axils, Indonesia	Thienemann 1934
'enchytraeid'	Tree holes, Germany	Röhnert 1950
ACANTHODRILIDAE		
Dichogaster annae	*Colocasia* axils, Indonesia	Thienemann 1934

LUMBRICIDAE		
'lumbricids'	Tree holes, Germany	Röhnert 1950
MEGASCOLECIDAE		
Perionyx excavatus	*Colocasia* axils, bamboos, tree holes, Indonesia (also in moss and elsewhere in Oriental region)	Thienemann 1934
Perionyx sp.	*Colocasia* axils, bamboos, Indonesia	Thienemann 1934
'oligochaetes'	Tree holes, bamboos, New Guinea	Kitching 1990
'oligochaetes'	Tree holes, Britain	Kitching 1969
'oligochaetes'	*Nepenthes* pitchers	Beaver 1983
'oligochaetes'	large axils, Uganda	Haddow 1948

Table A.2. *The Crustacea and their association with container habitats*

Taxon	Site	Authors
CLASS BRANCHIOPODA		
CLADOCERA		
CHYDORIDAE		
Biapertura affinis	Tree holes, UK	Gurney 1920
Chydorus ovalis	Tree holes, UK	Gurney 1920
C. spaericus	Tree holes, UK	Gurney 1920
CLASS MAXILLIPODA		
Subclass COPEPODA		
'copepod'	Bamboos, Sulawesi	Kitching 1987b
CALANOIDA		
'calanoid'	Tree holes, bamboos, New Guinea	Kitching 1990
CYCLOPOIDA		
CYCLOPIDAE		
Ectocyclops medius	Colocasia axils, Indonesia	Thienemann 1934
E. phaleratus		Picado 1913, Laessle 1961
Bryocyclops bogoriensis	Colocasia, Pandanus axils, bromeliads, Indonesia	Thienemann 1934
B. musicola	Colocasia, Pandanus axils, Indonesia	Thienemann 1934
B. chappuisi	Crinum, Hymenocallis axils, Indonesia; bromeliads, Puerto Rico	Thienemann 1934, Maguire 1970
B. anninae	Bromeliads, Java, Puerto Rico	Thienemann 1934, Maguire 1970
Tropocyclops jamaicensis	Bromeliads, Jamaica	Reid & Janetsky, 1996
Tropocyclops prasinus	Bromeliads, Jamaica	Laessle 1961
Diacyclops bisetosus	Tree holes, Europe	Wesenberg-Lund 1920–21

Taxon	Habitat, location	Reference
Diacyclops pulchellus	Tree holes, Europe	Wesenberg-Lund 1920–21
Cyclops sp.	Bromeliads, Argentina	Torales *et al.* 1972
HARPACTICOIDA		
'harpacticoid'	*Heliconia* axils, Costa Rica	Naeem 1990
PHYLLOGNATHOPIDAE		
Phyllognathopus nr. *coecus*	Bromeliads, Costa Rica, Jamaica	Menzel 1922
P. viguieri[a]	Bromeliads, Java	Thienemann 1934
	Zingiber, Colocasia, Cyrtandra axils, bromeliads, Indonesia	
	Bromeliads, Puerto Rico; *Nepenthes ampullaria*, Sumatra	Maguire 1970, Thienemann 1932
Tachidius brevicornis	Tree holes, Europe (usually a brackish water species!)	Gurney 1920
CANTHOCAMPTIDAE		
Atheyella ruttueri	*Colocasia* axils, Indonesia	Thienemann 1934
A. inopinata	*Cyrtandra* axils, Indonesia	Thienemann 1934
Canthocamptus sp.	Bromeliads, Indonesia	Menzel 1922
Elaphoidella bromeliaecola	*Colocasia, Pandanus* axils, bromeliads, Indonesia	Thienemann 1934
E. elegans	*Colocasia* axils, Indonesia	Thienemann 1934
E. malayica	Bromeliads, Java	Thienemann 1934
E. cornuta	*Cyrtandra* axils, Indonesia	Thienemann 1934
E. thienemanni	*Cyrtandra* axils, Indonesia	Thienemann 1934
E. sewelli	Bromeliads, Puerto Rico, Jamaica	Laessle 1961, Maguire 1970
Epactophanes richardi	*Nepenthes ampullaria*, Sumatra; tree holes, Europe	Thienemann 1932, 1934
	Zingiber, Colocasia, Cyrtandra axils, Indonesia	Thienemann 1934
Bryocamptus pygmaeus	Tree holes, Europe (cosmopolitan)	Gurney 1920
Moraria arboricola	Tree holes, UK	Scourfield 1915
PARASTENOCARIDAE		
Parastenocaris sp.	*Nepenthes ampullaria*, Sumatra	Thienemann 1932
'harpacticoid'	Tree holes, New Guinea	Kitching 1990

Table A.2. (*cont.*)

Taxon	Site	Authors
CLASS OSTRACODA		
PODOCOPIDA		
CYPRIDIDAE		
Candona pratensis	Tree holes, Europe	Thienemann 1934
C. compressa	Tree holes, Europe	Thienemann 1934
Candona sp.	Bromeliads, Costa Rica	Picado 1913
Condonopsis anisitsi	Bromeliads, Jamaica	Laessle 1961
C. kingsleyi	Bromeliads, Puerto Rico	Maguire 1970
CYTHERIDAE		
Metacypris bromeliarum	Bromeliads, Brazil	Müller 1879
M. laesslei	Bromeliads, Jamaica	Laessle 1961
M. maracaoensis	Bromeliads, Puerto Rico, Florida	Maguire 1970, Fish 1976
Cyclocypris laevis	Tree holes, Europe	Thienemann 1934
Metacypris sp.	Bromeliads, Costa Rica	Picado 1913
Prionocypris sp.	Bromeliads, Puerto Rico	Maguire 1970
LIMNOCYTHERIDAE		
Elpidium spp.	Bromeliads, Jamaica	Little & Hebert 1996
CANDONIDAE		
Candanopsis spp.	Bromeliads, Jamaica	Little & Hebert 1996
'ostracod'	Tree holes, New Guinea	Kitching 1990
CLASS MALACOSTRACA		
DECAPODA		

GRAPSIDAE		
Sesarma nodulifera	*Colocasia* axils, Indonesia	Thienemann 1934
S. angustipes	Bromeliads, Trinidad, Brazil	Sattler & Sattler 1965
Metopaulias depressus	Bromeliads, Jamaica	Laessle 1961

[a] *Viguierella coeca* (name used in Thienemann 1934).

and diatoms. I suspect most other copepods in these situations are, similarly, microsaprophages. However, some may feed on mosquito larvae and, indeed, such an interaction has been the basis for the attempted introduction of *Mesocyclops leuckarti* into container habitats as a possible control mechanism for vector mosquitoes (Marten 1984, B. Kay pers. comm.).

The remaining phytotelm copepods are Harpacticoida which, as a group, feed on detritus. They are not filter feeders and actively comminute particulate material (Dussart 1967). Harpacticoids are recorded from the full range of phytotelmata including pitchers of *Nepenthes* (see Table A.2).

Ostracoda

The tiny shelled ostracods are ubiquitous in aquatic habitats and are common in phytotelmata. According to Henderson (1990) they are particle feeders kicking up material using their antennal claw before filtering it through their carapace cavity. They appear to be omnivorous, feeding on detritus, diatoms, filamentous algae and even other animals. Some species are known to prey on protozoans, rotifers, oligochaetes, *Daphnia* or fish larvae. Henderson (1990) describes their habit of attacking prey communally and speculates on the sociobiological aspects of this behaviour for these parthenogenetically breeding organisms.

A range of species representing four families has been recorded in bromeliads and tree holes (Table A.2). I suggest that these species are saprophages specialising in fine particulate matter.

Decapoda

Crabs are much celebrated inhabitants of some leaf axils, bromeliads and even bamboo stumps. Three species have been recorded and these are undoubtedly generalist detritivores dissecting particulate matter. It seems likely that many of these are using plant containers merely for temporary shelter, although the specialist bromeliad species has been considered an integral part of the container community (Laessle 1961).

INSECTA

The great Class Insecta provides most of the larger organisms in phytotelmata, and hence those that are most readily observed and best recorded. A wide range of taxa spread over ten Orders has been recorded from container habitats, although seven are but minor elements of the bestiary (see accounts below). The remaining three, the Odonata, Coleoptera and Diptera, provide the most prominent and best known members of the fauna. Again, of these three the Diptera predominate, being represented by no fewer than twenty-seven families (cf. six families of Odonata and five of Coleoptera).

Ephemeroptera

The mayflies are an Order of insects with exclusively aquatic immature stages. These occur across a range of freshwater habitats from lakes and rivers, to temporary ponds and streams. In general they are associated with flowing water but there are exceptions (Peters & Campbell 1991). The larvae are largely detrital and algal feeders, scraping encrusting material of stones and leaves. A very few species are predacious.

The occurrence of Ephemeroptera in phytotelmata is almost certainly accidental. The instance known to me is of a larva of a leptophlebiid that we encountered recently in a water-filled tree hole in lowland rainforest near Bellenden Ker in North Queensland. This was reared and proved to be a species of *Atelophlebia*, a genus of catholic breeding habits with nymphs that feed on decaying leaves and wood. The site in which it was found was riparian and it may be that the nymph concerned became 'marooned' in the tree hole by retreating flood waters. This phenomenon is discussed further in the 'Odonata' section below.

Odonata

The dragonflies and damselflies are generally highly visible, often highly coloured insects associated with water bodies of all sizes and flow regimes. In general the larvae of odonates are aquatic. The small group of species that are terrestrial as larvae detracts from this generalisation but not from the further observation that all odonate larvae are quintessentially predatory. Their specially adapted extensible labia make them highly efficient sit-and-wait predators on organisms from oligochaete worms to immature fish and frogs. Accordingly they sit at the tops of food webs and, as we see in Chapter 7 (*et seq.*), confer structural complexity wherever they exist in food webs. Excellent general accounts of the systematics and biology of the Order are provided by Tillyard (1917) and Corbet (1963a).

A small but diverse group of species spanning the two principal suborders have been successful in exploiting phytotelm habitats in moist tropical environments. In these situations they prey upon dipterous larvae, annelids, tadpoles and smaller individuals of their own kind. Corbet (1983) provides an authoritative and exhaustive review of odonates recorded from phytotelmata to that time. Table A.3 summarises the records he presents, together with the few additional ones that have come to my notice since that time.

Representatives of three families of Zygoptera (the damselflies) and two

Table A.3. *The Odonata and their association with container habitats (based in part on Corbet 1983)*

Taxon	Site	Authors
MEGAPODAGRIONIDAE		
Coryphagrion grandis	Tree holes, Kenya	Lounibos 1980
Podopteryx selysi	Tree holes, Australia	Watson & Dyce 1978
PSEUDOSTIGMATIDAE		
Mecistogaster jocaste	Tree holes, Brazil, Peru	Machado & Martinez 1982; Louton *et al.*1996
M. modestus	Bromeliads, Costa Rica, Mexico	Calvert 1910, 1911
M. linearis	Tree holes, Panama, Peru	Fincke 1984,1992a,b; Louton *et al.* 1996
M. ornata	Tree holes, Panama	Fincke 1984,1992a,b
M. sp.	Bamboo internodes, Peru	Louton *et al.* 1996
Megaloprepus coerulatus	Tree hole, Costa Rica, Panama	Young 1980, Fincke 1984, 1992a,b
Microstigma rotundatum	Tree holes, Peru	Louton *et al.* 1996
Microstigma sp.	Fallen fruit	Santos 1981
Pseudostigma accedens	Tree holes, Panama	Fincke 1984,1992a,b
PROTONEURIDAE		
Roppaneura beckeri	*Eryngium* axils, Brazil	Machado 1976
COENAGRIONIDAE		
Amphicnemis spp.	*Pandanus* leaf axils	Lieftinck 1954
Diceratobasis macrogaster	Bromeliads, Jamaica	Laessle 1961
Diceratobasis worki	Fossil from Dominican amber, ?bromeliads	Poinar, 1996
Leptagrion andromache	Bromeliads, Brazil	Santos 1966a
L. bocainense	Bromeliads, Brazil	Santos 1979
L. dardanoi	Bromeliads, Brazil	Santos 1968a
L. elongatum	Bromeliads, Brazil	Santos 1966b
L. macrurum	Bromeliads, Brazil	Santos 1966a

L. perlongum	Bromeliads, Brazil	Santos 1962, 1966a
L. siqueirai	Bromeliads, Brazil	Santos 1968b, Lounibos et al. 1987a
L. vriesianum	Bromeliads, Brazil	Santos 1978
Megalagrion amaurodytum	*Freycinetia, Astelia* leaf axils, Hawaii	Williams 1936
M. asteliae	*Freycinetia, Astelia* leaf axils, Hawaii	Williams 1936
M. koelense	*Freycinetia, Astelia* leaf axils, Hawaii	Williams 1936
Pericnemis stictica	Bamboos, Malaysia	Lieftinck 1954, Kovac & Streit 1996
Teinobasis ariel	*Freycinetia* leaf axils?	Lieftinck 1954
'agrionid'	Bamboos, Indonesia	Thienemann 1934

ANISOPTERA

AESHNIDAE
Gynacantha membranalis	Tree holes, Panama	Fincke 1992a,b
Gynacantha sp.	Fallen tree hole, Costa Rica	Paulson in Corbet 1983
Indaeshna grubaueri	Tree holes, Borneo	Orr 1994

LIBELLULIDAE
Erythrodiplax sp.	Bromeliads, Jamaica	Laessle 1961
Hadrothemis camarensis	Bamboo sections, Uganda	Corbet 1961
H. scabrifrons	Tree holes, East Africa	Corbet & McCrae 1981
Lyriothemis cleis	Tree holes, Sulawesi, Borneo	Kitching 1986a, Orr 1994
L. magnificata	Bamboos, tree holes, Malaysia	Lieftinck 1954
L. salva	Old *Nepenthes* pitchers, Malaysia	Lieftinck 1954
L. tricolor	Bamboos, tree holes, Taiwan	Lien & Matsuki 1979
'libellulid'	Bamboos, Indonesia	Thienemann 1934
'libellulid'	Tree holes, North Australia	Kitching, Juniper & Mitchell unpublished
'libellulid'	*Sarracenia* pitcher, USA	Bradshaw in Corbet 1983

Based in part on Corbet (1983).

of Anisoptera (the dragonflies) are known from phytotelmata. These records range over all five major classes of phytotelm habitats. As Corbet (1983) points out, it is not surprising that the majority of records are of Zygoptera which, in general, are smaller as larvae than are the Anisoptera. Restrictions in physical access to phytotelmata by Odonata may represent barriers to ovipositing females. The records we have of odonates from smaller, leaf-axil phytotelmata are of Zygoptera whereas both Zygoptera and Anisoptera have been recorded from larger phytotelmata such as tree holes, bamboo sections and the larger pitcher plants and bromeliads.

In tropical water-filled tree holes whole guilds of odonates occur, presumably subdividing the available tree holes and the food resources they contain. Fincke (1984) has drawn attention to the group of five odonates that share the tree hole habitat in the tropical forest of Costa Rica. Giant damselflies (Pseudostigmatidae) of the genera *Mecistogaster* and *Megaloprepus* co-occur with a species of aeshnid dragonfly in these habitats. The males of one species of *Megaloprepus* actually establish territories around particular tree holes. The larvae of these species feed on mosquito larvae and tadpoles, and may be cannibalistic (Fincke 1984, 1992a,b). For water-filled tree holes in northern Borneo, Orr (1994) has described the co-occurrence of the aeshnid *Indaeschna grubaueri*, the libellulid *Lyriothemis cleis*, and the Zygopteran *Pericnemis triangularis* (see also Kitching & Orr 1996).

Corbet (1983), with Calvert (1911), speculates that odonates may have first invaded phytotelmata from other more usual aquatic habitats in situations where periodic inundation connected ground water to pools held in plants. This is borne out, circumstantially at least, by our recent records of a hereto unidentified libellulid from water-filled tree holes in the rainforest of North Queensland which is restricted in occurrence to riparian situations. This occasional connection between phytotelm situations and other water bodies following flooding may account, also, for the occasional records of other aquatic insect Orders in phytotelmata (see accounts under Ephemeroptera, Trichoptera, Plecoptera and Megaloptera).

Corbet (1983) also suggests that the phytotelm 'habit' in odonates would have arisen more in situations of general scarcity of aquatic habitats such as the seasonally dry forests of Kenya. As a generality, I find this a much less convincing argument and the records my colleagues and I have obtained, and odonates we continue to discover, in tree holes in Sulawesi, Brunei and North Queensland are all from perhumid or close to perhumid environments with an abundance of alternative aquatic environments. It seems to me more likely that, in situations wherein phytotelmata are abundant and more or less continuously water filled, odonate species have taken advantage of a situation offering, not only predictable food supplies, but also escape from the depredations of other species of predators.

Megapodagrionidae The Megapodagrionidae are generally considered to be a family showing primitive characteristics, allied to the presumed stem group of Odonata. They occur in all biological regions of the world except the Holarctic. As a family they breed in bogs and seepages but two species have been recorded from water-filled tree holes in Australia and Kenya respectively. Both appear to be obligate tree-hole breeders.

Pseudostigmatidae The Pseudostigmatidae are an exclusively Neotropical family comprising five or six genera only. Of these, four have been recorded from container habitats. Records exist from water-filled tree holes, bamboo internodes, bromeliads and water-filled fallen fruits (Table A.3, Corbet 1983).

Protoneuridae The Protoneuridae are a widespread family of damselflies occurring in all biological regions except the Holarctic (although poorly represented in Africa). They are, in general, stream and river dwellers. Only one protoneurid has been reported in phytotelmata to date: *Roppaneura beckeri* from the leaf axils of *Eryngium* (Umbelliferae) in Brazil (Machado 1976).

Coenagrionidae The Coenagrionidae are a large and cosmopolitan family specialising in still or slow moving water from lake margins to desert waterholes (Watson & O'Farrell 1991). This lentic habit, according to Corbet (1983) may well be the reason why more coenagrionids than any other odonate family have evolved to exploit phytotelm water bodies. Six genera are implicated to date in the Oriental, Neotropical and Oceanic regions. The genus *Amphinemis* in South-east Asia is reputed to be a specialist in utilising the axil waters of a variety of plants such as *Pandanus* (Lieftinck 1962). Corbet (1983) suggests that as many as seventeen species may be involved. Also in South-east Asia *Pericnemis strictica* is recorded from bamboo stumps and, in Micronesia, *Teinobasis ariel* is known from *Freycinetia* axils. In the Neotropics eight species of the genus *Leptagrion* have become specialised users of bromeliad tanks from Brazil to Jamaica (references in Table A.3). Laessle (1961) records larvae of *Diceratobasis macrogaster* from the same habitat. Williams (1936) in his review of the aquatic insects of Hawaii, recorded three species of *Megalagrion* from water-filled leaf axils of *Freycinetia* and *Astelia*. The antiquity of the phytotelm breeding habit in this family is indicated by Poinar (1996) who suggests that the species *Diceratobasis worki*, which he describes from Dominican amber, was in all probability a bromeliad breeder.

Aeshnidae The Aeshnidae are the quintessential dragonflies: large, colourful and predatory. The family is extensive and widely distributed across all zoo-geographical regions. They inhabit a range of larval habitats from streams to rivers and lakes. The larvae are (relatively) massive and voracious and have been recorded feeding on tadpoles and even fish as well as a wide range of aquatic invertebrates.

The aeshnids are perhaps the least likely group one might have imagined invading container habitats, and yet two undoubted records exist. Paulson (ref. in Corbet 1983) notes *Gynacantha* sp. breeding in a water-filled log hole (potentially the largest phytotelms known!). More impressive are the large aeshnids, *Indaeshna grubaueri*, reared from water-filled tree holes in Brunei by Dr A. G. Orr (Orr 1994, pers. comm.). Larvae of this very large and colour-ful species are commonly encountered in pans in the buttress roots of a num-ber of species of tree in the dipterocarp forests of the Sultanate. They have also been reared from temporary pools on the forest floor.

Libellulidae A widespread family of broad-bodied, smaller dragonflies, the Libellulidae are cosmopolitan in distribution and catholic in breeding habit. Larvae are recorded from bogs and seepages, through to streams, rivers and lakes. The larvae have a characteristic 'wide-bodied' appearance and are gen-eralist predators.

I am aware of records of nine species of libellulid from phytotelmata, prin-cipally from the Old-World tropics but including *Erythrodiplax* sp. from tanks of *Aechmea* bromeliads in Jamaica (Laessle 1961) (and one casual record from North America – see below). The Old-World species belong to the gen-era *Lyriothemis*, *Hadrothemis* and *Camacinia*. Four species of *Lyriothemis* are recorded from plant containers: from tree holes, old *Nepenthes* pitchers and bamboo stumps in Malaysia, Taiwan, Brunei and Sulawesi; two species of *Hadrothemis* occur in bamboos and tree holes in East Africa; and a sin-gle species of *Camacinia, harteri*, has been observed ovipositing in tree holes in Malaysia (see references in Table A.3). In addition larvae of a libellulid occur in water-filled tree holes in lowland riparian rainforests in North Queensland. We have not succeeded in breeding this to adulthood to date. Corbet notes a record of an unidentified 'libellulid' from pitchers of *Sar-racenia* in Maine (Bradshaw 1980). It seems likely that this last record is of an accidental occurrence as other authors make no mention of odonates from within the well-known New World pitcher plants.

Plecoptera

The stoneflies, like the Ephemeroptera, are an exclusively aquatic group with larvae which may be saprophages, herbivores or predators depending on the subfamily concerned. Their occurrence in phytotelmata is probably accidental. Picado (1913) recorded a species of *Perla* from bromeliads in Costa Rica. Members of this genus are generally thought of as being omnivorous, feeding both as predators and herbivores (Hynes 1958).

Hemiptera

The true bugs come in a great variety of shapes, sizes and habits. They may be terrestrial, amphibious or aquatic; plant-parasitic, herbivorous or predatory. Only four groups have been recorded as inhabitants of plant containers: water-bugs belonging to the families Veliidae, Hydrometridae and Gerridae respectively, and spittle bugs belonging to the superfamily Cercopoidea.

Veliidae The veliids, or water-crickets (not to be confused with the Tetrigidae – see below), are generalist predators in both nymphal and adult stages. They readily pass through the surface film of the water bodies they inhabit, foraging on moist areas both above and beneath the water surface.

Members of the genera *Microvelia* and *Velia* occur in bromeliad water bodies in Mexico, Trinidad, Jamaica, Peru, Panama, Venezuela and Brazil. Polhemus and Polhemus (1991) have reviewed the veliid fauna of bromeliads. They draw heavily, in this review, upon their own work and that of C. J. Drake and his co-workers (e.g. Drake & Maldonado 1952, Drake & Hussey 1954, Drake and Harris 1935, Drake & Chapman 1954). Kovac & Streit (1996) note a species of *Microvelia* in split bamboo internodes and bamboo stumps in Malaysia.

An unidentified species of veliid is also a common component of the fauna of some riparian water-filled tree holes in the lowland rainforest of North Queensland, Australia.

Hydrometridae The Hydrometridae are the familiar water-measurers of still waters. Surface dwellers, they prey upon soft-bodied invertebrates through the water film. Kovac & Streit (1996) record *Hydrometra* sp., like *Microvelia*, from split bamboo internodes and bamboo stumps in Malaysia.

Gerridae Gerrids are the familiar long-legged pond-skaters that forage for prey on the surface of ponds, lakes and rivers. These seem intrinsically unlikely candidates as fauna of phytotelmata but Kovac (1994) announced his discovery of a species of this family which lives in water-filled bamboo internodes in peninsular Malaysia. This species preys upon small insects on the water surface and, like its relatives inhabiting larger water bodies, individuals are territorial, defending their foraging area against other individuals.

Cercopidae/Aphrophoridae The spittle-bugs are herbivorous, sucking phloem contents from their host plants. Fish (1977) records an immature 'cercopid' from the bract axils of *Heliconia* species in Costa Rica that remains submerged in the water body, head downwards, with the ventral tube opening above the surface. Laird (1988) describes exactly similar behaviour on the part of an *Aphrophora* sp.[1] in the pitchers of *Sarracenia purpurea* in Newfoundland peat bogs. Judd (1959) records unidentified cercopids from nine of 489 pitchers examined in Ontario. These homopterans are presumably acting as true herbivores, feeding on the soft tissues of the plants that contain them and not reliant, like the vast majority of other container fauna, on detritus and its products that have entered the habitat allochthonously.

Orthoptera

The Orthoptera as an Order are not generally associated with water bodies. The single exceptional family, the Tetrigidae, also provide the only phytotelm record for the Order. Some tetrigids swim freely and may spend time underwater. Their feeding habits are varied but include dead vegetable matter and encrusting algae and other cryptogams (Rentz 1991). A species of tetrigid frequents the water/bark/air interface in riparian tree holes in lowland North Queensland. The species appears to spend most of its time foraging around the water line but enters the water and swims freely especially when disturbed.

Megaloptera

The alderflies are an exclusively aquatic group associated with streams. The larvae are predators upon a wide variety of small aquatic invertebrates (Theischinger 1991). Like members of the other 'aquatic' orders (Ephemeroptera, Plecoptera, Trichoptera) they are almost certainly rare accidental inhabitants of phytotelmata. Mather (1981) and Rymal & Folkerts (1982) draw attention to the occurrence of individuals of *Sialis joppa* in pitchers of *Sarracenia purpurea* in New Jersey. Pittman *et al.* (1996) extend this observation by noting *S. joppa* larvae in the same plant species in West Virginia. Fashing (1994b)

---

[1] Generally considered as a member of the Aphrophoridiae but designated by Laird as a 'cercopid spittle-bug or froghopper', reflecting a broader concept of the family Cercopidae.

recorded larvae of *Chauliodes pectinicornis* from tree holes in Virginia. As in other cases (see above) the presence of Megaloptera in ground pitchers and tree holes may be the result of earlier flooding of the sites.

Diptera

It is among the true flies that we find the great majority of species of phytotelm insects. Representatives of many families have been recorded at one time or another from plant containers. Many occur repeatedly in such situations: others are more generalist, exploiting the organically rich aquatic habitats offered by phytotelmata, as one of several habitat types that they use. Members of the families Culicidae, Ceratopogonidae, Chironomidae, Psychodidae and Syrphidae provide the greater part of the community of most phytotelmata, although members of many other families are recorded and some are undoubtedly phytotelm specialists.

Tipulidae The crane-flies (Figure A.1) are generally regarded as retaining features which are primitive within the Order Diptera. The larvae are unspecialised with the sclerotised head capsule, typical of the Nematocera, half concealed by the prothorax. The complex arrangement of lobes surrounding the posterior spiracles is characteristic of larvae of this family. The larvae occupy a variety of aquatic, semi-aquatic and damp-terrestrial situations and include the familiar 'leatherjackets' of soil and compost.

In container habitats they are 'edge' species occurring in crevices on the edges of tree holes or at the water/air interface in axil waters and bromeliads, reflecting their need for access to air and the maintenance of a respiratory plastron about the posterior spiracles. Trophically, in tree holes, they are macrodecomposers ('shredders' or 'collector-gatherers' according to Merritt & Cummins 1984), processing plant debris through their bodies. Fish (1983), however, records some tipulid larvae belonging to the genus *Sigmatomera* as predators. Alexander (1929) describes larvae of *S. shannoniana* feeding on mosquito larvae within bromeliads in Brazil and Fish himself describes unidentified but predatory tipulid larvae from bromeliads in Costa Rica and Brazil. Five individuals of the widespread neotropical species *S. occulta*, are recorded by Reidel (1921) (as *S. flavipennis*) from water-filled

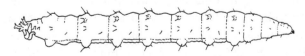

Fig. A.1. Larva of the tipulid fly (lateral view) (after Brauns 1954).

tree holes in Paraguay, all bred from tree holes. Lastly, Fish draws our attention to the record of Snow (1949) who described a species of *Sigmatomera* in tree holes in Central America which was also predatory upon mosquito larvae.

A number of saprophagous species have been recorded from bromeliads. Picado (1913) found larvae of the genus *Mongoma* in Costa Rica (Alexander 1915). *Trentophila bromeliadicola* occurs widely in the American tropics and *T. leucoxena* occurs in Mexico (Alexander 1920). Laessle (1961) recorded an unidentified species from Jamaica.

We found the barrage-balloon-like larvae of *Tipulodina* (s.l.) from tree holes and bamboo internodes in New Guinea and from tree holes in Sulawesi. Thienemann (1934) records *Lipnotes?* sp. and *Ctenacroscelis* sp. from bamboo stumps, and *Tipulodina pedata* from tree holes, all in Indonesia. Snow (1958) encountered *Teucholabis immaculata* in water-filled tree holes in Illinois. Röhnert (1950) found *Ctenophora pectinicornis* in north German tree holes but dismissed these as 'dendrolimnetophils' with catholic tastes across a range of semi-aquatic environments. Macfie & Ingram (1923) record an unidentified 'tipulid' from West African tree holes. Louton *et al.* (1996) noted larvae of *Trentepohlia* sp. and *Helius* sp. in their extensive studies of communities in bamboo internodes in Peru.

Plant-axil records for this family include those of Alexander (1931) of a species of *Lipnotes?* and *Limonia stantoni* from axils of *Colocasia antiquorum*. Both this author and Thienemann (1934) refer to larvae that may be *Gnathomyia* species from the axils of *Musa* spp. Alexander (1920) describes the occurrence of larvae of *Gnophomyia rufa* from 'the semi-liquid organic detritus' that accumulates in the leaf bases of the New Zealand epiphyte *Astelia solandri*. Lastly I encountered a small (probably limoniine) species in the water-filled bracts of *Curcuma australasica* inflorescences in New Guinea, often associated with the drying, later stages of development of the axil community.

It is likely that tipulids are generalist species exploiting the semi-aquatic, detritus-rich milieu of phytotelmata only facultatively, although the range and degree of specialisation of the bromeliad species remains to be established.

Chaoboridae The Chaoboridae (Figure A.2), together with the Dixidae and the true mosquitoes, the Culicidae, make up a well-defined superfamily of Diptera, the Culicoidea. According to Colless (1986) the Chaboridae are a heterogeneous group defined by difference. That is: when the distinctive groups Dixidae and Culicidae are removed from the Culicoidea, the Chaoboridae are what's left!

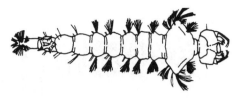

Fig. A.2. Larva of the chaoborid fly *Eucorethrella underwoodi* (after Pennak 1953).

Colless (1986) recognises four subfamilies (arguing that two of these at least might justify separate family status – but for the maxim that small, monogeneric families merely inflate and confuse subordinal taxonomy). Of these four subfamilies only one, the Corethrellinae, is recorded from phytotelmata and, hence, concerns us here.

The larvae of the only genus in the subfamily, *Corethrella*, are voracious predators feeding upon other insect larvae and microcrustaceans.

In water-filled tree holes, *Corethrella appendiculata* is recorded from Florida (Bradshaw & Holzapfel 1984). In pitchers of *Nepenthes* spp., Barr & Chellapah (1963) and Beaver (1983) record *Corethrella calathicola* from *N. ampullaria* and Clarke & Kitching (1993) record a species of *Corethrella* from *Nepenthes bicalcarata* and *N. ampullaria*. Finally, Frank & Curtis (1981) record fourteen species of *Corethrella* from water bodies held in various bromeliads from Panama, Brazil and Trinidad.

Culicidae The mosquitoes (Figures A.3, A.4), more than any other group of organisms, have become identified as *the* faunal elements of importance in virtually every phytotelm community that has been examined. Indeed the original interest that developed in phytotelmata within the English- and French-speaking scientific world seems to have been because tropical container habitats were identified as the sources of vector mosquitoes, in particular of the yellow fever mosquito, *Aedes aegypti*, and various vectors of malaria. Parenthetically, and in contrast, the German scientific interest in phytotelmata emerged as part of the general development of modern limnology under the direction of Albrecht Thienemann.

The Culicidae are an immense family of about two and a half thousand species usually classified into three subfamilies (when the Chaoboridae and Dixidae are considered as separate families, as in this treatment), and between 30 and 40 genera according to taste. General treatments of the biology of mosquitoes are few but the now somewhat elderly accounts of Bates (1949) and Mattingly (1969) remain excellent value. Horsfall (1955) is a mine of information about larval habitats and Laird's (1988) book on larval mosquito

Fig. A.3. Larva of the culicid fly *Tripteroides ceylonensis* (a composite after Mattingly 1981).

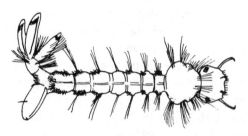

Fig. A.4. Larva of the predatory culicid fly *Toxorhynchites* (after Howard *et al.* 1913).

habitats also contains useful information. Service (1976) provides a comprehensive view of mosquito ecology, primarily from a methodological point of view, and Lounibos *et al.* (1985) is an edited compendium of many aspects of the ecology of mosquitoes. This last work includes a number of contributions dealing with phytotelm species. Lastly the much more recent volume of Clements (1992), the first of two projected volumes, is a general account of the biology of mosquitoes, although, on the author's own admission, the work takes a physiological viewpoint.

Within phytotelm communities it is the larvae of mosquitoes that are active: feeding, developing and pupating within the water bodies. Merritt *et al.* (1992) have provided an exhaustive review of larval feeding mechanisms and nutrition for mosquitoes in general.

Trophically, phytotelm larvae are diverse. Some are obligate predators. These include the massive larvae of *Toxorhynchites* spp., and the somewhat smaller but still formidable larvae of the genus *Topomyia*, the peculiar Malayan *Zeugnomyia gracilis* that lives in water-filled fallen leaves in the most deeply shaded parts of the rainforest (Horsfall 1955), and the subgenus *Culex (Lutzia)*. These predatory species feed upon other larvae within the phytotelm communities including, in many cases, smaller individuals of their own species. A more diverse group are particle feeders that may lapse into facultative predation from time to time. The American *Anopheles barberi* (Petersen *et al.* 1969) and the African genus *Eratmopodites* (Lounibos 1980, Corbet & Griffiths 1963) are cases in point. Yet other species, such as the greater part of the great genus *Culex*, browse on detritus particles. Many of these browsing species are certainly feeding actively upon micro-organisms such as protozoans and rotifers, as has been clearly demonstrated by the work of Addicott (1974) on the pitcher-plant mosquito, *Wyeomyia smithii*. Other species of mosquitoes, such as many of the anophelines, are true filter feeders living, more than is the case with other species, in the water column and at the water surface. Lastly many aedines combine browsing and filter feeding and have mouthbrush setae that appear adapted to allow this facultative switching between one mode and the other (Mattingly 1969).

The assignment of mosquito species into trophic positions within food webs is complicated by two factors. First, as with many phytotelm organisms, it is far from easy to designate organisms unequivocally as microsaprophages, feeding on minute detritus particles, or as micro-organism feeders, removing tiny living organisms by browsing them off detritus particles or even digesting them from among ingested detritus particles. Second, the feeding preferences of mosquito larvae undoubtedly will change as they grow and move from one instar to the next, as has been shown clearly for the related Chaoboridae by Moore (1988). Aedine larvae may be exclusively filter feeders as first instars less than a millimetre in length, but browse upon (relatively) large detritus particles when full-grown third instars some eight or nine millimetres long.

Table A.4 summarises the classification of the Culicidae and those genera and/or subgenera known to occur in containers. The table is largely based on a similar compendium in Horsfall (1955) with a number of additions based on subsequent investigations. Mattingly (1969) points out that of the subfamily Culicinae all of the large tribe Sabethini are obligate container dwellers, as are about 40% of the Culicini. All of the subfamily Toxorhynchitinae are similarly associated with containers. The remaining subfamily, the Anophelinae, although predominantly of non-container breeders, nevertheless has many species that occur in natural and artificial containers, especially within the genus *Anopheles* which has one large subgenus, *Kerteszia*, a container specialist. Mattingly and a number of later authors (e.g. Laird 1988) suggest that the ancestral habit of the Culicidae in general may well have been container-breeding and that the invasion of other breeding places is a more recently acquired characteristic.

A sequence of authors has attempted synoptic classifications of mosquito breeding places, within virtually all of which container habitats figure

Table A.4. *The Culicidae (mosquitoes) and their association with container habitats*

Taxon	Association with containers
ANOPHELINAE	
Anopheles (Anopheles)	Rot holes, bamboos, Oriental region
Anopheles (Kerteszia)	Bromeliads, bamboos, Neotropics
Anopheles (Myzomyia)	One record from Philippine rot hole
TOXORHYNCHITINAE	*All container breeders*
Toxorhynchites	Genus known from bromeliads, tree holes, bamboos, pitcher
CULICINAE	
Sabethini	All container breeders
Sabethes	Rot holes, palm axils, bamboos, Neotropics
Wyeomyia (Phoniomyia)	Bromeliads, Neotropics
Wyeomyia (Wyeomyia)	Bromeliads, bamboos, *Heliconia* axils, rot holes, Neotropics
Wyeomyia (Dendromyia)	Bromeliads, bamboos, *Heliconia*, *Colocasia* and other axils, Neotropics
Wyeomyia, other subgenera	Bromeliads, pitcher plants, palm and *Typha* axils, Neotropics
Limatus	Coconut shells, bamboos, Neotropics
Trichoprosopon	Bromeliads, bamboos, Neotropics
Tripteroides	Tree holes, coconut shells, pitcher plants, bamboos, banana and taro axils, Old World
Topomyia	Bamboos, *Colocasia* axils, South-east Asia
Culicini	
Culex (Anoedioporpa)	Bamboos, Peru
Culex (Isotomyia)	Rot holes, bamboos, Panama
Culex (Carrollia)	Tree holes, bromeliads, bamboos, Brazil, Ecuador, Peru, Central Africa
Culex (Lutzia)	Tree holes, Sulawesi (Kitching 1987b)
Culex (Lophoceratomyia)	Rot holes, bamboos, bromeliads, Asia, Central America
Culex (Culiciomyia)	Rot holes, *Pandanus*, banana axils, bamboos, Africa, Asia

Taxon	Habitat / Distribution
Culex (Culex)	Tree holes, coconuts, banana and *Colocasia* axils, cosmopolitan
Culex (Microculex)	Bromeliads, bamboos, Brazil, Costa Rica
Uranotaenia	Rot holes, *Curcuma*, taro, *Pandanus* axils, bamboos, *Nepenthes* pitchers, cup fungi, reed stems, cosmopolitan
Hodgesia	Tree holes, Australia (Kitching, Juniper and Mitchell unpubl.)
Zeugnomyia	Fallen leaf pools, Malaya
Orthopodomyia	Tree holes, bamboos, bromeliads, cosmopolitan
Ficalbia	Pitcher plants, leaf axils, Madagascar (Paulian 1961)
Eratmopodites	Leaf axils, bamboos, pools in fallen leaves, Africa
Armigeres	Rot holes, bamboos, *Curcuma* axils, *Nepenthes* pitchers, coconuts, Asia
Heizmannia	Rot holes, bamboos, Asia
Haemagogus	Rot holes, bamboos, coconuts, Central America
Aedes (Christophersomyia)	Rot holes, Asia
Aedes (Dunnius)	Rot holes, bamboos, Africa
Aedes (Leptosomatomyia)	Coconuts, rot holes, New Guinea
Aedes (Macleaya)	Rot holes, Australasia
Aedes (Ochlerotatus)	A few species in rot holes, Central Asia, Eastern Europe
Aedes (Finlaya)	Rot holes, bamboos, axil waters of all kinds, *Nepenthes* pitchers, fallen leaf pools, cosmopolitan
Aedes (Howardina)	Bromeliads, rot holes, axil waters, Central America
Aedes (Skusia)	Palm axils and stumps, bamboos, South-east Asia, New Guinea
Aedes (Stegomyia)	Rot holes, *Dracaena*, *Strelitzia* and sago axils, bamboos, Old World
Aedes (Aedimorphus)	Rot holes, banana and palm axils
Aedes (Diceromyia)	Rot holes, bamboos, Africa and Asia
Culiseta	A few species in rot holes, Africa

Based upon Horsfall 1955, unless otherwise indicated.
Genera and subgenera not recorded from containers excluded.

prominently. Laird (1988) reviews each of these in detail in a sequential jus-
tification for his habitat classification which is the basis for his book *The Natural History of Larval Mosquito Habitats*. In fact, beyond minor matters of
omission, there are few differences among the dozen or so schemes that he
reviews. Finally he suggests a simple arrangement in which mosquito habitats are divided, first, into above-ground and subterranean waters. Above-
ground waters are then classified into nine categories ranging from flowing
streams through lake edges and ephemeral puddles, to natural and artificial
containers. Laird (1988) devotes considerable space (his pp. 47–77) to a useful discussion of the role of containers in the evolution of the Culicidae, based
on analyses of levels of specialisation within container type. In this analysis
he draws heavily on the work of Horsfall (1955). He concludes that the more
generalised container habitats, tree holes and bamboos, are more likely to be
sources of evolutionary and ecological radiation both into other more restrictive phytotelmata and, presumably, beyond. He speculates, with Bates (1949),
that the mosquitoes probably arose in the early Tertiary when the 'global
forests of conifers, seed-ferns, ginkos, cycads and true ferns ... clearly
favoured ... a prodigious availability of tree hole habitats'. He backs up this
contention by drawing attention to the frequency of tree-hole habitats in the
eroded stumps of modern tree ferns.

The long-standing interest in phytotelm mosquitoes, motivated by epi-
demiological concerns, has led to numerous studies on the faunistics of mosquito assemblages in container habitats from many parts of the world. Well-
known 'western' studies such as that of Jenkins & Carpenter (1946) (nearctic
tree holes), Service (1965) (northern Nigerian tree holes), Haddow (1948)
(Ugandan plant axils) and Corbet (1964) (Ugandan bamboo cups) are complemented by much less well known studies such as Chow (1949) (plant containers in Yunnan), Miyagi & Toma (1980) (forest mosquitoes from Iriomote
Island, Japan), Machado-Allison *et al.* (1983) (Venezuelan phytotelmata), and
Lu *et al.* (1980) (Chinese bamboo breeders). In addition a number of authors
have provided invaluable overviews of mosquitoes from particular habitat
types. Frank & Curtis (1981) record 214 species of mosquito, the immature
stages of which are known from bromeliads, and Beaver (1983) lists 94
species from *Nepenthes* pitcher plants.

Lastly in this brief review of a vast literature, some species-level studies
must be noticed. A small subset of the container-breeding mosquitoes have
been selected as experimental organisms. The selection of these species
reflects a number of factors, most important of which appears to be accessi-
bility and roles as disease vectors. Accordingly, the tree-hole-breeding species
Aedes triseriatus (Table A.5) and *Toxorhynchites rutilus* (Table A.6),

Wyeomyia smithii from pitchers of *Sarracenia* (Table A.7), and *W. vanduzeei* from bromeliads (see references in Table 2.1), have received substantial scientific attention. *Ae. triseriatus* is a major vector of eastern equine encephalitis, endemic to the eastern and southern USA. The second division, so to speak, is occupied by the western American *Aedes sierrensis* (Hawley 1985a,b; Woodward *et al.* 1988), the European *Aedes geniculatus* (Yates 1979, Bradshaw & Holzapfel 1986b) and *Anopheles plumbeus* (Blacklock & Carter 1920), and the Japanese *Aedes albopictus* and *Ae. riversi* (Eshita & Kurihara 1979, Sota & Mogi 1992, Hawley *et al.* 1989). All of these species have received repeated attention in the literature and many aspects of the biology and ecology of each have been examined. Beyond this select group of species the biological literature becomes very scattered. By way of example: Lounibos (1980) discussed three container-breeding species of *Eratmopodites* from Kenya; Chapman (1964) studied the biology and ecology of *Orthopodomyia californica* in California, and Galindo (1958) described the bionomics of the tree-hole-breeding *Sabethes chloropterus* from central America.

Chironomidae After the Culicidae, the non-biting midges (Figure A.5) probably represent the most familiar and widespread animals found within phytotelmata. This having been said, however, it must be noted that they come only a very distant second in terms of the amount of scientific attention that has been devoted to them in this context.

In modern usage the Chironomidae are usually divided into ten subfamilies (Cranston 1995), only four of which have been recorded from phytotelmata. Of these four, one, the Podonominae, has been collected very recently from water-filled leaf axils of the curious alpine Tasmanian giant epacrid *Richea pandanifolia* (Cranston & Kitching 1994) but is otherwise unknown from plant container habitats. It is the Chironominae, Orthocladiinae and Tanypodinae that contain by far the greater part of phytotelm chironomids. These subfamilies also happen to be the richest in species within the Chironomidae overall (Oliver 1971). Table A.8 reviews the literature on the occurrence of chironomids in phytotelm habitats. In compiling this list I have

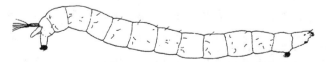

Fig. A.5 Larva of the chironomid midge *Parochlus bassianus* (after Cranston & Kitching 1994).

Table A.5. *Key works on the biology of the North American tree-hole mosquito, Aedes triseriatus*

Author(s)	Year	Subject matter
Bradshaw & Holzapfel	1983	Ecology within larger mosquito assemblage in Florida
Bradshaw & Holzapfel	1984	Seasonal development in relation to weather and predation
Bradshaw & Holzapfel	1985	Seasonal occurrence within a larger mosquito assemblage
Chambers	1985	Competitive and predatory interactions with other mosquitoes
Clay & Venard	1971	Diapause termination by moulting hormone
Clay & Venard	1972	Induction of larval diapause: effects of diet and temperature
Copeland & Craig	1990	Cold-hardiness of larvae in Great Lakes Region
Copeland & Craig	1992	Interactions with *Aedes hendersoni* in artificial tree holes
Craig	1983	Factors affecting control
Fish & Carpenter	1982	Leaf litter and larval mosquito dynamics
Grimstad *et al.*	1974	*A. triseriatus* and *A. hendersoni* in Wisconsin
Grimstad *et al.*	1977	Geographical variation invector abilities for LaCrosse virus
Hayes & Morland	1957	Egg incubation and colonisation
Ho *et al.*	1989	Competition with *A. aegypti* and *A. albopictus* in culture
Holzapfel & Bradshaw	1981	Geography of larval dormancy
Jalil	1974	Fecundity vs blood meal size, mating status, body weight etc.
Jenkins & Carpenter	1946	General account of North American tree-hole mosquitoes
Juliano & Reminger	1992	Vulnerability to predation and larval behaviour
Kappus & Venard	1967	Induction of diapause by photoperiod and temperature
Livdahl	1982	Intraspecific competition within and between hatching cohorts
Livdahl & Willey	1991	Competition with the introduced species, *Aedes albopictus*
Lounibos	1983	Occurrence and species interactions in subtropical Florida
Love & Whelchel	1955	Photoperiodism and development
Merritt & Craig	1987	Feeding mechanisms of larvae
Paulson	1984	Oviposition and egg diapause, *A. triseriatus* and *A. hendersoni*
Saul *et al.*	1977	Separation of sibling species *Aedes hendersoni*

Author	Year	Title
Saul *et al.*	1978	Genetic differences between adults from different tree species
Scholl & DeFoliart	1977	Vertical and temporal distribution of eggs
Scholl & DeFoliart	1978	Sex ratios and voltinism
Shroyer & Craig	1980	Egg hatching and diapause: effects of light and temperature
Shroyer & Craig	1981	Seasonal variation in sex ratios
Shroyer & Craig	1983	Egg diapause: geographical patterns
Sims	1982	Induction and intensity of larval diapause
Sinsko & Craig	1979	Adult population size in an isolated population
Sinsko & Grimstad	1977	Differentiation in oviposition behaviour from *A. hendersoni*
Truman & Craig	1968	Hybridisation with *A. hendersoni*
Vavra	1969	Gregarine parasites of *A. triseriatus*
Walker & Merritt	1988	Interaction between leaf detritus and larval productivity
Walker & Merritt	1991	Larval behaviour
Watts *et al.*	1972	Transmission of LaCrosse virus
Wright & Venard	1971	Diapause induction by photoperiod

Table A.6. *Key works on the biology of Toxorhynchites spp. in phytotelmata*

Author(s)	Year	Subject matter
Basham *et al.*	1947	Biology and distribution in Florida
Bonnet & Hu	1951	Introduction into Hawaii
Bradshaw & Holzapfel	1975	Photoperiodic control of development in *T. rutilus*
Bradshaw & Holzapfel	1977	Impact of photoperiod, temperature and chilling on larvae
Breland	1949	General biology
Corbet	1963b	Observations on *T. brevipalpis* in Uganda
Focks *et al.*	1979	Field performance of laboratory-reared *T. rutilus*
Focks *et al.*	1980	Interactions between *T. rutilus* and *Ae. aegypti*
Gerberg & Visser	1978	Biocontrol of *Ae. aegypti* by *T. rutilus*
Hemmerlein & Crans	1968	Review of biology of *T. rutilus*; the species in New Jersey
Holzapfel & Bradshaw	1976	Rearing, pre-diapause and pupal development of *T. rutilus*
Jenner & McCrary	1964	Photoperiodic control of larval diapause in *T. rutilus*
Juliano & Remiger	1992	Interaction between *T. rutilus* and *Ae. triseriatus*
Lamb & Smith	1980	Life history characteristics of *T. rutilus* and *T. brevipalpis*
Lounibos	1979b	Biology of *T. brevipalpis* in Kenya
Muspratt	1951	Bionomics of African species and biocontrol potential
Paine	1934	*T. splendens* introduced to Fiji for control of *Ae. variegatus*
Rubio *et al.*	1980	Behaviour of *T. theobaldi*
Russo	1983	Functional response in *T. rutilus*
Russo	1986	Predatory behaviour in five species of *Toxorhynchites*
Steffan	1975	Systematics and biocontrol potential
Steffan & Evenhuis	1981	General review of biology of genus
Steffan *et al.*	1980	Annotated bibliography of *Toxorhynchites*
Trimble	1979	Laboratory observations on *T. rutilus*
Trimble & Smith	1975	Bibliography of *T. rutilus*
Trimble & Smith	1978	Geographic variation, development, predation in *T. rutilus*

Trimble & Smith	1979	Geographic variation, temperature and photoperiod effects
Trpis	1972	Development and predatory behaviour in *T. brevipalpis*
Trpis	1972	Interaction between *T. brevipalpis* and *Ae. aegypti*
Vongtangswad & Trpis	1980	Prediction of pupation in *T. brevipalpis*
Watts & Smith	1978	Oogenesis in *T. rutilus*

Table A.7. *Key works on the biology of pitcher-plant mosquitoes*

Author(s)	Year	Subject matter
Addicott	1974	Impact of mosquito larvae on the protozoan assemblages
Bradshaw	1976	Geography of photoperiodic dormancy control
Bradshaw	1980	Blood feeding and adult fecundity in *Wyeomyia smithii*
Bradshaw	1983	*Wyeomyia smithii* and *Metriocnemus knabi* interaction
Bradshaw & Holzapfel	1983	Life cycle strategies in *Wyeomyia smithii*
Bradshaw & Holzapfel	1986a	Density-dependent selection in *Wyeomyia smithii*
Bradshaw & Lounibos	1972	Photoperiodic control of development in *Wyeomyia smithii*
Bradshaw & Lounibos	1977	Evolution of dormancy in *Wyeomyia smithii*
Buffington	1970	*Wyeomyia smithii* and *Metriocnemus knabi* interaction
Evans & Brust	1972	Diapause induction and termination in *Wyeomyia smithii*
Grjebine	1979	Mosquitoes in *Nepenthes* in Madagascar
Istock *et al.*	1983	Habitat selection by *Wyeomyia smithii*: behaviour and genetics
Istock *et al.*	1975	Population dynamics and laboratory trials on *Wyeomyia smithii*
Istock *et al.*	1976a	Resource tracking by *Wyeomyia smithii*
Istock *et al.*	1976b	Fitness factors in *Wyeomyia smithii*
Lang	1978	Fecundity and nutrition in *Wyeomyia smithii*
Lever	1950	Mosquitoes from *Nepenthes* in the Cameron Highlands, Malaya
Lounibos & Bradshaw	1974	A second diapause in *Wyeomyia smithii*
Moeur & Istock	1980	Larval impacts on adult fitness in *Wyeomyia smithii*
Price	1958	Biology and laboratory culturing of *Wyeomyia smithii*
Smith	1902	Life history of *Aedes* (=*Wyeomyia*) *smithii*
Smith & Brust	1971	Photoperiodic control of larval diapause in *Wyeomyia smithii*

drawn heavily upon the works of Fish (1983), Beaver (1983), Frank (1983) and Thienemann (1954) and personal communications from Dr Peter Cranston.

Chironomids in general are aquatic in the larval stages, inhabiting the decaying vegetation and other debris that accumulates in water bodies. All chironomines and some orthocladiines build a tube made of silk-like threads produced by the larval salivary glands. Organic and inorganic particles adhere to this silk forming the dwelling tube. Members of these two subfamilies are predominantly particle-feeding decomposers, processing organic and inorganic debris and the associated micro-algae (particularly diatoms). The feeding preferences and mechanisms of the Orthocladiinae and Chironominae are discussed by Oliver (1971). This author suggests that the orthocladiines probably feed on particulate matter directly, whereas the chironomines either collect food on extruded saliva, or filter particles from the water using 'nets' spun across currents or using feeding brushes present on the mouthparts of some species.

Of phytotelm species, only the feeding habits of *Metriocnemus knabi* from *Sarracenia* pitchers have been studied directly. Paterson & Cameron (1982) record larvae of this species ingesting particulate organic matter derived from drowned victims of the pitcher, often burrowing into the corpses themselves. The feeding habits of other phytotelm species we infer to be similar to those of related species. Members of the subfamily Tanypodinae are usually regarded as predatory although Oliver (1971) considers this a too sweeping view, suggesting instead that many tanypodines are only facultative predators which also ingest algae and detritus (references in Oliver 1971). My observations of species of the tanypodine genera '*Anatopynnia*' and *Paramerina* from water-filled tree holes in Australia and New Guinea indicate that these species at least are voracious predators which, certainly, may ingest surrounding particulate matter but give every impression of being obligate, generalist sit-and-wait predators which engulf smaller prey tail first. Larger prey are penetrated and sucked out by the tanypodine larvae. Some non-tanypodine chironomids are recorded as predators of small oligochaetes (Loden 1974) and these include the genus *Cryptochironomus*, which has been recorded from water bodies in bromeliads (Table A.8).

Current views on the origin of the Chironomidae, based on analysis of the morphology, systematics and biogeography of the relictual genus *Archaeochlus* (Cranston *et al.* 1987) suggests that protochironomids arose at least as far back as the Jurassic, in surface films and interstitial water in moss and stream margins. Cranston *et al.* (1987) suggest that from these habitats chironomids invaded the cool, clear flowing waters of woodland streams

Table A.8. *The Chironomidae and their association with container habitats*

Taxon	Association with containers	Key reference(s)
"CHIRONOMIDAE"	*Nepenthes ampullaria*, Malaysia	Mogi & Yong 1992
PODONOMINAE		
Parochlus bassianus	*Richea* axils, Tasmania	Cranston & Kitching 1994
CHIRONOMINAE		
Chironomus palpalis	Bamboos, Indonesia	Thienemann 1934
Chironomus subrectus	Tree holes, Germany	Thienemann in Benick 1924
Chironomus sp.	Tree holes, Papua New Guinea	Kitching 1990
Chironomus sp.	Bromeliads, Costa Rica, Jamaica	Picado 1913; Laessle 1961
Cryptochironomus sp.	Bromeliads, Jamaica	Laessle 1961
Chirocladius pedipalpus	Bromeliads, Costa Rica	Picado 1913
Polypedilum convexum	*Nepenthes bicalcarata, N. ampullaria*, Borneo	Clarke & Kitching 1993
	Bamboos, Malaya	Kovac & Streit 1996
Polypedilum sp.	Tree holes, Australia	Jenkins & Kitching 1990
Polypedilum sp.	*Nepenthes ?villosa*, Sabah	Cranston & Judd 1987
Polypedilum sp.	*Colocasia* axils, Indonesia	Thienemann 1934
Tanytarsus sp.	Bromeliads, US Virgin Islands	Miller 1971
Tanytarsus sp.	Bamboos, Indonesia	Thienemann 1934
'tanytarsines' indet.	Bromeliads, Florida, Brazil	Fish 1976, Winder & Silva 1972
ORTHOCLADIINAE		
Metriocnemus cavicola (=*martinii*)	Tree holes, Europe	Röhnert 1951, Kitching 1971
Metriocnemus abdominoflavidus	Bromeliads, Costa Rica	Picado 1913
Metriocnemus edwardsi	*Darlingtonia* pitchers, California	Jones 1916
Metriocnemus hirticollis	*Dipsacus* axils, Europe	Zavrel 1941
Metriocnemus inopinatus	*Scirpus* axils, Europe	Strenzke 1950
Metriocnemus knabi	*Sarracenia* pitchers	Bradshaw & Creelman 1984

Species	Habitat, location	Reference
Metriocnemus scirpi	*Scirpus* axils, Europe	Strenzke 1950
Metriocnemus wittei	Banana axils, Uganda	Freeman 1956
Metriocnemus lobeliae	*Lobelia* and *Senecio* axils, Kenya	Freeman 1956, Cranston & Judd 1987
Metriocnemus canus	*Senecio* axils, Kenya	Cranston and Judd 1987
Metriocnemus sp.	*Nepenthes tentaculata, ?villosa*, Sabah	Cranston and Judd 1987
Metriocnemus sp.	*Dipsacus* axils, USA	Baumgartner 1986
Metriocnemus sp. nr. *knabi*	*Dipsacus* axils, Russia	Borobev 1960
Metriocnemus sp.	*Heliamphora* pitchers, Venezuela	Jaffé *et al.* 1992
Metriocnemus sp.	*Nepenthes* sp., Borneo	Beaver 1983
Metriocnemus sp.	Tree holes, Central America	Snow 1949
Orthocladius sp.	Bromeliads, Costa Rica	Picado 1913
Bryphaenocladius sp.	*Richea* axils, Tasmania	Cranston & Kitching 1994
Compterosmittia sp.	*Richea* axils, Tasmania	Cranston & Kitching 1994
Compterosmittia sp.	Bamboos, Malaya	Kovac & Streit 1996
TANYPODINAE		
Anatopynnia pennipes	Tree holes, Australia	Kitching & Callaghan 1982
Anatopynnia sp.	Tree holes, Sulawesi	Kitching 1987b
Thienemannimyia facilis	Tree holes, Indonesia	Zavrel 1933, Thienemann 1934
Pentaneura alterna	*Colocasia* axils, Indonesia	Thienemann 1934
Pentaneura monilis	Bamboos, Indonesia	Thienemann 1934
Pentaneura sp.	Tree holes, Australia	Kitching 1990
Pentaneura sp.	Bromeliads, Jamaica, US Virgin Is, Brazil	Laessle 1961, Miller 1971
Pentaneura sp.	*Heliconia* axils, Costa Rica	Naeem 1990
Paramerina ignobilis	Bamboos, tree holes, *Pandanus* axils, Sumatra, Java, Malaysia, Australia, Papua New Guinea	Kitching 1990, Thienemann 1934, P. S. Cranston pers. comm.
Monopelopia tillandsia	Bromeliads, Florida	Fish 1976a
Monopelopia sp.	Bromeliads, Brazil	Winder 1977, Winder & Silva 1972
Ablabesmyia costaricensis	Bromeliads, Costa Rica	Picado 1913
Genus nr. *Trissopelopia* sp.	Bamboos, Malaya	Kovac & Streit 1996

(previously considered the archaic habitat; Brundin 1966) from whence they invaded the very wide range of habitats from which we know the family today. If this scenario is accepted, then, in striking contrast with the Culicidae, chironomids of phytotelmata are to be regarded as highly specialised derivatives of parent groups associated with other habitat types. This view might well account for the relatively few genera of chironomids that have been found in container habitats (Table A.8).

Of particular note is the predominance of the orthocladiine genus *Metriocnemus* which provides most of the specialised chironomids in phytotelmata as diverse, biologically and geographically, as European tree holes and *Heliamphora* pitchers from Venezuela. Cranston & Judd (1987) review the occurrence of the genus in container habitats. A species of *Metriocnemus* (*cavicola* Kieffer = *martinii* Thienemann) is the most common inhabitant of tree holes in temperate Europe (Kitching 1972a) although the genus is absent from such habitats in temperate North America. Cranston and Judd speculate that this may be because of the presence there of predatory larvae of *Toxorhynchites*, absent from northern Europe. In North America, the genus is predominant in pitchers of *Sarracenia* (in the east) and *Darlingtonia* (in the west). The recent record of a *Metriocnemus* sp. from pitchers of *Heliamphora* (Jaffé *et al.* 1992) apparently parallels these records in northern South America. Cranston & Judd (1987) note additional species of *Metriocnemus* from *Nepenthes* pitchers in Sabah and from axil waters in East Africa and Europe and add these to well-known records from bromeliad tanks (Table A.8). The genus is absent from Australia.

Tanypodinae from tree holes and bromeliads have been recorded under the names *Anatopynnia, Pentaneura* and *Paramerina*. Although the first two of these genera are well defined on the basis of their type species from other habitats, the phytotelm species do not sit comfortably within these restricted generic concepts and revision is sorely needed (P. S. Cranston, pers. comm.). Within tree holes and bromeliads, they are generalist predators which may reach high levels of abundance in some instances. The additional tanypodine genera *Ablabesmyia* and *Monopelopia* occur in bromeliad water bodies within which they undoubtedly act as predators.

Ceratopogonidae The Ceratopogonidae (Figure A.6) are a large and ubiquitous family of Nematocera which, as adults, are known as 'biting midges'. As well as being widespread nuisance pests some species are dangerous and economically important vectors of human and animal diseases. They occupy a wide range of larval habitats from moist soil through the range of freshwater situations to intertidal pools. They are readily distinguished as larvae

Fig. A.6 Larva of the ceratopogonid midge *Culicoides annulatus* (dorsal view).

from the chironomids with which they so often occur as they lack the pro-
thoracic proleg characteristic of larval Chironomidae. The species that occur
in container habitats are streamlined, fast swimmers, travelling through the
liquid medium by throwing their bodies into rapid undulations.

The biting midges are common and widespread inhabitants of phytotelmata
of all kinds (Table A.9). Frequently overlooked, or judiciously ignored
because of the taxonomic intractability of some genera, they are probably
even more widespread in container habitats than existing records suggest.
Within phytotelm food webs they probably act as decomposers, processing
decaying animal and plant matters for nutrients and for the associated micro-
organisms. They will congregate around moribund insect larvae, oligochaetes
and nematodes, feeding readily on these dying organisms. This has led to
their being designated as predators in some studies (such as Kitching &
Callaghan 1982, for instance) but mature reflection suggests they are better
designated as generalist saprophages.

Taxonomically the Ceratopogonidae are dominated by a small number of
very large and uniform genera. So of the 87 different records of larvae in
phytotelmata that I have assembled in Table A.9 all but six belong to either
Dasyhelea (20+ species), *Culicoides* (about 29 species) or *Forcipomyia* (about
28 species). Groups within the large and poorly known genus *Dasyhelea*
appear to have specialised as container breeders. The genus occurs widely in
pitcher plants, tree holes, axil waters and bamboo internodes. There is a single record from a bromeliad tank. *Culicoides*, undoubtedly the best known
genus in the family because it contains many species that are nuisance pests
or vectors, has radiated widely in tree holes in the Americas and Australia
and is also known from bromeliad tanks, axil waters and *Heliamphora* pitchers in South America. Lastly the great genus *Forcipomyia* is, *inter alia*, a
bromeliad specialist throughout the southern USA and South America and
even has turned up in garden bromeliads in Africa and Asia. A few additional records of species in this genus are available from axil waters. The list
of records contains (by implication) some curious negative records: no *Culicoides* species seem to have been recorded from water-filled tree holes in
Europe, and no ceratopogonids at all are recorded from North American
pitcher plants.

Table A.9. *The Ceratopogonidae and their association with container habitats*

Taxon	Association with containers	Key reference(s)
DASYHELEA		
ampullariae	*Nepenthes ampullaria, N. gracilis*, Malaya	Beaver 1983
assimilis	Bamboo stumps, Indonesia, Malaysia	Johannsen 1931, Thienemann 1934, Kovac & Streit 1996
bilineata	*Dipsacus fullonum* axils, UK	Disney & Wirth 1982
biseriata	*Nepenthes ampullaria*, Malaya	Beaver 1983
confinis	*Nepenthes mirabilis*, Sumatra	Thienemann 1932, Beaver 1983
dufouri	Tree holes, Europe	Röhnert 1951; Kitching 1972b
grata	Bamboo stumps, Indonesia, Malaysia	Johannsen 1932, Thienemann 1934, Kovac & Streit 1996
hitchcocki	Tree holes, Tonga and Samoa	Wirth 1976
lignicola	Tree holes, Europe	Röhnert 1951
nepenthicola	*N. albomarginata, ampullaria, gracilis*, Malaya	Beaver 1983
oppressa	Tree holes, Alabama, Florida	Snow 1958, R. Kitching unpublished
subgrata	*Nepenthes mirabilis*, New Guinea	Beaver 1983
Dasyhelea sp.	Bromeliads, Brazil	Winder & Silva 1972
Dasyhelea sp.	*Nepenthes bicalcarata, rafflesiana, ampullaria, mirabilis, ampullaria, albomarginata*, Brunei	Clarke & Kitching 1993
Dasyhelea sp.	*Nepenthes ampullaria*, Malacca, Sarawak	Mogi & Yong 1992
Dasyhelea sp.	Bamboo internodes, Japan	Mogi & Suzuki, 1983
Dasyhelea sp.	*Cephalotus follicularis* pitchers, Australia	Clarke 1985
Dasyhelea sp.	Tree holes, South Pacific	Wirth 1976
Dasyhelea sp.	*Dipsacus* axils, Europe	Disney & Wirth 1982
Dasyhelea sp.	*Curcuma* bract axils, New Guinea	Kitching 1991
Dasyhelea sp.	*Zingiber, Commelina* axils, Indonesia	Thienemann 1934
Dasyhelea sp.	Occurrence in phytotelms	Zavrel 1935

CULICOIDES		
angularis	Tree holes, Australia	Kitching & Callaghan 1982
anophelis	*Colocasia* axils, Indonesia	Thienemann 1934
arboricola	Tree holes, central and southern USA	Snow 1958; Lamberson *et al.* 1992
bayano	Bromeliads, Panama	Vitale *et al.* 1981
beckae	Tree holes, Georgia	Lamberson *et al.* 1992
confinis	*Nepenthes mirabilis*, Sumatra	Oldroyd 1964
decor	Bromeliads, Trinidad	Williams 1964
elemae	Tree holes, Nebraska	Lamberson *et al.* 1992
flukei	Tree holes, New York	Lamberson *et al.* 1992
footei	Tree holes, central and southern USA	Lamberson *et al.* 1992
gutifer	Tree holes, Indonesia	Thienemann 1934
gutipennis	Tree holes, Kansas, Florida	Fashing 1975; R. Kitching, unpublished
hayesi	Bromeliads, Trinidad, Venezuela	Williams 1964; Wirth & Blanton 1968
heliconiae	*Heliconia* inflorescence bracts, Panama	Wirth *et al.* 1968
hinmani	Tree holes, Nebraska	Lamberson *et al.* 1992
lahillei	Tree holes, Georgia, Missouri	Lamberson *et al.* 1992
nanus	Tree holes, central and southern USA	Lamberson *et al.* 1992
nigrigenus	Bromeliads, Trinidad, Columbia, Panama, Mexico	Wirth & Blanton 1968
oklahomensis	Tree holes, southern USA	Lamberson *et al.* 1992
paraensis	Bromeliads, Panama; tree holes, southern USA	Vitale *et al.* 1981; Lamberson *et al.* 1992
snowei	Tree holes, southern USA	Lamberson *et al.* 1992
sumatrae	*Curcuma* inflorescence bracts, New Guinea	Kitching 1991
villosipennis	Tree holes, central and southern USA	Lamberson *et al.* 1992
sp. nr *debilipalpis*	Bromeliads, Brazil	Winder 1977
Culicoides sp.	*Heliamphora* pitchers, Venezuela	Jaffé *et al.* 1992
Culicoides sp.	Tree hole, India	Wirth & Hubert 1972
Culicoides sp.	*Calathea* leaf axils, Brazil, Colombia	Wirth & Soria 1981
Culicoides sp.	Bamboo internodes, Malaysia	Kovac & Streit 1996
Culicoides spp.	Bamboo internodes, Brazil	Wirth & Blanton 1968

Table A.9. (*cont.*)

Taxon	Association with containers	Key reference(s)
FORCIPOMYIA		
aeria	Bromeliads, Puerto Rico	Saunders 1956
antiguensis	Bromeliads, Trinidad	Meillon & Wirth 1979
bacoti	Bromeliads, Ghana	Saunders 1956, Meillon & Wirth 1979
bikanni	Bamboo internodes, Malaysia	Kovac & Streit 1996
brevis	Bromeliads, Hawaii	Meillon & Wirth 1979
bromeliae	Bromeliads, Brazil	Saunders 1956, Winder 1977
bromelicola	Bromeliads, Brazil	Meillon & Wirth 1979
caribbeana	Bromeliads, Trinidad, Brazil, Guyana	Fish & Soria 1978, Meillon & Wirth 1979
sp. nr. *cinctipes*	Bromeliads, Brazil	Winder 1977
edwardsi	Bromeliads, Brazil, Trinidad	Saunders 1956
falcifera	Bromeliads, Antigua	Saunders 1959
fuliginosa	Bromeliads, Trinidad, Brazil	Wirth 1975, Winder 1977
genualis	Bromeliads, Brazil	Winder 1977
harpegonata	Bromeliads, Brazil	Winder 1977
hutsoni	*Pandanus* axils, Aldabra Island	Wirth & Ratanaworabhan 1976
jocosa	Bromeliads, Trinidad	Meillon & Wirth 1979
keilini	Bromeliads, Brazil	Meillon & Wirth 1979
magna	Bromeliads, Brazil	Meillon & Wirth 1979
nicopina	Bromeliads, Singapore	Meillon & Wirth 1979
oligarthra	Bromeliads, West Indies, Singapore, USA	Meillon & Wirth 1979
pictoni	Bromeliads, Brazil	Winder 1977
seminole	Bromeliads, Florida	Fish 1976
simulans	Bamboos, Indonesia, Malaysia	Thienemann 1934, Kovac & Streit 1996
swezeyanaadfinis	Bamboos, Malaysia	Kovac & Streit 1996
tuberculata	Bromeliads, Trinidad	Saunders 1956

Taxon	Host/locality	Reference
Forcipomyia sp.	Cacao pods, Brazil	Soria *et al.* 1978
Forcipomyia sp.	*Saccharum* axils, Brazil	Soria *et al.* 1978
Forcipomyia sp.	*Nepenthes ampullaria*, Malacca, Sarawak	Mogi & Yong 1992
LASIOHELEA sp.	Bromeliads, Trinidad	Williams 1964
BEZZIA		
Bezzia sp.	Bromeliads, Jamaica, Brazil	Laessle 1961, Winder 1977
Bezzia sp.	Tree holes, Europe	Röhnert 1950
APELMA		
Apelma canis	*Colocasia* axils, Indonesia	Thienemann 1934
Apelma sp.	*Colocasia* axils, Brazil	Thienemann 1934
JOHANNSENOMYIA prominens	Bamboos, Indonesia	Thienemann 1934
'ceratopogonine'	Bamboos, Peru	Louton *et al.* 1996
'forcipomyiine'	Bamboos, Peru	Louton *et al.* 1996
'ceratopogonid'	*Musa* axils, Brazil	Fish & Soria 1978

Fig. A.7 Larva of the psychodid fly *Telmatoscopus* sp. (dorsal view) (after Brauns 1954).

Fig. A.8 Larva of the anisopodid fly *Anisopus fenestralis* (dorsal view) (after Peterson 1951).

Psychodidae The moth flies or psychodids (Figure A.7) are a cosmopolitan family of very distinctive flies which include the phlebotomine vectors of leishmaniasis in South America. The remaining five subfamilies, the Bruchomyiinae, Trichomyiinae, Syncoracinae, Horaiellinae and Psychodinae, are innocuous although may become nuisance pests of sewage treatment systems, within which they also serve a valuable function as agents of particulate breakdown (e.g. Lloyd *et al.* 1940). General accounts of the family are presented by Quate & Vockeroth (1981) and Duckhouse & Lewis (1989). Vaillant (1971) provides a useful classification of larval habitats for the Palaearctic Psychodinae and includes phytotelmata and (separately) tree holes as categories in his scheme. The larvae of all psychodids, in general, are detritus feeders inhabiting moist areas around and within aquatic and semi-aquatic habitats, rotting wood, carrion, fungi and dung. Some are regular, specialist inhabitants of phytotelmata (Table A.10) having been recorded in water-filled tree holes, bromeliad tanks, bamboo internodes and a range of plant axils. There are no records of psychodids from pitcher plants.

Table A.10 lists the genera and species that are recorded in the literature as occurring in phytotelmata. All but one of the species that have been recorded are psychodines. Larvae of the family in general are characteristically nematoceran with a sclerotised head capsule and no legs or prolegs. Larvae of the Psychodinae predominate in container habitats and these have distinctive darker dorsal scutes and a terminal syphon on the abdomen. They reach very high densities in tree hole habitats in Australia.

Anisopodidae The anisopodids (Figure A.8) are a small family of Nematocera with larvae that inhabit decomposing organic matter feeding on, and helping break down, particulate detritus. Some species favour wetter situations and, accordingly, have occurred in phytotelmata of various kinds. The

Table A.10. *The Psychodidae and their association with container habitats*

Taxon	Association with containers	Key reference(s)
PSYCHODINAE		
Pericoma fagocavatica	Tree holes, Germany	Röhnert 1950
Pericoma spp.	*Musa* axils, Indonesia	Thienemann 1934
Pericoma sp.	Bamboos, Indonesia	Thienemann 1934
Brunettia nitida	Tree holes, USA	Snow 1958
Clogmia albipunctata	Tree holes, PNG	Kitching 1990
Clogmia sp.	Bamboos, *Curcuma* bracts, PNG	Kitching 1990
Clogmia sp.	Aroid axils	Mogi *et al.* 1985
Clogmia spp.	Tree holes, Australia	Kitching & Pimm 1986
Clogmia sp.	Bamboo internodes, Malaysia	Kovac & Streit 1996
Telmatoscopus albipunctatus	Tree holes, USA	Fashing 1975
Telmatoscopus nr *albipunctatus*	Bromeliads, Jamaica	Laessle 1961
Telmatoscopus rotschildi	Tree holes, Europe	Vaillant 1971
Tematoscopus superbus	Tree holes, USA	Fashing 1975, Snow 1958
Telmatoscopus sp.	Bromeliads, Brazil	Winder 1977
Alepia sp.	Bamboo internodes, Peru	Louton *et al.* 1996
Jungiella laurencei	Tree holes, Europe	Vaillard 1971
Clytocerus xylophilus	Tree holes, Europe	Vaillard 1971
Psychoda parsivena	*Curcuma* bracts, PNG	Kitching 1990
Psychoda savaiiensis	*Curcuma* bracts, PNG	Kitching 1990
Psychoda sp.	*Curcuma* bracts, PNG	Kitching 1990
Psychoda sp.	*Colocasia* axils, Indonesia	Thienemann 1934
Philosepedon fumata	Bromeliads, Florida	Fish 1976
Philosepedon sp.	Bamboo internodes, Malaysia	Kovac & Streit 1996
'Psychodinae'	Various plant axils, Uganda	Haddow 1948
TRICHOMYIINAE		
Trichomyia crucis	Tree holes, Australia	Duckhouse & Lewis 1989
UNPLACED		
Neurosystasis amplipenna	Bromeliads, Cuba	Knab 1913a
Neurosystasis sp.	Bromeliads, Mexico	Knab 1913a
'Psychodidae sp.'	*Scirpus* axil, Europe	Strenzke 1950
'psychodids'	Tree holes, Florida	Lounibos 1983
'psychodids'	*Curcuma* axils, Thailand	Mogi & Yamamura 1988

window fly *Anisopus fenestralis* has been recorded in water-filled tree holes in Britain (Kitching 1969, Colyer & Hammond 1951). Röhnert (1950) records the related *Mycetobia* sp. from holes in northern Germany. Both Kitching (1969) and Röhnert (1950) reject the anisopodids as specialist members of the tree-hole communities they were studying, accepting them only as accidental 'philic' species. Picado (1913) records larvae of *Anisopus picturatus* from bromeliads in Costa Rica.

Scatopsidae The Scatopsidae (Figure A.9) are a family of flies linked taxonomically to both the sciarids and cecidomyids (Oldroyd 1964). Keilin (1927) included a scatopsid species among the fauna he recorded associated with a tree hole in a horse-chestnut tree (*Aesculus hippocastanum*) near Cambridge, England. Wirth (1952) describes three Nearctic species of *Systenus* from tree-hole debris. For all of these scatopsid species I suggest the association with tree holes is more 'philic' than 'biotic' based upon a generalised attraction to the moisture and decay of these situation. Haenni and Vaillant (1994), however, record larvae of *Holoplagia richardi* and *Ectaetia platscelis* as being fully aquatic in the water body within maple tree holes near Grenoble, France. Laurence (1953) had earlier described the larvae of a British species of *Ectaetia* from 'moist tree holes'.

Cecidomyiidae The Cecidomyiidae (Figure A.10) are a large, relatively poorly known family. The majority of the family are gall formers (hence, 'gall-midges') but a significant minority have larvae that are scavengers, fungivores or predators. As plant parasites, members of the family are significant pests of agriculture (Colless & McAlpine 1991).

There are few records of cecidomyids from phytotelmata. However, mem-

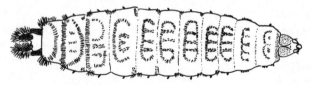

Fig. A.9 Larva of the scatopsid fly *Scatopse* sp. (dorsal view) (after Freeman 1983).

Fig. A.10 Larva of a cecidomyiid fly (lateral view) (after Brauns 1954).

bers of the genus *Lestodiplosis* occur commonly in *Nepenthes* pitcher plants in South-east Asia where they are predatory upon other members of the pitcher community. Beaver (1983) records *L. syringopais* from *Nepenthes albomarginata* and *N. gracilis* in West Malaysia and Sumatra. Beaver (1979a) observed *Lestodiplosis* feeding upon phorid larvae and sugested that ceratopogonid larvae were also likely prey. Clarke and Kitching (1993) record a species of *Lestodiplosis* from *N. gracilis* and *N. mirabilis* in Brunei, and Mogi and Yong (1992) encountered a similar insect in *N. ampullaria* in Malacca. Kovac and Streit (1996) record four undescribed species of this family from water-filled bamboo internodes in Malaysia (placed in *Clinodiplosis, Feltiella, Lestodiplosis* and *Xylodiplosis* respectively). These authors present remarkable observations of larvae of *Xylodiplosis* acting as a top predator within these internode food webs, preying upon tanypodine larvae and those of *Toxorhynchites*. Louton *et al.* (1996) record cecidomyid larvae 'infrequently' from Peruvian bamboo internodes and there is a record by Winder (1977) of an unidentified 'cecidomyid' larvae from bromeliad tanks in Brazil. In addition we have encountered what may be *Lestodiplosis* larvae in some water-filled tree holes in North Queensland, but these records need confirmation.

Sciaridae The sciarids (Figure A.11) are a large and diverse family of nematoceran flies closely related to (indeed, sometimes classified with) the fungus gnats (Mycetophilidae) (Freeman 1983). Their larvae commonly occur in rotting vegetable matter but some species have become pests of mushrooms and glasshouse plants (Colless & McAlpine 1991). Other species inhabitat dung and yet others are associated with bird and mammal nests. The general association with rotting vegetation predisposes the family to occur in phytotelmata.

Three records exist of sciarids in phytotelmata. A species of *Corynoptera* has been recorded by Fish (1976) from bromeliads in Florida. Fish himself (1983) urges caution in accepting it as a member of the aquatic fauna. Given the larval proclivities of most of the family it seems likely that it is a 'philic' species rather than a bromeliad specialist. The other records are both of larvae of *Bradysia* spp. *Bradysia macfarlanei* occurs as an obligate associate of *Sarracenia* pitchers. The larvae of this species 'can be found burrowing among the entrapped prey' (Rymal & Folkerts 1982). Presumably the larvae

Fig. A.11 Larva of the sciarid fly *Corynoptera* sp. (lateral view) (after Stehr 1987).

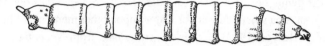

Fig. A.12 Larva of the tabanid fly *Tabanus* sp. (lateral view) (after McAlpine *et al.* 1981).

are detritivorous upon the drowned prey. In addition, Kovac and Streit (1996) note larvae of *Bradysia* sp. from bamboo internodes in Malaysia.

Tabanidae The horse-flies (Figure A.12) are a large and ubiquitous family of considerable medical importance. Their larval habitats range from floating vegetation to river mud to dry sand, in which situations they are generally assumed to be predatory on other, soft-bodied, larvae.

Tabanid larvae are not commonly encountered in phytotelmata but English *et al.* (1957) described a species of the tabanine *Chalybosoma casuarinae* from larvae collected in moist debris in a rot hole in *Casuarina cunninghamiana* in south-eastern, coastal Australia. Oldroyd (1957) recorded tabanids of the tribe Rhinomyzini which are from very similar habitats in Africa, and Burger (1977) found larvae of *Leucotabanus ambiguus* in 'rot holes of the willow ... sycamore ... and cottonwood ... trees in riparian habitats in southeastern Arizona'. In all these cases it seems the larvae are merely semi-aquatic, seeking moist rather than water-filled situations. Indeed when English *et al.* (1957) flooded the hole in which they found *C. casuarinae* it was subsequently invaded by species of mosquito and ceratopogonid midges and no further tabanids were found. Goodwin & Murdoch (1974) record larvae of *Stibasoma* spp. from bromeliads in Panama and Brazil and state, quite explicitly: '... spaces between the leaf bases which catch and hold water. It is in these spaces that larvae live'. Accordingly, in this case (and, probably, in this case alone) we may properly consider tabanid larvae as part of the aquatic community. Other tabanids, of course, are fully aquatic, with species occurring in both still and flowing waters.

Stratiomyidae The stratiomyids (Figure A.13) are a cosmopolitan family with larvae that range from terrestrial to fully aquatic. Rozkosny (1982) presents a complete account of the biology and systematics of the European species. Larval habitats include leaf litter and soil, compost, manure, under bark, in moss, and in both flowing and stagnant water. Some even occur in salt-springs and ditches. They are generally root- or detritus-feeders and some species are pests of pasture crops.

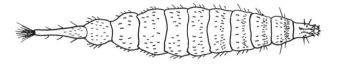

Fig. A.13 Larva of the stratiomyid fly *Stratiomys norma* (dorsal view) (after McAlpine *et al.* 1981).

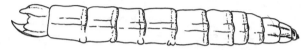

Fig. A.14 Larva of a dolichopodid fly (lateral view) (after Brauns 1954).

Like tabanids they are only occasional inhabitants of phytotelmata. English *et al.* (1957) describe larvae of *Syndipnomyia auricincta* co-occurring with the tabanid they described from rot holes in southern Australia (see above). In addition we have encountered occasional stratiomyid larvae in treehole samples from tropical north Queensland. These always occurred as singletons and, as in the case of the tabanids, I believe are not to be considered as a consistent member of the food webs. Seifert & Seifert (1976a) and Machado-Allison *et al.* (1983) recorded species of *Merosargus* from the water-filled bract axils of *Heliconia bihae, caribaea, imbricata* and *wagneriana* from Costa Rica and Venezuela. *Heliconia* axils in Costa Rica contained larvae of *Sargus fasciatus* (Naeem 1990). The stratiomyid larvae from *Heliconia* axils fed on floral parts and detritus. Small numbers of stratiomyid larvae were noted by Thienemann (1934) from bamboos in South-east Asia and by Louton *et al.* (1996) from similar habitats in Peru. Kovac & Streit (1996) encountered larvae of *Captopteromyia fractipennis* and *Ptecticus longipennis* in bamboo internodes in Malaysia.

Dolichopodidae The Dolichopodidae (Figure A.14) are a large family within the suborder Brachycera. Larvae are, in general, predacious although Oldroyd (1964) suggests dead insects and even plant material may be used by some species. The family is basically a terrestrial one but, as in some other dipterous families, a few genera have become secondarily aquatic. Johannsen (1935) records three genera in this category. Only one phytotelm record exists: that of an unidentified species from the pitchers of *Nepenthes ampullaria* from Sarawak (Mogi & Yong 1992). These authors denote the fly as a nepenthebiont but append a question mark to their designation!

Table A.11. *The Phoridae and their association with container habitats*

Taxon	Association with containers	Key reference(s)
METOPININAE		
Megaselia anomaloterga	Bamboos, Malaysia	Kovac & Streit 1996
Megaselia bivesicata	Bamboo cups, *Nepenthes ampullaria*, Java	Thienemann 1932, 1934
Megaselia cambodiae	*Nepenthes* sp., Cambodia	Disney 1994
Megaselia campylonympha	*Nepenthes ampullaria*, *N. mirabilis*, Brunei	Clarke & Kitching 1993
Megaselia decipiens	*Nepenthes gymnamphora*, Java	Thienemann 1932
Megaselia deningsi	*Nepenthes* spp., Brunei, Sri Lanka; bamboos, Malaysia	Clarke & Kitching 1993, Kovac & Streit 1996
Megaselia gombakensis	Bamboos, Malaysia	Kovac & Streit 1996
Megaselia gregalis	*Nepenthes gymnamphora*, Java; *N. distillatoria*, Sri Lanka	Thienemann 1932, Disney 1994
Megaselia humida	Bamboos, Malaysia	Kovac & Streit 1996
Megaselia kovaci	Bamboos, Malaysia	Kovac & Streit 1996
Megaselia nepenthina	*Nepenthes* spp., South-east Asia	Beaver 1979b
Megaselia orestes	*Darlingtonia californica* pitchers, Oregon	Fashing 1981
Megaselia scalaris	Bamboos, Malaysia	Disney 1994
Megaselia schuitemakeri	*Nepenthes* spp., South-east Asia	Beaver 1979b; Kovac & Streit 1996
Megaselia tobaica	*Nepenthes tobaica*, Indonesia	Thienemann 1932; Disney 1994
Megaselia sp.	Tree holes, Sulawesi	Kitching 1987
Megaselia sp.	*Nepenthes ampullaria*, Sarawak	Mogi & Yong 1992
Chonocephalus sp.	Tree holes, USA	Rettenmeyer & Akre 1968
Plastophorides gigantochloae	Bamboos, Malaysia	Disney 1994
PHORINAE		
Dohrniphora sp.	Pitchers of *Sarracenia flava*, South Carolina	Jones 1918, Rymal & Folkerts 1982
Dohrniphora sp.	Bromeliad tanks, Brazil	Winder 1977

Fig. A.15 Larva of the phorid fly *Megaselia scalaris* (lateral view) (after Ferrar 1987).

Phoridae The Phoridae, or scuttleflies (Figure A.15), are one of the largest and most complex families of Diptera (Disney 1983, 1994; Ferrar 1987). They are cosmopolitan in distribution and occupy a wide range of larval habitats. They include true parasites, parasitoids, predators, fungivores and generalist saprophages (Disney 1994). Many live in specialised associations with social Hymenoptera and Isoptera (Disney 1986, 1994). Disney's (1994) general account provides an extended treatment of the life cycles, ecology and taxonomy of the group and is without parallel as an introduction to this important and highly speciose family. Ferrar (1987) provides a genus by genus summary of the known larval habits.

Disney (1994) recognises five subfamilies within the Phoridae of which three are specialist associates of social insects. The remaining two subfamilies, the Metopininae and the Phorinae, are large and heterogeneous in habit (Ferrar 1987). Representatives of both these subfamilies are recorded as regular inhabitants of phytotelmata, occurring in pitcher plants, bromeliad tanks, bamboo internodes and/or tree holes. Within these communities they act as decomposers, feeding on dead animal and plant material. Table A.11 summarises the phytotelm records for this family.

Within the Metopininae all but two of the records are of members of the immense genus *Megaselia* (including *Endonepenthia* erected for a group of *Nepenthes*-dwelling species but now (Disney 1981) synonymised with the larger genus).

The phorine genus *Dohrniphora* is recorded from *Sarracenia flava* pitchers from South Carolina (Jones 1918, Rymal & Folkerts 1982) and from bromeliad tanks in Brazil by Winder (1977).

Syrphidae The hover flies (Figure A.16) are familiar elements of the flower-visiting fauna, feeding on nectar and pollen. As larvae they are remarkably diverse in habit from free-living terrestrial predatory forms, through to aquatic decomposers. Ferrar (1987) notes that modern classifications recognise three subfamilies, the Syrphinae, Milesiinae and the Microdontinae, although F. Gilbert informs me (pers. comm.) that the Milesiinae are a heterogeneous set of 'grade groups' rather than a monophyletic taxon. The Syrphinae, usually,

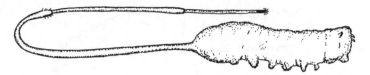

Fig. A.16 Larva of the syrphid fly *Eristalis bastardi* (lateral view) (modified from Johannsen 1935).

have slug-like predatory larvae, familiar as natural enemies and biocontrol agents of aphids and other Homoptera. The Microdontinae are predators on ant-brood within ant nests. The Milesiinae, for the most part, have saprophagous larvae associated with organically enriched habitats, and decaying wood. All but one of the species that have been recorded from phytotelmata fall within this subfamily (Table A.12).

One ecological group, the larvae of which are referred to as 'rat-tailed maggots', inhabit a variety of phytotelmata and, to the naturalist at least, are among the best known of the fauna of plant containers. The long telescopic respiratory processes attached to the posterior of these larvae may be as much as 30 centimetres in length when extended and enable the larvae to live in substrates that are putrid and oxygen-starved. Rat-tailed maggots of the tribe Eristalini are filter feeders on suspended detritus, although it has been suggested that some species 'rasp' surfaces, including that of plants containing phytotelmata (Hartley 1963).

Larvae of *Eumerus* sp. recorded from axils of *Colocasia* in Java by Sack (1931) belong to a genus which, in western Europe, has phytophagous larvae (Ferrar 1987). In Africa members of this genus have radiated to exploit aquatic habitats in which the larvae have become filter feeders. The trophic role of the species encountered by Sack (1931) is accordingly uncertain. Elsewhere I have speculated (Kitching 1990) that the 'eristaline' larvae found in the bract axils of *Curcuma australasica* were basically phytophagous, scraping the plant epidermis and beginning the plant-detritus-based food web found in those axils (see Chapters 5, 10 and 13). Seifert & Seifert (1979a) recorded larvae of *Copestylum rovaina* feeding in a similar fashion in the bract axils of *Heliconia bihai* in Venezuela.

The only non-milesiine syrphid recorded from phytotelmata is an unnamed syrphine found by Francis Gilbert (pers. comm.) in the water-filled tanks of bromeliads in Venezuela. This species is long-tailed and has a ventral sucker which it uses to anchor itself to the bromeliad leaves.

Table A.12 summarises published records of syrphid larvae from container habitats and draws heavily on the analysis presented by Ferrar (1987) modified on advice from F. Gilbert (pers. comm.) and other members of the World

Fig. A.17 Larva of the lauxaniid fly *Homoneura* sp. (lateral view) (after Ferrar 1987).

Fig. A.18 Puparium of the richardiid fly *Epiplatea hondurana* (dorsal view) (after Ferrar 1987) (larval stages undescribed, Steyskal 1987).

Wide Web syrphid discussion group. Syrphid larvae have been recorded in the full range of phytotelmata although water-filled tree holes provide more than half of these records. Curiosities among the list include the records of *Copestylum* spp. which have been recorded in rot pockets from the cactus *Cereus* in the American desert (Maier 1982) and the four species of *Nepenthosyrphus* which have evolved as specialist inhabitants of *Nepenthes* pitchers in South-east Asia.

Lauxaniidae The Lauxaniidae (Figure A.17) are a large family of flies with recorded breeding media ranging from decomposing leaves, through fungus to animal carcasses (Ferrar 1987). Meijere (1910), according to Beaver (1983), recorded an unidentified species of this family from a species of *Nepenthes* from Java. Obviously this record needs checking and the exact relationship of the lauxaniid to the pitcher remains as uncertain as the trophic role such a larvae might have performed within the pitcher.

Richardiidae The Richardiidae (Figure A.18) are a Neotropical family of schizophoran flies. Machado-Allison *et al.* (1983) recorded a species of *Beebiomyia* from the bract containers of *Heliconia caribaea* in Venezuela. Siefert & Siefert (1976a,b, 1979a) recorded other species of this genus from *H. wagneriana, H. imbricata, H. latispatha* and *H. bihae*. Machado-Allison *et al.* (1983) comment that the larvae feed 'on floral parts, and nectar seems to be the main source of larval food'. Fish (1983) rightly points out that further investigation is needed before these species are admitted as part of the aquatic fauna of phytotelmata.

Table A.12. *The Syrphidae and their association with container habitats*

Taxon	Association with containers	Key reference(s)
Blera	Detritus in tree holes and stumps	Maier 1982
Brachyopa sp.	Tree holes	Speight 1974
Brachypalpoides bicolor	Tree rot hole	Heiss 1938
Callicera sp.	In pine rot holes, USA	Perry & Stubbs 1978
Ceriona sp.	Sap-soaked detritus in tree holes	Maier 1982
Chalcosyrphus spp.	Tree holes	Maier 1982
Copestylum ernesta	*Heliconia* bract axils	Seifert & Seifert 1976a
Copestylum spp.	Rot pockets in cactus	Maier 1982
Criorina umbratilis	Tree holes, USA	Johannsen 1935
Eristalis sp.	Bromeliads, Guyana, Jamaica	Smart 1938, Laessle 1961
Eristalis sp.	Bamboos, Japan	Kurihara 1983
Eristalis sp.	Bamboos, Indonesia	Thienemann 1934
Eristalis transversus	Tree holes, USA	Snow 1958
Eumerus parallelus	*Colocasia* axils, Java	Sack 1931, Thienemann 1934
Graptomyza longqishanica	Bamboo stems, China	Huang & Cheng 1993
Graptomyza sp.	Bamboo internodes, Malaysia	Kovac & Streit 1996
Leptomyia sp.	Bromeliads, Florida	Fish 1976
Mallota bautias	Tree holes, USA	Fashing 1975
Mallota cimbiciformis	Tree holes, Europe	Coe 1953
Mallota posticata	Tree holes, USA	Johannsen 1935, Fashing 1976
Meromacrus acutus	Tree holes, USA	Snow 1958
Meromacrus spp.	Bromeliads, Florida, Brazil	Fish 1976
Milesia virginiensis	Tree holes, USA	Snow 1958; Maier 1982
Myiatropa florea	Tree holes, Europe	Johannsen 1935, Kitching 1971
Myolepta nigra	Tree holes	Johannsen, 1935
N. malayanus	*Nepenthes* sp.	Beaver 1979a

N. oudemansi	*N. ampullaria, N. rafflesiana*	Beaver 1979a
N. venustus	*Nepenthes* sp.	Beaver 1979a
Nepenthosyrphus capitatus	*N. reinwardtiana, N. tobaica*	Thienemann 1932, Beaver 1979
Pocota personata	Tree holes	Johannsen 1935
Quichiana calathea	*Heliconia* axils, Costa Rica	Naeem 1990
Quichiana picadoi	Bromeliads, Costa Rica	Picado 1913, Knab 1913b
Quichiana sp.	*Heliconia* axils, Costa Rica	Seifert & Seifert 1976a
Somula sp.	Wet detritus in rot holes	Teskey 1976; Maier 1982
Spilomyia sp.	Wet detritus in rot holes	Maier 1982
Xylota nemorum	Tree holes, USA	Johannsen 1935
Xylota sp.	*N. ampullaria*	Beaver 1979a
Xylota sp.	Tree holes, USA	Hartley 1961; Perry and Stubbs 1978
Chrysogasterinae genus?	Bamboo internodes, Peru	Louton et al. 1996
Eristalinae genus?	Bamboo internodes, Peru	Louton et al. 1996
'eristaline' species	*Curcuma* bract axils, New Guinea	Kitching 1991
'eristaline' species	*Colocasia* axils, Java	Thienemann 1934
'microeristalis'	*Commelina* axils, Indonesia	Thienemann 1934

Modified from Ferrar (1987).
Specific records have not been placed within a higher classification of the family because of present taxonomic uncertainties (see text).

Fig. A.19 Larva of the micropezid fly *Badisis ambulans* (lateral view) (after Yeates 1992).

Fig. A.20 Larva of the sphaerocerid fly *Leptocera* sp. (lateral view) (after Ferrar 1987).

Micropezidae The Micropezidae (Figure A.19) are a small family of stilt-legged flies, the larvae of which generally inhabit rotting wood or other vegetable matter or fruit (Oldroyd 1964). Just one species is known from phytotelmata. The species *Badisis ambulans* of the subfamily Eurybatinae is an apparently obligate associate of the Western Australian pitcher plant *Cephalotus follicularis*. Yeates (1992) records the larvae feeding on the ant *Iridomyrmex conifer* species group, drowned in the pitchers. The larvae are apparently truly aquatic and have non-functional posterior spiracles. The species is recorded by Clarke (1985) as an early invader of pitchers and Yeates (1992) records up to 25 larvae in a single pitcher with larvae present 'in almost all mature pitchers'. The apterous adult fly mimics living individuals of the same species of ant.

Sphaeroceridae The Sphaeroceridae (Figure A.20) are a group of small flies (sometimes known as the Borboridae) whose larvae inhabit, generally, dung and fouled detritus. They are presumably microsaprophages, feeding upon smaller particles of detritus. *Leptocera (Limosina) bromeliarum* was recorded by Knab & Malloch (1912) in Mexican bromeliads and by Malloch (1914) from bract axils of *Heliconia* from Costa Rica. Thienemann (1934) found larvae of *Cypselosoma flavinotata* in *Musa* axils in South-east Asia. A further species of *Leptocera* has been noted as a regular inhabitant of the pitchers of the western American pitcher plant *Darlingtonia californica* (Szerlip 1975).

Drosophilidae A large and important family of flies, the drosophilids (Figure A.21) are largely fungivorous or frugivorous as larvae. Their occurrence in phytotelmata is somewhat problematical but Thienemann (1934), working in Indonesia, noted larvae of species of *Paradrosophila* from the leaf axils

Fig. A.21 Larva of the drosophilid fly *Drosophila melanogaster* (lateral view) (after Ferrar 1987).

Fig. A.22 Larva of the anthomyiid fly *Limnophora* sp. (lateral view) (after Pennak 1953).

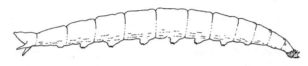

Fig. A.23 Larva of the neurochaetid fly *Neurochaeta inversa* (lateral view) (after Ferrar 1987).

of *Commelina obliqua, Pseudodrosophila curvicapillata* in banana axils, and *Acanthophila hypocausta* in bamboos. It seems unlikely that these are phytotelm specialists.

Anthomyiidae The Anthomyiidae (Figure A.22) are a small family of 'flower-flies' with largely herbivorous larvae. I am aware of only one phytotelm record: that of *Atherigona excisa* within axils of *Colocasia indica* in Sumatra (Thienemann 1934). It is likely that this occurrence is also fortuitous, implying no specialisation on the part of the fly species.

Neurochaetidae The Neurochaetidae (Figure A.23) are distinguished among the Insecta as the only family which appears to be composed entirely of phytotelm specialists. McAlpine erected the family in 1978 to contain species he referred to as 'upside down flies' reflecting the adult habit of perching and running while inverted on their host plants (or supposed host plants in many cases). Originally erected for three species from Australia, Madagascar and Zimbabwe, respectively, the family has now had additional New Guinean (McAlpine 1988a), Malaysian (Woodley 1982) and Malagasy (McAlpine 1993) species added to it. McAlpine (1988b, 1993) describes aspects of the biology and evolution of the family.

Fig. A.24 Larva of the periscelid fly *Stenomicra orientalis* (dorsal view) (after Ferrar 1987).

Fig. A.25 Larva of the chloropid fly *Botanobia darlingtoniae* (lateral view) (after Johannsen 1935).

The larvae of *Neurochaeta inversa* are found associated with the water-filled axils of the inflorescence spathe in *Alocasia macrorrhizos* in Australia and, it is speculated, feed on the micro-organisms that also occur within the spathe axils (McAlpine 1978). Other adult neurochaetids have been found associated with the water-filled inflorescences of *Zingiber* in Malaysia, the water-filled axils of *Pandanus* and *Musa* in New Guinea, and the water-filled axils of *Strelitzia* and *Ravenala* in Madagascar. McAlpine (1993) infers, reasonably, that these phytotelmata represent the larval habitats for members of the family and discusses the likely coevolution of the neurochaetids and their host plants within the Zingiberales, Arecales and Pandanales (see Chapter 2, Table 2.5).

Periscelididae The Periscelididae (Figure A.24) are another small and little known family of flies, the larvae of which are best known as inhabitants of decomposing sap flows from tree wounds. One genus, however, *Stenomicra*, is restricted to phytotelmata (Khoo 1984). As summarised by Fish (1983), the genus, which he ascribes to the Aulacigastridae, is known to occur in bromeliads and in the axil waters of *Pandanus*, sugar cane (Williams 1936), the grass *Coix lacrymajobi* (Swezey 1932), *Musa* spp. (Fish & Soria 1978) and the aracean *Xanthosoma* sp. Fish (1976) observes that larvae of *Stenomicra* are predatory and records them feeding upon the larvae of the bromeliad mosquitoes, *Wyeomyia* spp.

Chloropidae The Chloropidae (Figure A.25) are a widespread and very large family, the best known member of which is *Oscinella frit*, a major pest of pastures and cereals. Chloropid larvae display the full range of life styles, acting as predators or even parasites, through herbivores, to fungivores and decomposers (well summarised by Ferrar 1987).

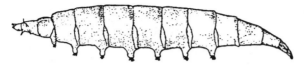

Fig. A.26 Larva of the muscid fly *Graphomya maculata* (lateral view) (after Ferrar 1987).

In phytotelmata two records exist, both from pitcher plants. Paulian (1961) records an unidentified species of chloropid from pitchers of *Nepenthes madagascarensis* in Madagascar. The trophic habits of this species are unknown. In addition, Rymal and Folkerts (1982) report a 'previously undescribed species' from pitchers of *Sarracenia leucophylla* on the Gulf Coast of the USA. These authors describe the larvae as burrowing within the prey mass inside the pitchers, presumably feeding as saprophages.

Muscidae The house-flies and their relatives typify the large fly family, Muscidae (Figure A.26), the habits and habitats of which are diverse. According to Skidmore (1985) the larvae of most species are at least facultatively carnivorous although many can be reared exclusively on vegetable material. Significantly Johannsen (1935) includes no muscids in his exhaustive review of the aquatic Diptera. The family is customarily divided into thirteen subfamilies but only the Coenosiinae and Phaoniinae contain species that have been recorded from phytotelmata.

From the Coenosiinae, Picado (1913) records a species of *Coenosia* from bromeliads in Costa Rica and Fish (1976) records one of *Neodexiopsis* from similar habitats in Florida. Based on the known habits of members of their respective genera (Skidmore 1985), both these species are likely to be predators upon other fly larvae. A species of the genus *Graphomya* was recorded by Kitching (1990) from water-filled bract axils of the inflorescences of *Curcuma australasica* in New Guinea. The larvae were undoubtedly predatory, feeding on all available dipterous larvae within the system including those of eristaline syrphids. Another species, *G. maculata*, has been collected from water-filled rot holes in trees in Great Britain where the larvae feed on syrphid larvae and puparia. Skidmore (1985) notes that this species is more likely to be found in any habitat in which syrphid (particularly eristaline) larvae occur, rather than being a specific associate of rot holes. Kovac *et al.* (1997) describe a further species from water-filled bamboo internodes in Indonesia. They record the species as preying upon a wide variety of the phytotelm inhabitants but feeding principally upon culicid larvae.

Within the Phaoniinae, Meijere (1910) records *Phaonia nepenthicola* from

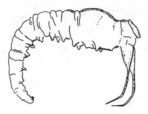

Fig. A.27 Larva of the calliphorid fly *Wilhemina nepenthicola* (lateral view) (after Johannsen 1935).

Nepenthes pitchers in Java, where the larvae prey upon other dipterous larvae breeding in the pitchers. Tate (1935) observed larvae of *P. exoleta* (recorded by him as *P. mirabilis*) feeding on mosquito larvae in rot holes in England.

Muscid larvae also occurred occasionally in samples taken from water-filled tree holes in tropical forests of North Queensland. Levels of occurrence and abundance were sufficiently low to discount the species as a regular member of the aquatic food webs involved.

Calliphoridae The Calliphoridae, or blowflies (Figure A.27), are an immense and cosmopolitan family. The larval habitats of most of its members are carrion, dung and rotting waste and they occur as parasites of both vertebrates and invertebrates. As with the Sarcophagidae (see below) the general association of the family with habitats rich in decomposing animal material appears to have set the scene for the invasion of the water bodies in pitcher plants by one or two highly specialised members of the family.

Two species of calliphorid have been described from *Nepenthes* pitcher plants in South-east Asia, feeding directly on the drowned prey within the pitchers. Schmitz & Villeneuve (1932) described the larvae of *Wilhelmina nepenthicola* from pitchers in Borneo. These are distinctive due to their possession of two ventrally directed caudal processes of unknown function. Clarke & Kitching (1993) also record this species from *N. bicalcarata* in Brunei. A second calliphorid, *Nepenthomyia malayana*, is described by Kurahashi & Beaver (1979) from pitchers of *N. ampullaria* in West Malaysia. The maggot-like larvae are more typically calliphorid, lacking the greatly extended caudal processes of *Wilhelmina*.

Sarcophagidae The Sarcophagidae (Figure A.28), perhaps inappropriately known as flesh-flies, are generally thought of as dung or carrion feeders and parasites. Their larvae generally have the advanced, 'maggot' form of the higher Diptera and most species are larviparous: that is, the eggs either hatch *in utero* or immediately upon laying. In either case species of this family are

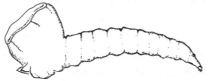

Fig. A.28 Larva of the sarcophagid fly *Sarcophaga dux* (lateral view) (after Johannsen 1935).

admirably pre-adapted to take advantage of the availability of food resources wherever and whenever they may occur.

Suprisingly a number of species have become highly specialised aquatic inhabitants of pitcher plants in both the Old and New Worlds. These species have evolved a circum-spiracular process resembling an 'inverted umbrella' (Fish 1983) which folds over the spiracles when the larvae submerge.

In pitchers of *Nepenthes* spp. four species have been recorded and all appear to be regular 'biontic' members of the infaunal community, acting as direct feeders on drowned carrion within the pitchers. Shinonaga & Beaver (1979) described *Pierettia urceola* as a new species from *N. albomarginata* in West Malaysia. Beaver (1979a) recorded the species as being cannibalistic (or at least fatally antagonistic to conspecifics) although *not* generally predatory. An earlier record of '*Sarcophaga*' sp. from *N. sanguinea* is due to Lever (1956). *Sarcosolomonia pauensis* is recorded by Yeates *et al.* (1989) from pitchers of *N. mirabilis* in Cape York, Australia. These authors record up to two or three larvae from single pitchers. The congeneric *S. carolinensis* was decribed by Souza-Lopez (1958) from *Nepenthes* sp. from the Micronesian island of Palau. Larvae recorded by Kitching & Pimm (1985) as 'syrphines' from pitchers of *N. mirabilis* from Cape York are undoubtedly *S. papuensis*. Similarly the 'syrphines' recorded by me from *N. maxima* from Sulawesi (Kitching 1987b) are probably sarcophagids.

Another suite of sarcophagid species act as generalist carrion feeders in North American pitchers of the genus *Sarracenia*. Aldrich (1916) records four species of *Fletcherimyia* from these situations, one of which, *fletcheri*, has been recorded many times by other authors (but usually under the generic designation *Blaesoxipha*) (Rymal & Folkarts 1982, Fish 1983, Laird 1988). Aldrich (1916) also records *Sarraceniomyia sarraceniae* and *Wohlfartiopsis utilis* from *Sarracenia* pitchers. As Fish (1983) indicates, the general host-plant relationships of these North American species are poorly understood. More than ten years later this appears still to be the case. The family has not been recorded, hereto, from the much less well studied American pitcher genera, *Darlingtonia* and *Heliamphora*.

Tachinidae The tachinids are virtually all parasitoids, infesting other arthropods of various kinds. Some records are available from phytotelmata but, as in the case of parasitoidal Hymenoptera (see below) it is almost certainly inappropriate to regard these as aquatic. *Succingulum fransseni* is recorded by Beaver (1983) from pitchers of *Nepenthes mirabilis* in Borneo. It seems likely that adult tachinids lay eggs above the level of the liquid to be encountered by some of the larger dipterous larvae as they leave the fluid to pupate.

Coleoptera

The vast Order Coleoptera has radiated to fill almost every conceivable ecological role with almost 170 families acting variously as parasitoids, predators, saprophages, xylophages, phytophages and fungivores. A few families are exclusively aquatic and others are amphibious. In terms of the numbers of inquiline species, the Order has not dominated phytotelm communities the way the Diptera has, although in some tree hole situations scirtid larvae are abundant and dominate the metazoan fauna in terms of biomass.

Records of five beetle families properly belong with the aquatic fauna of phytotelmata and these are collated in Table A.13. Other beetle families have members that are specifically associated with container plants, such as the hispine chrysomelids (*Cephaloleia* sp. and *Xenerascus* sp.) associated with *Heliconia* bract axils (Seifert & Seifert 1976a), but these can only be regarded as associated herbivores, not part of the aquatic food web within these phytotelmata.

Carabidae The ground beetles are among the most familiar and best known of beetles. More than 21 000 species of the family have been described (Kryzhanovsky 1976) spread across eight subfamilies and 78 tribes (Ball 1979). The vast majority of these species are predacious and wholly terrestrial. However, guilds of riparian species are well known (Spence 1979) and so the existence of facultatively aquatic species in bromeliad tanks is perhaps not surprising.

Laessle (1961) records three species of the carabine genus *Colpodes* from bromeliads in Jamaica. He noted both adults and larvae within water samples from these plants and observed them feeding on the larvae of scirtids within these communities. *Colpodes* is a large and difficult genus of principally tropical beetles within the tribe Agonini. Thiele (1977) reproduces Darlington's (1971) illustration (Figure A.29) of an adult of one of these species showing the extraordinary dorso-lateral flattening of the body: presumed to be an adaptation to interstitial life within the bromeliad plant.

Fig. A.29 The bromeliad-inhabiting carabid beetle *Colpodes* sp. (dorsal and lateral views) (from Darlington 1971).

Dytiscidae The dytiscids are the water beetles proper and as such comprise an enormous cosmopolitan family associated with water bodies of all kinds from water-filled wheel ruts to continental lakes. They are predatory both as larvae and adults, feeding on a wide variety of the invertebrates with which they live. The adults are highly mobile and this means that they may play a significant role within a particular aquatic community without, necessarily, breeding within that water body. I suspect that this may well be the case with the phytotelm records with which I am familiar.

Three species of dytiscids have been described from bromeliads in Trinidad and Guyana (see Table A.13) and other species are known from bromeliads in Brazil and Jamaica. We have encountered adult dytiscids frequently in water-filled tree holes in the lowland rainforests of North Queensland. As Laessle (1961) also suggests, I believe these species are unlikely to be phytotelm specialists, merely exploiting the container water bodies as a source of prey for these generalist predators.

Hydrophilidae This family is cosmopolitan and of moderate size as coleopteran families go. The worldwide fauna of about 2000 species is divided into about 125 genera. The hydrophilids are amphibious beetles occuring as both adults and larvae in freshwater habitats as well as moist vegetation, soil and even dung. Although generally herbivorous as adults, the larvae are all predacious upon other insects and annelids (Booth *et al.* 1990).

Table A.13 summarises 22 records of hydrophilids from phytotelmata representing eleven genera. Hydrophilids have been found in bromeliads, axil waters and tree holes, although they are rarely found in the latter habitat. We encountered them as rare members of the tree hole fauna only in the cool temperate rainforests of southern Australia.

Scirtidae The marsh beetles or Scirtidae (sometimes referred to as the Helodidae, Elodidae, Cyphonidae or Dascillidae) are a small cosmopolitan family of beetles that have characteristic aquatic larvae (Fig. A.30), distinguished by their long filamentous antennae (Arnett 1960). The larvae leave the water body to pupate and the adults are wholly terrestrial. The larvae are probably

Table A.13. *The Coleoptera and their association with container habitats*

Taxon	Association with containers	Key reference(s)
CARABIDAE		
Colpodes punctus	Bromeliads, Jamaica	Laessle 1961
C. bromeliarum	Bromeliads, Jamaica	Laessle 1961
C. darlingtoni	Bromeliads, Jamaica	Laessle 1961
DYTISCIDAE		
Aglymbus bromeliarum	Bromeliads, Trinidad	Scott 1912, Balfour Browne 1938
Copelatus cordylinoides	Bromeliads, Guyana	Balfour Browne 1938, Smart 1938
Copelatus fulviceps	Bromeliads, Guyana	Balfour Browne 1938, Smart 1938
Desmopachria sp. nr. *laevis*	Bromeliads, Jamaica	Laessle 1961
'dytiscid'	Bromeliads, Brazil	Fish & Soria 1978
'dytiscid'	Tree holes, north Queensland	Kitching, Juniper & Jenkins, unpublished
HYDROPHILIDAE		
Coelostoma coaptatum	*Colocasia* axils, Indonesia	Thienemann 1934
Coelostoma sp.	Bromeliads, Brazil	Winder 1977
Cercyon javanus	*Colocasia* axils, Indonesia	Thienemann 1934
C. udus	*Colocasia* axils, Indonesia	Thienemann 1934
C. trossulus	*Musa* axils, Indonesia	Thienemann 1934
C. fulvus	*Musa* axils, Indonesia	Thienemann 1934
C. madidus	*Musa* axils, Indonesia	Thienemann 1934
Dactylosternum hydrophiloides	*Musa* axils, Indonesia	Thienemann 1934
D. seriatum	*Musa* axils, Indonesia	Thienemann 1934
D. wagneri	*Musa* axils, Indonesia	Thienemann 1934
Dactylosternum sp.	Bromeliads, Jamaica	Laessle 1961
Lachnodachnum saundersi	Bromeliads, Brazil	d'Orchymont 1937
?Perochthes sp.	Bromeliads, Costa Rica	Champion 1913
Phaenonotum tarsale	Bromeliads, Costa Rica	Champion 1913

Phaenonotum uncatum	Bromeliads, Brazil	d'Orchymont 1937
Psilodacnum urichi	Bromeliads, Trinidad	d'Orchymont 1937
Pelosoma sumatrense	*Musa* axils, Indonesia	Thienemann 1934
Omicrogiton insularis	*Musa* axils, Indonesia	Thienemann 1934
Noteropagus politus	*Musa* axils, Indonesia	Thienemann 1934
N. occlusus	*Musa* axils, Indonesia	Thienemann 1934
'hydrophilid'	Bromeliads, Brazil	Fish and Soria 1978
'hydrophilid'	Tree holes, southern Australia	Kitching, Jackson & Jenkins, unpublished
SCIRTIDAE (=HELODIDAE)		
Cyphon bromelius	Bromeliads, Costa Rica	Klausnitzer 1980
Cyphon spp.	Bromeliads, Jamaica	Laessle 1961
Prionocyphon serricornis	Tree holes, Europe	Röhnert 1950, Kitching 1971
Prionocyphon niger	Tree holes, Australia and New Guinea	Kitching & Allsopp 1987
Prionocyphon discoideus	Tree holes, USA	Petersen 1953
Elodes pulchella	Tree holes, Kansas	Fashing, 1975
Elodes thoracica	Tree holes, Kansas	Fashing, 1975
Elodes sp.	Bromeliads, Mexico	Zaragoza 1974
Flavohelodes thoracica	Tree holes, Maryland	Stribling & Young 1990
Helodes pulchella/P. discoideus	Tree holes, Pennsylvania	Paradise & Dunson 1997
Pentameria bromeliarum	Bromeliads, Brazil	Friedenreich 1883
Scirtes championi	Bromeliads, Costa Rica	Picado 1913
Scirtes sp.	Tree holes, bamboos, Papua New Guinea	Kitching 1990
Scirtes spp. ('A', 'B' and 'C')	Bamboo internodes, Malaysia	Kovac & Streit 1996
'scirtid'	Bromeliads, Brazil	Fish and Soria 1970
'scirtid'	Bamboos, Papua New Guinea	Kitching 1990
'scirtid'	Tree holes, Sulawesi	Kitching 1987
'cyphonid'	Bamboos, Indonesia	Thienemann, 1934
'cyphonid'	Tree holes, Indonesia	Thienemann, 1934
Scirtidae, genus?	Bamboo internodes, Peru	Louton *et al.* 1996

Fig. A.30 Larva of the scirtid beetle *Prionocyphon* sp. (lateral view) (from Arnett 1960).

saprophagous, using their highly specialised mouthparts to sweep particles of detritus from their substrate. These are filtered and compressed using the complex mouth structures (Beier 1952, Kitching & Allsopp 1987).

Several genera of scirtids are highly characteristic members of bromeliad, bamboo and tree-hole communities (Table A.13) and they may achieve very high levels of larval abundance in some sites and seasons. The scirtids are in sore need of generic revision and the phytotelm species of the complex *Cyphon/Prionocyphon/Elodes/Scirtes* are assigned to these genera out of historical necessity only!

Trichoptera

The caddisflies are an excusively aquatic Order occurring in a very wide range of aquatic habitats, from moist 'splash-zones' through streams, rivers, seepage pools, ponds and lakes. The larvae show a variety of feeding habits from filter feeders upon suspended detritus, through detritivores and herbivores, to free-living predators. They are rare inhabitants of phytotelmata but one species has been recorded consistently from some bromeliads. The calamoceratid *Phylloicus bromeliarum* was described by Müller from bromeliads in Brazil (Müller 1878, Scott 1914). Other members of the family are known to be detritus feeders as larvae (Neboiss 1991). The only other record I am aware of is that of a limnephilid (another saprophagous family), *Frenesia difficilis*, from pitchers of *Sarracenia purpurea* in the USA (Brower & Brower 1970). Rymal and Folkerts (1982) imply that this occurrence was accidental and that the primary habitat of the species is outside the pitchers. Günther's (1913) record of a trichopteran from Sri Lankan pitchers in fact refers to a psychid moth (see below).

Lepidoptera

The general biology and taxonomy of the order Lepidoptera have received two excellent treatments in recent years, by Common (1990) and Scoble (1992), and my general remarks in this section are based on these accounts. As an Order they are less diverse, ecologically, than the beetles and the majority of species are plant feeders as both larvae and adults. The larvae of some species, however, feed on fungi and/or leaf litter, some on animal detritus and faeces, and a few are predatory.

The larvae of a number of species, spread across several families, are aquatic or semi-aquatic as larvae. The essential problem with such a life style for a terrestrial group such as the Lepidoptera is maintaining gaseous exchange while feeding underwater. Scoble (1992) reviews the mechanisms used by lepidopterous larvae to circumvent this problem. Some carry bubbles or layers of air with them, returning to the surface to renew these (or sometimes using bubbles generated by their food plants), others employ direct oxygen exchange across their cuticles, and a third category has gills or gill-like structures like many other aquatic insects. Only the nymphuline pyralids have become fully aquatic and it is species of this subfamily that have developed gill-like structures. Nymphulines are not recorded from phytotelmata.

Larvae from four lepidopteran families are known from phytotelmata although many other species live on the foliage of the container plants, even within pitchers in some cases. The records described here are only of those species that are known or suspected to spend a major part of their larval existence actually in the water bodies themselves. All of the Lepidoptera recorded from phytotemata actually feed on parts of the container plants and, as such, may be regarded as inimicable to the infaunal community overall. Once a hole is established in the container its liquid contents will be reduced or even eliminated, to the obvious disadvantage of the aquatic organisms within the container.

Psychidae The Psychidae are a family of tineoid moths distinguished, in part, by the larval habit of the secretion of a silken bag around them which then becomes covered in particles of vegetation, sand or other material and within which the larvae live. Many species feed on lichen although others are more conventional chewers upon vegetation (Scoble 1992).

Günther (1913) describes a new genus and species of case-bearing moth, *Nepenthophilus tigrinus,* from pitchers of the Sri Lankan *Nepenthes distillatoria.* Günther in fact described the species as a trichopteran and presents a photograph of the 'case' of the species composed of arthropod skeletal components, presumably gleaned from the prey mass within the pitchers. It is not entirely clear what the feeding habits of this species are but it seems most likely to be eating the inner surfaces of the pitchers, as do some other moths elsewhere in the range of *Nepenthes* (see below). Erber (1979) expresses doubts as to the habitat specificity of this species.

Acrolophidae The Acrolophidae are an exclusively Neotropical group of tineoid moths that exhibit a range of 'unusual' habits as larvae. Most feed on plant roots and tunnel in soil to achieve this end. Others eat decaying plant material and even faeces in rodent and reptile burrows.

Beutelspacher (1972) described *Acrolophus vigia* from larvae collected on the bromeliads *Aechmea bracteata*, *A. mexicana*, *Vriesia gladiolifolia* and *V. chiapensis* in various parts of Mexico. He indicates that the larvae feed on the bromeliad leaves and are 'semi-aquatic'. In the same paper he draws attention to a number of other bromeliad-feeding Lepidoptera but these are conventionally terrestrial in habit.

Yponomeutidae The yponomeutids, in general, feed on leaf tissues, often spinning silken webs over their host plants and skeletonising the leaves beneath. Other species feed upon buds, stems, and plant storage bodies.

Very recently, Neilsen and McQuillan have discovered an extraordinary giant species of this family living in the water-filled leaf axils of the alpine Tasmanian epacrid, *Richea pandaniifolius*, feeding on the living meristems of the plant and, indeed, causing the characteristic branching commonly seen in large specimens of this very long-lived, slow-growing plant. The species is in process of being described in a new genus (E. S. Neilsen and P. McQuillan, pers. comm).

Noctuidae The Noctuidae are an enormous cosmopolitan family of macro-moths. Within the subfamily Acontiinae the large genus *Eublemma* is widespread in South-east Asia. As larvae, many feed conventionally on plant materials but some prey upon scale insects and other Homoptera, and members of a closely related genus feed on detritus in spider webs. One species of *Eublemma, radda*, feeds inside the pitchers of *Nepenthes bicalcarata, mirabilis* and *rafflesiana* in Thailand, peninsular Malaysia, Borneo and Sumatra (Beaver 1983). In our studies of *N. mirabilis* in Brunei a high proportion of pitchers in some locations contained single larvae of *Eublemma* feeding on the green tissue of the pitcher. In North America the genus *Exyra* displays a similar relationship with pitcher plants of the genus *Sarracenia* (Folkerts & Folkerts 1996).

Gracillariidae The Gracillariidae are a world-wide family of small moths which, for the most part, are leaf-miners as larvae. Hering (1931) described the species *Phyllocnistis nepenthae* from pitchers of *Nepenthes tobaica* in Sumatra. The mines apparently extend beneath the surface of the liquid within the pitchers. Beaver (1983) assigns this species to a separate family, the Phyllocnistidae – a distinction described by Scoble (1992) as unnecessary.

Hymenoptera

The ants, bees and wasps are quintessentially terrestrial and hence seem unlikely candidates for membership in the aquatic food webs of phytotelmata.

Beaver (1983) presents a short list of parasitoidal species which have been bred from Diptera and Arachnida associated with pitchers of *Nepenthes* spp. However, Beaver (1983) specifically designates these as terrestrial organisms, which presumably attack the prepupae or pupae of the aquatic Diptera after they leave the aquatic medium.

There remains one exception to this terrestrial 'rule'. Beccari (1884) described the curious association between the Bornean pitcher plant *Nepenthes bicalcarata* and an inquiline ant later described by Schuitemaker & Stärke (1933) as *Camponotus (Colobopsis) schmitzi*. The ants invariably build a nest (or subnest) within the swollen tendril of the upper pitchers and are to be found sheltering beneath the peristome within the pitchers. Many species of ant forage over the external surfaces of the pitchers, seeking the extra-floral nectaries that are scattered over the surface. For their pains, many end up as part of the prey mass within the pitcher fluid.

In general it was supposed that a mutualistic interaction existed between *C. schmitzi* and *N. bicalcarata* – an idea predicated on the provision by the plant of an ant domicile. The benefit to the plant was assumed to be one of protection by the ants of the pitcher tissue against herbivory. As recently as 1990, Hölldobler and Wilson repeated this idea in their masterwork on *Ants*. It was left to Charles Clarke to demonstrate the real nature of the relationship. In fact the ants forage nowhere but within the fluid-body of the pitcher, removing oversized prey items and from time to time preying upon the dipterous larvae which live within the pitcher fluid. The ants actually swim in the fluid, acting co-operatively to remove prey items from the liquid back to the peristome where they are broken up. The comminutive activities of the ants appear to permit digestion by the plant of prey items which otherwise would have been overlarge and lead to putrescence and, even, pitcher death (Clarke 1992, 1997; Clarke & Kitching 1993, 1995). *C. schmitzi* forms a part of the *aquatic* food web within *N. bicalcarata* acting as a top predator in these situations (see Chapters 5, 7).

ARACHNIDA

Acari

The mites are perhaps the least well known component of the phylum Arthropoda. The group as a whole is extremely diverse in morphology and ecological role and is ubiquitous across terrestrial and freshwater ecosystems. It has been suggested that, like the Nematoda and the Fungi, the mites represent one of the great unknowns in terms of biodiversity with a suspected richness that may rival that of the Insecta (May 1990, Hammond 1992).

The mites present challenges to the ecologist for several reasons, not least among which is the fact that they transcend the lower limits of size of most of the other arthropods and accordingly are often overlooked and/or ignored in community studies. This is unfortunate given the potential dynamic importance of mites as decomposers, herbivores, parasites and predators. Most studies of phytotelm communities have noted the presence of mites within containers but then have largely ignored them in favour of the other arthropods present. However, a reasonably complete picture of the range of forms which may be expected in these habitat types has begun to emerge largely due to the systematic and ecological work of Barry O'Connor and Norman Fashing. Table A.14 summarises known records and was compiled with the assistance of Dr Fashing.

Perhaps suprisingly the great range of families within the Acarina as a whole has not translated into a great diversity of phytotelm records. Only six of the dozens of known mite families are recorded from phytotelmata. Three of these are in the Order Astigmata, two the Prostigmata, and but one from the Mesostigmata. Records of at least twenty species exist and, undoubtedly, many more species await discovery especially within tropical tree holes which have been little studied.

O'Connor (1994) provides a review of the life histories of astigmatid species, in general and, in the same volume, Fashing (1994a) focuses on the astigmatid inhabitants of tree holes. Among other notable general treatments, Vitzhum (1931) treats the mites from *Nepenthes* species in South-east Asia.

Histiostigmatidae The astigmatid Histiostigmatidae (previously known as the Anoetidae) are one of a group of families which prefer 'moist habitats' (Evans 1992) ranging from the interior of cells in bee nests, through carrion to genuinely aquatic situations. Fully half of all mites recorded from phytotelmata are histiostigmatids. As a group they appear to be microsaprophages living on the dead and decaying vegetable and animal matter within containers, or the associated micro-organisms. The species *Homosianoetus mallotae*, according to Fashing (1994a), is a filter feeder, straining micro-organisms from the water of tree holes using its specially modified mouthparts.

Records of histiostomatids exist from *Nepenthes* and *Sarracenia* pitchers, water-filled tree holes and bromeliads. Representatives of the genera *Creutzeria, Zwickia* and *Sarraceniopsis* appear to be present in virtually every pitcher examined. Like other mites the histiostigmatids face an ecological problem because of their flightlessness when coupled with the discrete and scattered nature of phytotelm habitat units. Within this family the deutonymphs attach to the legs of emerging Diptera and, in this fashion, are carried between habitat units. The precise relationships involved are not well known in general but Fashing (1973) has described *H. mallotae* using adults of the syrphid species *Mallota* as a dispersal agent. It shares this habit with co-occuring acarid and algophagid species (see below). Again like these species, *H. mallotae* is also larviparous. Fashing (1975) suggests this is an adaptation to their tree hole habitats within which saprophagous species of insect are abundant and in which mite eggs would, accordingly, be at substantial risk.

Acaridae The Acaridae are an astigmatid family, the members of which are saprophagous, fungivorous, phytophagous and graminivorous (Evans 1992) but, although often occuring in moist places, they are not generally aquatic (Fashing 1975). Fashing (1975) drew attention to two tree-hole-inhabiting species from the eastern USA which *are* fully aquatic. He erected a new sub-family, the Naiadacarinae, to accommodate these two species of *Naiadacarus* (Fashing 1974). Subsequently Nesbitt (1985) described an additional acarid, *Bromelioglyphus monteverdensis* from Neotropical bromeliad containers. Species of *Naiadacarus* appear to be fully saprophagous, actively skeleton-ising decaying leaves. Deutonymphs of this species 'hitchhike' upon syrphids to move between tree holes (see Chapter 6).

Algophagidae The Algophagidae (earlier referred to as the Hyadesiidae) complete the set of astigmatid families known from phytotelmata. Three tree-hole-inhabiting species are known to date: one from the USA and two undescribed species from Australia. The American species is relatively well known. Fashing & Campbell (1992) have studied the feeding behaviour of this and other tree-hole mites (see comments above). The algophagid *Algophagus pennsylvaticus* apparently is a grazer, feeding upon decaying vegetable matter within the tree-hole environment. Again the species is known to use syrphid flies as agents of dispersal.

Arrhenuridae The arrhenurids are one of two families of Prostigmata which have been encountered in phytotelmata. Members of the family are the highly distinctive predatory water mites of ponds and lake margins, readily recognised by their bright red, near spherical facies. Arrhenurids are predacious upon other arthropods and, in water-filled tree holes at least, appear to be generalists, feeding upon anything small enough for them to subdue. Unlike the astigmatids, nymphal arrhenurids are obligate ectoparasites of arthropods and, in this fashion, achieve transportation from habitat unit to habitat unit. Known hosts include mosquitoes and chironomid midges.

Species of *Arrhenurus* are recorded from water-filled tree holes in Australia and from bromeliads in Jamaica (Kitching & Callaghan 1982, Laessle 1961).

Table A.14. *The Acari (mites) and their association with container habitats*

Taxon	Site	Authors
ASTIGMATA		
HISTIOSTIGMATIDAE		
Anoetus sp.	Bromeliads, Florida	Fish 1976
Creutzeria seychellensis	*Nepenthes pervillei*, Seychelles	Nesbitt 1979
Creutzeria tobaica	*Nepenthes*, South-east Asia, Madagascar	Oudemans 1932, Paulian 1961
Hormosianoetus mallotae	Water-filled tree holes, USA	Fashing 1973
Hormosianoetus sp.	Bromeliads, Americas	O'Connor 1994
Sarraceniopus gibsoni	*Sarracenia purpurea*, USA	Nesbitt 1954, Rymal & Folkerts 1982
Sarraceniopsis hughesi		Hunter & Hunter 1964
S. darlingtoniae	*Darlingtonia* pitchers, USA	Fashing & O'Connor 1984
Zwickia guentheri	*Nepenthes* spp., Java, Sumatra, Sri Lanka	Vitzhum 1931, Thienemann 1932
Zwickia nepenthesiana	*Nepenthes* spp., South-east Asia generally	Hirst 1928, Vitzhum 1931, Thienemann 1932
Zwickia spp.	*Nepenthes*, 6 species, Borneo	Clarke & Kitching 1993
Tyroglyphus sp.	Bromeliads, Costa Rica	Picado 1913
Histiostoma 2 spp.	*Heliconia* axils, Costa Rica	Naeem 1990
ACARIDAE		
Naiadacarus arboricola	Water-filled tree holes, USA	Fashing 1975
N. oregonensis	Water-filled tree holes, USA	Fashing 1975
Bromelioglyphus monteverdensis	Bromeliads, Costa Rica	Nesbitt 1985
Naiacus muertensis	Bromeliads, Costa Rica	Nesbitt 1990
Rhizoglyphus sp.	Bromeliads, Americas	O'Connor 1994
Schwiebia sp.	Bromeliads, Americas	O'Connor 1994
ALGOPHAGIDAE		
Algophagus pennsylvaticus	Water-filled tree holes, USA	Fashing & Wiseman 1980
Fusohericia sp.	*Heliconia* axils, Costa Rica	Naeem 1990

'hyadesiid sp.' (2 species)	Water-filled tree holes, Australia	Kitching & Callaghan 1982
PROSTIGMATA		
ARRHENURIDAE		
Arrhenurus kitchingi	Tree holes, Australia	Kitching & Callaghan 1982;
Arrhenurus sp.	Bromeliads, Jamaica	Laessle 1961
ANISITSIELLIDAE		
Anisitsiella sp.	*Richea* axils, Tasmania	M. S. Harvey and R. L. Kitching, unpublished
MESOSTIGMATA		
ASCIDAE		
Cheiroseius sp.	Water-filled tree holes, Australia	Kitching & Callaghan 1982

Anisitsiellidae The Anisitsiellidae are a family of prostigmatid mites with a distinctly Gondwanan distribution. Members of the family are generally interstitial and, like arrhenurids, adults are predators on other arthropods. Harvey (1990) has revised the Australian members of the family.

In recent studies of the water-filled axils of the extraordinary giant epacrid *Richea pandanifolia* from the alpine zone of Tasmania, I discovered a predatory mite in abundance, together with several species of chironomid (see Table A.8 above). This species is almost undoubtedly ectoparasitic as a nymph upon adults of the chironomid *Parochlus bassianus* with which it invariably co-occurs.

Ascidae The Ascidae is the only mesostigmatid group recorded from phytotelmata. A species of the predatory genus *Cheiroseius* occurs in water-filled tree holes in the subtropical rainforest of south-east Queensland but may be an accidental in this situation.

Phylum: Mollusca

The molluscs are a vast phylum, ubiquitous in marine, freshwater and terrestrial habitats. There are several records of slugs and snails sheltering around the rims of tree holes and bromeliads and in moist associated leaf litter. In addition I found many individuals of an aquatic gastropod, *Gyraulus* sp. nr. *convexiosculus* (Planorbidae), in the water of tree holes in northern Sulawesi. This detritus-feeding species occurred in two very large water-filled holes: one in a log-hole next to a river, the other in an enormous rot hole in the stump of a tree.

Phylum: Chordata

Vertebrates of various kinds make use of phytotelmata either as sources of water or shelter. Picado (1913) records fer-de-lance as an occasional temporary resident of bromeliad tanks and I encountered a brown tree snake (*Boiga irregularis*) in a tree hole in New Guinea, cooling off and escaping biting flies coiled tight beneath the water surface. Birds use tree holes for bathing and as a source of drinking water (Clarke 1997). *Nepenthes* pitcher plants are colloquially referred to as 'monkey cups' in Malaya and 'the places birds drink' in Sulawesi. Many frogs have been recorded resting as adults in bromeliads, tree holes and leaf axils (see for example Haddow 1948).

This last section of my bestiary, however, concerns only those vertebrates that are known (or strongly suspected) to breed in phytotelmata: a far more select list, entirely of frogs.

ANURA

Table A.15 is a compilation of breeding records of frogs in phytotelmata. The list draws heavily on the earlier review of Lannoo *et al.* (1987) and was assembled with the aid of Dr John Cadle of Harvard's Museum of Comparative Zoology and Stephen Richards of James Cook University. This account has also benefited greatly from discussions with Dr Cadle.

Members of ten families have been recorded breeding in phytotelmata (treating the Madagascan Mantellidae as a separate family rather than as a subfamily of the Ranidae). Many of these also breed in other forest pools but a fascinating subset are phytotelm specialists. Many more occurrences will no doubt come to light as our knowledge of the tropical herpetofauna increases. There are many records of phytotelm-breeding frogs from Central and South America, from Madagascar and from South-east Asia. In contrast records from continental Africa and Australia are few. Of course additional records may well come to light as these last two faunas become better known but, in preparing Table A.15, I have worked through several major monographs on, for instance, the amphibians of the Congo, Malawi, Botswana and southern Africa. That, in most instances, these are not referenced is because they added no further examples to the list. The fauna of Australia is also relatively well known (although this is not the case with New Guinea). Accordingly it may well be that the phytotelm-breeding habit is somewhat restricted geographically.

Frog tadpoles present a very varied set of feeding strategies including filter feeding, detritus feeding, herbivory and predation. In many species the tadpoles may not feed at all and, in others, cannibalism is commonplace. The females of some species of *Dendrobates* that breed in bromeliads deposit single fertile eggs in the water bodies and lay infertile trophic eggs in the same locations to act as food for the larvae. Recent observations suggest that some such tadpoles are also predatory on other members of the bromeliad fauna. Other authors have postulated that the same or similar habits may occur in tree-hole and axil-breeding species (see, for examples, Glaw & Vences 1992, Wassersug *et al.* 1981, Wilson *et al.* 1985). All this having been said it is still true to say that the diets of most tadpoles are unknown (J. Cadle *in litt.*). The placing of tadpoles, where they occur, within phytotelm food webs must be done with great care and, ideally, only after study of the biology of the particular species encountered. In the case of those species with non-feeding tadpoles, or those that rely in large part on conspecific trophic eggs for their food supply, the species may be trophically unlinked to the rest of the phytotelm community, although remaining an integral part of that community through less direct interactions.

Table A.15. *The frogs and their association with container habitats*

Taxon	Association with containers	Key reference(s)
MYOBATRACHIDAE		
Lechriodus fletcheri	Water-filled tree holes, Australia	Kitching & Callaghan 1982
LEPTODACTYLIDAE		
Crossodactylodes bokermanni/izecksoni	Bromeliads, Brazil	Peixota 1981, 1983
Eleutherodactylus jamaicensis	Bromeliads, Jamaica	Dunn 1926, Laessle 1961
E. jasperi	Bromeliads, Puerto Rico	Drewry & Jones 1976
Leptodactylus sp. (*wagneri* grp.)	Log holes, Peru	Hoogmoed & Cadle 1991
DENDROBATIDAE		
Colostethus bromelicola	Bromeliads, Venezuela	Dixon & Rivero-Blanco 1985
Dendrobates arboreus	Bromeliads, Panama	Myers et al. 1984
D. auratus	Bromeliads, tree holes, Costa Rica, Panama	McDiarmid & Foster 1975, Silverstone 1975, Zimmermann 1974, E. Rand pers. comm.
D. histrionicus	Bromeliads, Colombia	Silverstone 1973, 1975, Zimmermann & Zimmermann 1985
D. leucomelass	Bromeliads, tree holes, Colombia	Zimmermann & Zimmermann 1980
D. pumilio	Bromeliads, aroid axils, Costa Rica	Starrett 1960, Young 1979, Weygoldt 1980
D. quinquevittatus	Bromeliads, Neotropics	Zimmermann & Zimmermann 1984
D. reticulatus	Bromeliads, Neotropics	Zimmermann & Zimmermann 1984
D. speciosus	Bromeliads, Neotropics	Jungfer 1985
Phyllobates lugubris	Tree holes, bromeliads, Neotropics	Silverstone 1976, Zimmermann 1982
Phyllobates lugubris	Tree holes, bromeliads, Neotropics	Silverstone 1976, Zimmermann 1982
HYLIDAE		
Agalychius calcarifer	Log holes, Costa Rica, Panama	Duellman 1970, Wilson et al. 1985
A. craspedopus	Log holes, Peru	Hoogmoed & Cadle 1991

Species	Habitat	Reference
A. spurelli	Log holes, Panama	Duellman 1970
Anotheca spinosa	Bromeliads, tree holes, Costa Rica	Picado 1913, Taylor 1954, Wilson *et al.* 1985
Calyptohyla crucialis	Bromeliads, Jamaica	Dunn 1926
Hyla bromeliacia	Bromeliads, Honduras, Guatemala	Schmidt 1933, Stuart 1948
H. bromeliana	Bromeliads, Mexico	Taylor 1939
H. brunnea	Bromeliads, Jamaica	Dunn 1926
H. dendroscarta	Bromeliads, Mexico	Taylor 1940, Smith 1941, Duellman 1970
H. lichenata	Bromeliads, Jamaica	Dunn 1926
H. marianae	Bromeliads, Jamaica	Dunn 1926
H. miliaria	Tree holes (probably), Central America	Duellman 1979
H. picadoi	Bromeliads, Costa Rica	Robinson 1977, Duellman 1979
H. salvaje	Tree holes, Honduras	Wilson *et al.* 1985
H. smaragdina	Bromeliads, Mexico	Taylor 1940
H. wilderi	Bromeliads, Jamaica	Dunn 1926, Laessle 1961
H. zetecki	Bromeliads, Costa Rica (probably)	Picado 1913, Dunn 1937, Duellman 1970
Litoria sp.	Tree holes, New Guinea	S. Richards, pers. comm.
Nyctimantis rugiceps	Bamboo stumps, Ecuador	Duellman & Trueb 1976
Ololygon perpusilla	Bromeliads, Brazil	Lutz 1973
Osteocephalus buckleyi	Tree holes, Brazil	Zimmerman & Rodriguez 1990
Osteocephalus sp.	Bromeliad tanks, Brazil	Hero 1990
Osteopilus brunneus	Bromeliads, Jamaica	Dunn 1926, Orton 1944, Lannoo *et al.* 1987
Phyllodytes auratus	Bromeliads, Trinidad	Kenny 1969
P. luteolus	Bromeliads, Brazil	Bokermann 1966
P. tuberculosus	Bromeliads, Brazil	Bokermann 1966
Phrynohyas venulosa	Tree holes, bromeliads, bamboos, S. America	Duellman 1971
P. resinifictrix	Tree holes, Brazil	Zimmerman & Rodriguez 1990, Hero 1990

RHACOPHORIDAE

Species	Habitat	Reference
Edwardtayloria spinosa	Tree holes, Philippines	Brown & Alcala 1983
Nyctixalus spinosus	Tree holes, Borneo, Philippines	Inger 1966, Wassersug *et al.* 1981
Nyctixalus pictus	Tree holes, Thailand	Taylor 1962, Alcala & Brown 1982

Table A.15. (*cont.*)

Taxon	Association with containers	Key reference(s)
P. lissobrachius	Fern and *Pandanus* leaf axils, Philippines	Brown & Alcala 1983
Philautus schmackeri	*Alocasia* axils, Philippines	Inger 1954, Liem 1970
Philautus sp.	*Pandanus* axils, Philippines	Brown & Alcala 1983
Philautus sp.	Tree holes, Thailand	Wassersug *et al.* 1981
Philautus spp.	Pitcher plants, Sabah	Inger & Steubing 1989
Rhacophorus colleti	Bamboos, Indonesia	Thienemann, 1934
R. harissoni	Tree holes, Borneo	Inger 1966, Wassersug *et al.* 1981
Theloderma gordoni	Tree holes, Indochina	Duellman & Trueb 1986
T. horridum	Tree holes, Thailand	Boulenger 1903, Wassersug *et al.* 1981
T. stellatum	Tree holes, Thailand	Wassersug *et al.* 1981
BUFONIDAE		
Bufo sp.	Log hole, Borneo	Inger 1985
Mertensophryne micranotis	Tree holes	Grandison 1980, Grandison & Ashe 1983
Pelophryne brevipes	Leaf axils, Philippines, Borneo	Brown & Alcala 1983, Inger 1985
Stephopaedes anotis	Tree holes, Zimbabwe, Mozambique	Poynton & Broadley 1988
RANIDAE		
Rana palavanensis	Tree holes, Borneo	Inger 1985
HYPEROLIIDAE		
Acanthixalus spinosus	Tree holes, Cameroon	Savage 1952, Inger 1966, Perret 1962, 1966,
Callixalus pictus	Bamboos, Central Africa	Laurent 1964
MANTELLIDAE		
Mantella laevigata	Tree holes, Madagascar	Glaw & Vences 1992
Mantidactylus bicalcaratus	Tree holes, Madagascar	Blommers-Schlösser 1979, Razahelisoa 1974
M. pulcher	*Pandanus* axils, Madagascar	Glaw & Vences 1992

M. bicalcaratus	*Pandanus* axils, Madagascar	Glaw & Vences 1992
M. punctatus	*Pandanus* axils, Madagascar	Glaw & Vences 1992
M. flavobrunneus	*Pandanus* axils, Madagascar	Glaw & Vences 1992
MICROHYLIDAE		
Anodonthyla boulengeri	Tree holes, *Ravenala* axils, Madagascar	Glaw & Vences 1992
A. nigrogularis	Tree holes, Madagascar	Glaw & Vences 1992
A. rouxi	Bamboos, Madagascar	Glaw & Vences 1992
Chaperina fusca	Log holes, Borneo, Philippines	Inger 1956, 1966, Brown & Alcala 1983, Parker 1934
Cophyla phyllodactyla	Bamboos, Madagascar	Glaw & Vences 1992
Hoplophryne rogersi	Bamboos, banana axils	Noble 1929
Hoplophryne uluguruensis	Bamboos, banana axils	Noble 1929
Kalophrynus pleurostigma	Log holes, Sabah; bamboos, Malaysia	Inger & Steubing 1989, Kovac & Streit 1996
Microhyla borneensis	Log holes, Borneo	Inger 1985
M. perparva	Log holes, Borneo	Inger 1985
M. petrigena	Log holes, Borneo	Inger 1985
Metaphrynella sundana	Tree holes, Borneo	Inger 1966, Inger & Steubing 1989
Paracophyla tuberculata	Tree holes, Madagascar	J. Cadle, pers. comm.
Platypelis grandis	Tree holes, Madagascar	Glaw & Vences 1992
P. barbouri	Tree holes, Madagascar	Glaw & Vences 1992
P. tuberifera	*Pandanus* axils, Madagascar	Glaw & Vences 1992
P. milloti	*Ravenala* axils, Madagascar	Glaw & Vences 1992
P. pollicaris	Tree holes, Madagascar	J. Cadle, pers. comm.
Plethodontohyla notosticta	Tree holes, Madagascar	Glaw & Vences 1992
P. inguinalis	Tree holes, Madagascar	J. Cadle, pers. comm.
Ramanella triangularis	Tree holes, India	Inger *et al.* 1985

General comments upon the habits of the families given in the following account are based on those of Duellman & Trueb (1986) for the first six families, and upon Glaw & Vences (1992) for the Mantellidae and Microhylidae.

Myobatrachidae

The Myobatrachidae are an Australasian family of ground-dwelling frogs with a wide range of life histories. Some lay eggs on land which then produce aquatic tadpoles, others have tadpoles which develop wholly in foam nests and yet others are brooded in pouches or even within the alimentary canals of the adult frogs. The family is combined with the Leptodactylidae by some authors.

Leptodactylidae

The Leptodactylidae are a New World tropical family that have radiated to fill many ecological roles in the region. The more primitive termatobiines are fully aquatic whereas the more advanced leptodactylines make foam nests on vegetation or have terrestrial eggs which release tadpoles into adjacent water bodies.

Dendrobatidae

The Dendrobatidae are an exclusively Neotropical family of four genera and about 120 species. They are sometimes combined with the Ranidae. There appears to be some measure of parental care in many species and adults will remain with or return to eggs. The resulting tadpoles may be carried to water on the backs of the adults.

Hylidae

The Hylidae are tree frogs which occur throughout the Americas, southern Europe, Asia and Australasia. Duellman & Trueb (1986) recognise four subfamilies. The hylines and pelodryadines, in general, have aquatic eggs and tadpoles. A few species from these subfamilies and all of the Phyllomedusinae lay eggs in foam nests from whence tadpoles drop into water bodies. Members of the fourth subfamily, the Hemiphractinae, brood their eggs on their backs.

Most of the phytotelm hylids are true tree frogs belonging to the Hylinae but three pelodryadine genera are known to breed in water-filled tree holes, and one in bromeliads (Hero 1990).

Rhacophoridae

The Rhacophoridae are principally an Oriental family of arboreal frogs. They generally build foam nests in which eggs are deposited and may develop. In some cases the tadpoles may leave the foam nest and drop into water beneath to complete their development. Madagascan species are fully aquatic (Glaw & Vences 1992). Eleven South-east Asian species are known from phytotelmata (Table A.15).

Bufonidae

The bufonids are the true toads and their close relatives. The family is a large one with a natural distribution encompassing all continents except Australia and Antarctica. In general, toads lay strings of eggs in still water bodies. Tadpoles of *Pelophryne*, however, have an abbreviated life history, surviving solely on yolk. Three records of phytotelm bufonids are available: two from Southeast Asian phytotelms, and one from southern African tree holes. These include a species of *Pelophyrne* from moist leaf axils in Borneo and the Philippines.

Ranidae

The ranids are another large family of 'typical' frogs distributed worldwide, although entering only the northern parts of South America and Australia. Of the four subfamilies recognised (if the Mantellidae of Madagascar are accorded family status), three are restricted to the African continent. Only the Raninae are widely distributed. Like the bufonids, the ranids, in general, have life cycles associated with still or slow-moving water bodies. A small group of Asiatic species, however, have direct development and these include *Platymantis* spp. The records of this genus from leaf and fern axils in the Philippines (Brown & Alcala 1983) are unlikely to be associated with water bodies (S. Richards, pers. comm.). *Rana palavanensis*, however, is recorded from water-filled tree holes in Borneo by Inger (1985).

Hyperoliidae

The Hyperoliidae are an African and Malagasy family which may have 'normal' aquatic eggs or may use arboreal foam nests or container habitats as breeding sites. Two species are recorded from phytotelmata – tree holes and bamboos – in East and Central Africa.

Mantellidae

This family, closely related to the Ranidae and Rhacophoridae, is endemic to Madagascar and contains three genera, two of which are commonly phytotelm specialists. Mantellids in general may be arboreal or terrestrial but, in any case, never lay eggs directly into the water bodies in which their tadpoles live (Glaw & Vences 1992). One species of *Mantella* is known from tree holes and four species of *Mantidactylus* from water-filled *Pandanus* axils. Adults of other species of *Mantidactylus* often rest in *Pandanus* axils, although they may breed elsewhere.

Microhylidae

The microhylids are a widely distributed pan-Tropical family that, in Madagascar at least, have adopted phytotelmata as a common breeding site. Other species of the family are terrestrial and yet others develop totally in foam nests in vegetation. Glaw & Vences (1992) record species breeding in tree holes, bamboos and plant axils (Table A.15).

References

Adam, P. (1992). *Australian Rainforests*. Clarendon Press, Oxford.

Addicott, J. F. (1974). Predation and prey community structure: an experimental study of the effect of mosquito larvae on the protozoan communities of pitcher plants. *Ecology* **55**, 475–492.

Alcala, A. C. & Brown, W. C. (1982). Reproductive biology of some species of *Philautus* (Rhacophoridae) and other Philippine anurans. *Philippine Journal of Biology* **11**, 203–220.

Aldrich, J. M. (1916). Sarcophaga *and allies in North America*. Thomas Say Foundation, Lafayette.

Alexander, C. P. (1915). A second bromeliad-inhabiting crane-fly (Tipulidae, Diptera). *Entomological News* **26**, 29–30.

Alexander, C. P. (1920). The crane-flies of New York II. Biology and phylogeny. *Cornell University Agricultural Experimental Station Memoirs* **38**, 691–1133.

Alexander, C. P. (1929). The genus *Sigmatomera* Osten Sacken with observations on the biology by Raymond C. Shannon. *Encyclopedie Entomologique, Serie B*, **5**, 155–162.

Alexander, C. P. (1931). Deutsche Limnologische Sunda-expedition. The crane-flies (Tipulidae). *Archiv für Hydrobiologie, Supplementumband* **9**, 135–191.

Alpatoff, W. W. (1922). Epiphytengewässer und ihren fauna. *Russkii Gidrobiologicheskii Zhurnal* **1**, 164–166.

Arnett, R. H. (1960). *The Beetles of the United States*. Catholic University of America Press, Washington.

Balfour-Browne, J. (1938). On two new species of bromeliadicolous *Copelatus* (col. Dytiscidae). *Entomologist's Monthly Magazine*, **74**, 100–102.

Ball, G. E. (1979). Conspectus of carabid classification: history, holomorphology, and higher taxa. In: *Carabid Beetles: their evolution, natural history and classification*, (eds. T. L. Erwin, G. E. Ball, D. R. Whitehead & A. L. Halpern), Junk, the Hague, pp. 61–111.

Barr, A. R. & Chellapah, W. T. (1963). *The mosquito fauna of pitcher plants in Singapore*. Singapore Medical Journal **4**, 184–185.

Barrera, R., Fish, D. & Machado-Allison, C. E. (1989). Ecological patterns of aquatic insect communities in two *Heliamphora* pitcher-plant species of the Venezuelan highlands. *Ecotropicos* **2**, 31–44.

Basham, E. H., Mulrennan, J. A. & Obermuller, A. J. (1947). The biology and distribution of *Megarhinus* Robineau-Desvoidy in Florida. *Mosquito News* **7**, 64–66.

385

Bates, M. (1949). *Natural History of Mosquitoes*. Macmillan, New York.

Baumgartner, D. L. (1986). Failure of mosquitoes to colonize teasel axils in Illinois. *Journal of the American Mosquito Control Association* **2**, 371–373.

Bazzaz, F. A. (1979). The physiological ecology of plant succession. *Annual Review of Ecology and Systematics*, **10**, 351–371.

Beattie, M. V. F. & Howland, L. J. (1929). The bionomics of some tree-hole mosquitoes. *Bulletin of Entomological Research* **20**, 45–58.

Beaver, R. A. (1972). Ecological studies on Diptera breeding in dead snails. I. Biology of the species found in *Cepaea nemoralis* (L.). *Entomologist* **105**, 41–52.

Beaver, R. A. (1973). The effects of larval competition on puparial size in *Sarcophaga* spp. (Diptera: Sarcophagidae) breeding in dead snails. *Journal of Entomology* (A) **48**, 1–9.

Beaver, R. A. (1979a). Biological studies of the fauna of pitcher plants (*Nepenthes*) in West Malaysia. *Annales de la Société entomologiques de France* (N. S.) **15**, 3–17.

Beaver, R. A. (1979b). Description of the male and larva of *Endonepenthia schuitemakeri* Schmitz from *Nepenthes* pitchers (Diptera, Phoridae). *Annales de la Société entomologiques de France* (N. S.) **15**, 19–25.

Beaver, R. A. (1980). Fauna and food webs of pitcher plants in West Malaysia. *Malayan Nature Journal* **33**, 1–10.

Beaver, R. A. (1983). The communities living in *Nepenthes* pitcher plants: fauna and food webs. In: *Phytotelmata: Terrestrial Plants as Hosts of Aquatic Insect Communities* (eds. J. H. Frank & L. P. Lounibos), Plexus, Medford, pp. 129–159.

Beaver, R. A. (1985). Geographical variation in food web structure in *Nepenthes* pitcher plants. *Ecological Entomology* **10**, 241–248.

Beccari, O. (1884). Piante ospitatrice, ossia piante formicarie della Malesia e della Papuasia. *Malesia* (Genoa) **2**, 1–340.

Beccari, O. (1904). *Wanderings in the Great Forests of Borneo*. Archibald Constable, London.

Begon, M., Harper, J. L. & Townsend, C. R. (1990). *Ecology – Individuals, Populations, Communities*, 2nd Edn., Blackwell Scientific Publications, Oxford.

Beier, M. (1952). Bau und Funktion der Mundwerkzeuge bei den Helodiden-Larven (Col.) *Transactions of the Ninth International Congress of Entomology* **1**, 135–138.

Benick L, (1924). Zur Biologie Der Käferfamilie Helodidae (mit einer Übersicht der Baumhöhlenfauna von Prof. Dr. A. Thienemann) *Mitteliegung Geographische Gesellschaft Naturihistoriche Muzeum von Lübeck*, **29**, 47–78.

Bennett, I. (1966). *The Fringe of the Sea*. Rigby, Adelaide.

Benzing, D. H., Derr, J. A. & Titus, J. E. (1972). The water chemistry of microcosms associated with the bromeliad *Aechmea bracteata*. *American Midland Naturalist* **87**, 60–70.

Bernardi, R. de, Giussani, G. & Manca, M. (1987). Cladocera: predators and prey. *Hydrobiologia* **145**, 225–243.

Beutelspacher, C. R. (1971). Una bromeliácea como ecosistema. *Biologia* **2**, 82–87.

Beutelspacher R. C. (1972). Some observations on the Lepidoptera of bromeliads. *Journal of the Lepidopterists' Society* **26**, 133–137.

Blacklock, B. & Carter, H. F. (1920). Observations on *Anopheles (Coelodiazesis) plumbeus* Stephens with special reference to breeding places, occurrence in the

Liverpool district and possible connection with the spread of malaria. *Annals of Tropical Medicine and Parasitology* **13**, 421–444.

Blommers-Schlösser, R. M. A. (1979). Biosystematics of the Malagasy frogs. I. Mantellinae (Ranidae). *Beaufortia* **29**, 1–77.

Bockermann, W. C. A. (1966). O gênero *Phyllodotes* Wagler, 1830 (Anura, Hylidae). *Anais de Academia Brasileira de Ciências* **38**, 335–344.

Bonnet, D. D. & Hu, S. M. K. (1951). The Introduction of *Toxorhynchites brevipalpis* Theobald into the Territory of Hawaii. *Proceedings of the Hawaiian Entomological Society* **14**, 237–243.

Booth, R. G., Cox, M. L. & Madge, R. B. (1990). *IIE Guides to Insects of Importance to Man 3. Coleoptera*. Natural History Museum, London.

Borobev, B. A. (1960). Dipterous larvae inhabiting water lying in the leaf axils of the teasel. *Entomological Review* **39**, 579–580.

Boulenger, G. A. (1903). Report on the batrachians and reptiles. *Fascicula Malayana, Zoology* **1**, 131–176.

Bradshaw, W. E. (1976). Geography of photoperiodic response in a diapausing mosquito. *Nature,* **262**, 384–386.

Bradshaw, W. E. (1980). Thermoperiodism and the thermal environment of the pitcher-plant mosquito, *Wyeomyia smithii*. *Oecologia (Berlin)* **46**, 13–17.

Bradshaw, W. E. (1983). Interaction between the mosquito, *Wyeomyia smithii*, the midge, *Metriocnemus knabi*, and their carnivorous host, *Sarracenia purpurea*. In: *Phytotelmata: Terrestrial Plants as Hosts of Aquatic Insect Communities* (eds. J. H. Frank & L. P. Lounibos), Plexus, Medford, pp. 161–189.

Bradshaw, W. E. & Creelman, R. A. (1984). Mutualism between the carnivorous purple pitcher plant and its inhabitants. *American Midland Naturalist* **112**, 294–304.

Bradshaw, W. E. & Holzapfel, C. M. (1975). Biology of tree-hole mosquitoes: photoperiodic control of development in northern *Toxorhynchites rutilus* (Coq.) *Canadian Journal of Zoology* **53**, 889–893.

Bradshaw, W. E. & Holzapfel, C. M. (1977). Interaction between photoperiod, temperature and chilling in dormant larvae of the tree-hole mosquito *Toxorhyncites rutilus* Coq. *Biological Bulletin* **152**, 147–158.

Bradshaw, W. E. & Holzapfel, C. M. (1983). Predator-mediated non-equilibrium coexistence of tree-hole mosquitoes in south-eastern North America. *Oecologia (Berlin)* **57**, 239–256.

Bradshaw, W. E. & Holzapfel, C. M. (1984). Seasonal development of treehole mosquitoes and chaoborids in relation to weather and predation. *Journal of Medical Entomology* **21**, 366–378.

Bradshaw, W. E. & Holzapfel, C. M. (1985). The distribution and abundance of tree-hole mosquitoes in eastern North America: perspectives from north Florida. In: *Ecology of Mosquitoes: Proceedings of a Workshop* (eds. L. P. Lounibos, J. R. Rey & J. H. Frank), Florida Medical Entomology Laboratory, Vero Beach, pp. 3–23.

Bradshaw, W. E. & Holzapfel, C. M. (1986a). Geography of density-dependent selection in pitcher-plant mosquitoes. In: *The Evolution of Insect Life-cycles* (eds. F. Taylor & F. Karban), Springer-Verlag, New York, pp. 48–65.

Bradshaw, W. E. & Holzapfel, C. M. (1986b). Habitat segregation among European tree-hole mosquitoes. *National Geographic Research* **2**, 167–178.

Bradshaw, W. E. & Lounibos, L. P. (1972). Photoperiodic control of development in the pitcher-plant mosquito, *Wyeomyia smithii*. *Canadian Journal of Zoology* **50**, 13–719.

Bradshaw, W. E. & Lounibos, L. P. (1977). Evolution of dormancy and its photoperiodic control in pitcher-plant mosquitoes. *Evolution* **21**, 546–567.

Brandt, A. von (1934). Untersuchungen in Baumhöhlengewässern auf *Fagus sylvatica. Archiv für Hydrobiologie* **27**, 546–563.

Brauns, A. (1954). *Terricole Dipterenlarven.* Musterschmidt Wissenschaftliche Verlag, Göttingen.

Brehm, V. (1925). Hängenden Aquarien in der Pflanzenwelt. *Mikrokosmos* **19**, 1–6.

Breland, O. P. (1949). The biology and immature stages of the mosquito, *Megarhinus septentrionalis* Dyar & Knab. *Annals of the Entomological Society of America* **42**, 38–47.

Briand, F. & Cohen, J. E. (1984). Community food webs have scale-invariant structure. *Nature (London)* **307**, 264–266.

Briand, F. & Cohen, J. E. (1987). Environmental correlates of food chain length. *Science (New York)* **238**, 956–960.

Brower, J. H. & Brower, A.E. (1970). Notes on the history and distribution of moths associated with the pitcher-plant in Maine. *Proceedings of the Entomological Society of Ontario* **101**, 79–83.

Brown, W. C. & Alcala, A. C. (1983). Modes of reproduction of Philippine anurans. In: *Advances in Herpetology and Evolutionary Biology* (eds. A. G. J. Rhodin & K. Miyata), Museum of Comparative Zoology, Cambridge, Massachusetts, pp. 416–428.

Brundin, L. (1966). Transantarctic relationships and their significance, as evidenced by chironomid midges with a monograph of the subfamilies Podonominae and Aphroteniinae and the austral Heptaginae. *Kunglica Svenska Vetenskapsakademiens Handlingar* **11**, 1–472.

Buffington, J. D. (1970). Ecological considerations of the cohabitation of the pitcher plant by *Wyeomyia smithii* and *Metriocnemus knabi. Mosquito News* **30**, 89–90.

Burger, J. F. (1977). The biosystematics of immature Arizona Tabanidae (Diptera). *Transactions of the American Entomological Society* **103**, 145–258.

Calvert, P. P. (1910). Plant-dwelling odonate larvae. *Entomological News* **22**, 365–366.

Calvert, P. P. (1911). Studies on Costa Rican Odonata. II. – The habits of the plant-dwelling larva of *Mecistogaster modestus. Entomological News* **22**, 402–411.

Cameron, C. J., Donald, G. L. & Paterson, C. G. (1977). Oxygen-fauna relationships in the pitcher plant *Sarracenia purpurea* L. with reference to the chironomid *Metriocnemus knabi* Coq. *Canadian Journal of Zoology* **55**, 2018–2023.

Carlisle, A., Brown, A. H. F. & White, E. J. (1966). The organic matter and nutrient elements in the precipitation beneath a sessile oak (*Quercus petraea*) canopy. *Journal of Ecology* **54**, 87–98.

Carpenter, S. R. (1982). Stemflow chemistry: effects on population dynamics of detritivorous mosquitoes in tree-hole ecosystems. *Oecologia (Berlin)* **53**, 1–6.

Carpenter, S. R. (1983). Resource limitation of larval tree hole mosquitoes subsisting on beech detritus. *Ecology* **64**, 219–223.

Carpenter, S. R., Kitchell, J. F. & Hodgson, J. R. (1985). Cascading trophic interactions and lake productivity. *BioScience* **35**, 634–639.

Carpenter, S. R., Kitchell, J. F., Hodgson, J. R., Cochran, P. A., Elses, J. J., Elser, M,. M., Lodge, D. M., Kretchmer, D., He, X. & von Ende, C. N. (1987). Regulation of lake primary productivity by food web structure. *Ecology* **68**, 1863–1876.

Carpenter, S. R., Kraft, C. E, Wright, R., He, X., Soranno, P. A. & Hodgson, J. R. (1992). Resilience and resistance of a lake phosphorus cycle before and after food web manipulation. *American Naturalist* **140**, 781–798.

Carpenter, S. R., Frost, T. M., Ives, A. R., Kitchell, J. F. & Kratz, T. K. (1994). Complexity, cascades and compensation in ecosystems. In: *Biodiversity: its Complexity and Rôle* (eds M. Yasuno & M. M. Watanabe). Global Environmental Forum, Tokyo, pp. 197–207.

Chambers, R. C. (1985). Competition and predation among larvae of three species of treehole breeding mosquitoes. In: *Ecology of Mosquitoes: Proceedings of a Workshop* (eds. L. P. Lounibos, J. R. Rey & J. H. Frank), Florida Medical entomology Laboratory, Vero Beach, pp. 25–54.

Champion, G. C. (1913). Coleoptera, &c. in bromeliads. *Entomologist's Monthly Magazine*, **49**, 2–7.

Chapman, H. C.(1964). Observations on the biology and ecology of *Orthopodomyia californica* Bohart (Diptera: Culicidae). *Mosquito News* **24**, 432–439.

Chow, C. Y. (1949). Observations on mosquitoes breeding in plant containers in Yunnan. *Annals of the Entomological Society of America* **42**, 465–470

Clarke, C. M. (1992). *The ecology of metazoan communities in* Nepenthes *pitcher plants in Borneo, with special reference to* Nepenthes bicalcarata *Hook. f.* PhD Thesis, University of New England, Armidale.

Clarke, C. M. (1997). *The* Nepenthes *of Borneo*. Natural History Publications, Kota Kinabalu.

Clarke, C. M. (1998). A re-examination of geographical variation in *Nepenthes* foodwebs. *Ecography* **21**, 430–436.

Clarke, C. M. & Kitching, R. L. (1993). The metazoan food webs of six Bornean *Nepenthes* species. *Ecological Entomology* **18**, 7–16.

Clarke, C. M. & Kitching, R. L. (1995). Swimming ants and pitcher plants: a unique ant-plant interaction from Borneo. *Journal of Tropical Ecology* **11**, 589–602.

Clarke, S. A. (1985). *Demographic aspects of the pitcher of* Cephalotus follicularis *(Labill.) and development of the contained community*. PhD Thesis, University of Western Australia, Perth.

Clay, M. E. & Venard, C. E. (1971). Diapause in *Aedes triseriaus* (Diptera: Culicidae) larvae terminated by molting hormones. *Annals of the Entomological Society of America* **64**, 968–970.

Clay, M. E. & Venard, C. E. (1972). Larval diapause in the mosquito *Aedes triseriatus:* effects of diet and temperature on photoperiodic induction. *Journal of Insect Physiology* **18**, 1441–1446.

Clements, A. N. (1992). *The Biology of Mosquitoes Vol. 1. Development, Nutrition and Reproduction*. Chapman and Hall, London.

Clements, F. E. (1916). *Plant Succession: an Analysis of the Development of Vegetation*. Carnegie Institution Publication 242, Washington, D.C.

Closs, G. (1991). Multiple definitions of food web statistics – an unnecessary problem for food web research. *Australian Journal of Ecology* **16**, 413–415.

Coe, R. L. (1953). Diptera: Syrphidae. *Handbooks for the Identification of British Insects,* **10**, 1–98.

Cohen, J. E. (1978). *Food Webs and Niche Space*, Princeton University Press, Princeton.

Cohen, J. E. & Newman, C. M. (1985). A stochastic theory of community food webs I. Models and aggregated data. *Proceedings of the Royal Society of London B* **224**, 421–461.

Cohen, J. E., Newman, C. M. & Briand, F. (1985). A stochastic theory of community food webs II. Individual webs. *Proceedings of the Royal Society of London* B **224**, 449–461.

Cohen, J. E., Briand, F. & Newman, C. M. (1986). A stochastic theory of community food webs I. Predicted and observed lengths of food chains. *Proceedings of the Royal Society of London* B **228**, 317–353.

Cohen, J. E., Beaver, R. A., Cousins, S. H., DeAngelis, D. L., Goldwasser, L., Heong, K. L., Holt, R. D., Kohn, A. J., Lawton, J. H., Martinez, N., O'Malley, R., Page, L. M., Patten, B. C., Pimm, S. L., Polis, G. A., Rejmanek, M., Scoener, T. W., Scoenly, K., Sprules, W. G., Teal, J. M., Ulanowicz, R. E., Warren, P. H., Wilbur, H. M. & Yodzis, P. (1993). Improving food webs. *Ecology* **74**, 252–258.

Colinvaux, P. (1978). *Why Big, Fierce Animals are Rare*. Princeton University Press, Princeton.

Colless, D. H. (1986). The Australian Chaoboridae (Diptera). *Australian Journal of Zoology Supplement* **124**, 1–66.

Colless, D. H. & McAlpine, D. K. (1991). Diptera (Flies). In: *Insects of Australia*, 2nd Edition, (ed. I. Naumann), Melbourne University Press, Melbourne, pp. 717–786.

Colyer, C. N. & Hammond, C.O. (1951). *Flies of the British Isles*. Frederick Warne & Co., London.

Common, I. (1990). *The Moths of Australia*. University of Melbourne Press, Melbourne.

Connell, J. H. (1980). Diversity and coevolution of competitiors, or the ghost of competition past. *Oikos* **35**, 131–138.

Connell, J. H. & Slatyer, R. O. (1977). Mechanisms of succession in natural communities and their role in community stability and organisation. *American Naturalist* **111**, 1119–1144.

Conover, A. (1994). Bamboo and katydid: their surprising 'inside' story. *Smithsonian* **25**, 120–128.

Copeland, R. S. & Craig, G. B. (1990). Habitat segregation among treehole mosquitoes (Diptera: Culicidae) in the Great Lakes region of the United States. *Annals of the Entomological Society of America* **83**, 1063–1073.

Copeland, R. S. & Craig, G. B. (1992). Interspecific competition, parasitism and predation affect development of *Aedes hendersoni* and *A. triseriatus* in artificial treeholes. *Annals of the Entomological Society of America* **85**, 154–163.

Corbet, P. S. (1961). Entomological studies from a high tower in Mpanga Forest, Uganda. XII. Observations on Ephemeroptera, Odonata and some other orders. *Transactions of the Royal Entomological Society of London* **113**, 356–361.

Corbet, P. S. (1963a). *The Biology of Dragonflies*. Witherby, London.

Corbet, P. S. (1963b). Observations on *Toxorhynchites brevipalpis conradti* Gruub. (Diptera: Culicidae) in Uganda. *Bulletin of Entomological Research* **54**, 9–17.

Corbet P. S. (1964). Observations on mosquitoes ovipositing in small bamboo containers in Zika forest, Uganda. *Journal of Animal Ecology* **33**, 141–164.

Corbet, P. S. (1983). Odonata in phytotelmata. In: *Phytotelmata: Terrestrial Plants as Hosts of Aquatic Insect Communities* (eds. J. H. Frank & L. P. Lounibos), Plexus, Medford, pp. 29–54.

Corbet, P. S. & Griffiths A. (1963). Observations on the aquatic stages of two species of *Toxorhynchites* (Diptera: Culicidae) in Uganda. *Proceedings of the Royal Entomological Society of London (A)*, **38**, 125–135.

Corbet, P. S. & McCrae, A. W. R. (1981). Larvae of *Hadrothemis scabrifrons* (Ris) in a tree cavity in East Africa (Anisoptera: Libellulidae). *Odonatolgica* **10**, 311–317.

Cotgreave, P., Hill, M. J. & Middleton, D. A. J. (1993). The relationship between body size and population size in bromeliad tank faunas. *Biological Journal of the Linnaean Society* **49**, 367–380.

Cousins, S. H. (1987). The decline of the trophic level concept. *Trends in Ecology and Evolution* **2**, 215–245.

Craig, G. B. (1983). Biology of *Aedes triseriatus:* some factors affecting control. In: *California Serogroup Viruses* (eds. C. H. Calsiher & W. H. Thomson), Liss, New York, pp. 329–341.

Cranston, P. S. (1995). Systematics. In: *The Chironomidae: the Biology and Ecology of Non-biting Midges* (eds. P. Armitage, P. S. Cranston & L. C. V. Pinder), Chapman & Hall, London, pp. 31–61.

Cranston, P. S. & Judd, D. D. (1987). *Metriocnemus* (Diptera: Chironomidae) – An ecological survey and description of a new species. *Journal of the New York Entomological Society* **95**, 534–546.

Cranston, P. S. & Kitching, R. L. (1994). The Chironomidae of Austro-oriental phytotelmata (plant-held waters): *Richea pandanifolia* Hook. f. In: *Chironomids from Genes to Ecosystems* (ed. P. Cranston), CSIRO, Melbourne, pp. 225–231.

Cranston, P. S., Edward, D. H. D. and Colless, D. H. (1987). *Archaeochlus* Brundin: a midge out of time (Diptera: Chironomidae). *Systematic Entomology* **12**, 313–334.

Cuenot, L. (1949). Les Tardigrades. In: *Traité de Zoologie*, Volume 6, (ed. P. P. Grassé), pp. 39–59.

Darlington, P. J. (1971). The carabid beetles of New Guinea. Part IV. General considerations, analysis and history of fauna. Taxonomic Supplement. *Bulletin of the Museum of Comparative Zoology* **142**, 129–337.

DeAngelis, D. L. (1992). *Dynamics of Nutrient Cycling and Food Webs.* Chapman & Hall, London.

Disney, R. H. L. (1981). A new species of *Megaselia* from *Nepenthes* in Hong Kong, with reevaluation of genus *Endonepenthia* (Diptera: Phoridae). *Oriental Insects* **12**, 201–206.

Disney, R. H. L. (1982). A new species of *Megaselia* (Diptera: Phoridae) which breeds in pitchers of *Nepenthes* in Sri Lanka. *Ceylon Journal of Science (Biological Science)* **14**, 89–101.

Disney, R. H. L. (1983). Scuttle Flies Diptera, Phoridae (except *Megaselia*). *Royal Entomological Society of London* **10**, 1–39.

Disney, R. H. L. (1986). Two remarkable new species of scuttle-fly (Diptera: Phoridae) that parasitise termites (Isoptera) in Sulawesi. *Systematic Entomology* **11**, 413–422.

Disney, R. H. L. (1994). *Scuttle Flies: the Phoridae.* Chapman and Hall, London.

Disney, R. H. L. & Wirth, W. W. (1982). A midge (Dipt., Ceratopogonidae) new to Britain from teasel in Suffolk. *Entomologist's Monthly Magazine* **18**, 233–234.

Dixon, J. R. & Rivero-Blanco, C. (1985). A new dendrobatid frog (*Colostethus*) from Venezuela with notes on its natural history and that of related species. *Journal of Herpetology* **19**, 177–184.

D'Orchymont, A. (1937). Sphaeridiini broméliocoles nouveaux (Coleoptera: Hydrophilidae, Sphaeridiinae). *Annals and Magazine of Natural History* (10 ser.) **20**, 127–140.

Downs, W. G. & Pittendrigh, C. S. (1946). Bromeliad malaria in Trinidad, British West Indies. *American Journal of Tropical Medicine and Hygiene* **26**, 47–66.

Downs, W. G., Gillette, H. P. S. & Shannon, R. C. (1943). A malaria survey of Trinidad and Tobago. *Journal of the National Malaria Society* **2**, 3–44.

Drake C. J. & Chapman H. C. (1954). New American waterstriders (Hemiptera). *Florida Entomologist* **37**, 151–155.

Drake C. J. & Harris H. M. (1935). New Veliidae (Hemiptera) from central America. *Proceedings of the Biological Society of Washington* **48**, 191–194.

Drake C. J. & Hussey R. F. (1954). Notes on some American Veliidae (Hemiptera), with the description of two new *Microvelias* from Jamaica. *Florida Entomologist* **37**, 133–138.

Drake C. J. & Maldonado C. J. (1952). Water-striders from Territorio Amazonas of Venezuela (Hemiptera: Hydrometridae, Veliidae). *Great Basin Naturalist* **12**, 47–54.

Drewry, G. E. & Jones, K. L. (1976). A new ovoviviparous frog, *Eleutherodactylus jasperi* (Amphibia, Anura, Leptodactylidae) from Puerto Rico. *Journal of Herpetology* **10**, 161–165.

Duckhouse, D. A. & Lewis, D. J. (1989). Superfamily Psychodoidea. 15. Family Psychodidae. In: *Catalog of the Diptera of the Australian and Oceanian Regions* (ed. N. L. Evenhuis), Bishop Museum, Honolulu, pp. 166–181.

Duellman, W. E. (1970). The hylid frogs of Middle America. *Monograph of the Museum of Natural History, University of Kansas* **1**, 1–753.

Duellman, W. E. (1971). A taxonomic review of South American hylid frogs, genus *Phrynohyas*. *Occasional Papers of the Museum of Natural History, University of Kansas* **4**, 1–21.

Duellman, W. E. (ed.) (1979). The South American herpetofauna: its origin, evolution and dispersal. *Monograph of the Museum of Natural History, University of Kansas* **7**, 1–485.

Duellman, W. E. & Trueb, L. (1976). The systematic status and relationships of the hylid frog *Nyctimantes rugiceps* Boulenger. *Occasional Papers of the Museum of Natural History, University of Kansas*, **58**, 1–14.

Duellman, W. E. & Trueb, L. (1986). *Biology of Amphibians*. McGraw Hill, New York.

Dunn, E. R. (1926). The frogs of Jamaica. *Proceedings of the Boston Natural History Society* **38**, 111–130.

Dunn, E. R. (1937). The amphibian and reptile fauna of bromeliads in Costa Rica and Panama. *Copeia* **1937**, 163–167.

Dussart, B. (1967). *Les Copépodes des Eaux continentales d'Europe occidentale. I. Calanoïdes et Harpacticoïdes*. Editions N. Boubée, Paris.

Edgerly, J. S., Willey, M. S. & Livdahl, T. (1999). Intraguild competition among larval treehole mosquitoes, *Aedes albopictus, Ae. aegypti* and *Ae. triseriatus* (Diptera: Culicidae), in laboratory microcosms. *Journal of Medical Entomology* **36**, 394–399.

Elton, C. S. (1966). *The Pattern of Animal Communities*. Methuen, London.

English, K. M. I., Mackerras, I. M. & Dyce, A. L. (1957). Notes on the morphology and biology of a new species of *Chalybosoma* (Diptera, Tabanidae). *Proceedings of the Linnean Society of New South Wales* **82**, 289–296.

Erber, D. (1979). Untersuchungen zur Biozönose und Nekrozönose in Kannenpflanzen auf Sumatra. *Archiv fur Hydrobiologie* **87**, 37–48.

Erickson, R. (1968). *Plants of Prey*. Lamb, Perth.

Eshita, Y. & Kurihara, T. (1979). Studies on the habitats of *Aedes albopictus* and *Ae. riversi* in the south western part of Japan. *Japanese Journal of Zoology* **30**, 181–185.

Evans, G. O. (1992). *Principles of Acarology*. CAB International, London.

Evans, K. H. & Brust, R. A. (1972). Induction and termination of diapause in *Wyeomyia smithii* (Diptera: Culicidae), and larval survival studies at low and subzero temperatures. *Canadian Entomologist*, **104**, 1937–1950.

Fashing, N. J. (1973). The post-embryonic stages of a new species of *Mauduyta* (Acarina: Anoetidae). *Journal of the Kansas Entomological Society* **46**, 454–468.

Fashing, N. J. (1974). A new subfamily of Acaridae, the Naiadacarinae, from water-filled treeholes (Acarina: Acaridae) *Acarologia*, **16**, 166–181.

Fashing, N. J. (1975). Life history and general biology of *Naiadacarus arboricola* Fashing, a mite inhabiting water-filled treeholes (Acarina: Acaridae). *Journal of Natural History* **9**, 413–424.

Fashing, N. J. (1976). The evolutionary modification of dispersal in *Naiadacarus arboricola* Fashing, a mite restricted to water-filled tree-holes (Acarina: Acaridae). *American Midland Naturalist* **95**, 337–346.

Fashing, N. J. (1981). Arthropod associates of the cobra lily (*Darlingtonia californica*). *Virginia Journal of Science* **32**, 92.

Fashing, N. J. (1994a). Life-history patterns of astigmatid inhabitants of water-filled treeholes. In: *Mites: Ecological and Evolutionary Analyses of Life-History Patterns* (ed. M. A. Houck), Chapman and Hall, New York, pp.160–185.

Fashing, N. J. (1994b). A novel habitat for larvae of the fishfly *Chauliodes pectinicornis* (Megaloptera: Corydalidae). *Banisteria* **3**, 25–26.

Fashing, N. J. & Campbell, D. M. (1992). Observations on the feeding biology of *Algophagus pennsylvanicaus* (Astigmata: Algophagidae), a mite restricted to water-filled treeholes. *International Journal of Acarology* **18**, 77–81.

Fashing, N. J. & O'Connor, B. M. (1984). *Sarraceniopus* – A new genus for histiostomatid mites inhabiting the pitchers of the Sarraceniaceae (Astigmata: Histiostomatidae), *International Journal of Acarology* **10**, 217–227.

Fashing, N. J. & Wiseman, L. L. (1980). *Algophagus pennsylvanicus* – a new species of Hyadesiidae from water-filled treeholes. *International Journal of Acarology* **6**, 79–84.

Ferrar, P. (1987). *A Guide to the Breeding Habits and Immature Stages of Diptera Cyclorrhapha*. Entomonograph **8**, Brill, Leiden & Copenhagen.

Fincke, O. M. (1984). Giant damselflies in a tropical forest: reproductive biology of *Megaloprepus coerulatus* with notes on *Mecistogaster* (Zygoptera: Pseudostigmatidae), *Advances in Odonatology* **2**, 13–27.

Fincke, O. M. (1992a). Interspecific competition in tree holes: consequences for mating systems and coexistence of Neotropical damselflies. *American Naturalist* **139**, 80–101.

Fincke, O. M. (1992b). Consequences of larval ecology for territoriality and reproductive success of male dragonflies. *Ecology* **73**, 449–462.

Fincke, O. M. (1994). Population regulation of a tropical damselfly in the larval stage by food limitation, cannibalism, intraguild predation and habitat drying. *Oecologia (Berlin)* **100**, 118–127.

Fincke, O. M. (1999). Organization of predator assemblages in Neotropical tree holes: effects of abiotic factors and priority. *Ecological Entomology* **24**, 13–23.

Fish, D. (1976). Insect-plant relationships of the insectivorous pitcher-plant *Sarracenia minor*. *Florida Entomologist* **59**, 199–203.

Fish, D. (1977). An aquatic spittle bug (Homoptera; Cercopidae) from a *Heliconia* flower bract in southern Costa Rica. *Entomological News,* **88,** 10–12.

Fish, D. (1983). Phytotelmata: flora and fauna. In: *Phytotelmata: Terrestrial Plants as Hosts of Aquatic Insect Communities* (eds. J. H. Frank & L. P. Lounibos), Plexus, Medford, pp. 161–190.

Fish, D. & Beaver, R. A. (1978). A bibliography of the aquatic fauna inhabiting bromeliads (Bromeliaceae) and pitcher plants (Nepenthaceae and Sararceniaceae). *Proceedings of the Florida Anti-mosquito Association* **49,** 11–19.

Fish, D. & Carpenter, S. L. (1982). Leaf litter and larval mosquito dynamics in tree-hole ecosystems. *Ecology* **63,** 283–288.

Fish, D. & Hall, D. W. (1978). Succession and stratification of aquatic insects inhabiting the leaves of the insectivorous pitcher plant, *Sarracenia purpurea. American Midland Naturalist* **99,** 172–183.

Fish, D. & Soria, S. de J. (1978). Water-holding plants (phytotelmata) as larval habitats or ceratopogonid pollinators of Cacao in Bahia, Brazil. *Revista Theobroma* **8,** 133–146.

Floyd, A. G. (1990). *Australian Rainforests in New South Wales.* Vol. 2., Surrey Beatty, Chipping Norton, NSW.

Focks, D. A. Seawright, J. A., & Hall, D. W. (1979). Field survival, migration and ovipositional characteristics of laboratory-reared *Toxorhynchites rutilus rutilus. Journal of Medical Entomology* **16,** 121–127.

Focks, D. A., Dame, D. A., Cameron, A. L. & Boston, M. D. (1980). Predator-prey interaction between insular populations of *Toxorhynchites rutilus rutilus* and *Aedes aegypti. Environmental Entomology* **9,** 37–42.

Folkerts, D. R & Folkerts, G. W. (1996). Aids for field identification of pitcher plant moths of the genus *Exyra* (Lepidoptera: Noctuidae). *Entomological News* **197,** 128–136.

Frank, J. H. (1983). Bromeliad phytotelmata and their biota, especially mosquitoes. In: *Phytotelmata: Terrestrial Plants as Hosts of Aquatic Insect Communities* (eds. J. H. Frank & L. P. Lounibos), Plexus, Medford, pp. 101–128.

Frank, J. H. & Curtis, G. A. (1977). On the bionomics of bromeliad-inhabiting mosquitoes. VI. A review of bromeliad-inhabiting species. *Journal of the Florida Anti-mosquito Association* **52,** 4–23.

Frank, J. H. & Curtis, G. A. (1981). On the bionomics of bromeliad-inhabiting mosquitoes. VI. A review of the bromeliad-inhabiting species. *Journal of the Florida Anti-Mosquito Association* **52,** 4–23.

Frank, J. H. & Lounibos, L. P. (1983). *Phytotelmata: Terrestrial Plants as Hosts for Aquatic Insect Communities.* Plexus, Medford, N. J.

Frank, J. H. & O'Meara, G. F. (1985). Influence of micro- and macrohabitat on distribution of some bromeliad-inhabiting mosquitoes. *Entomologia experimentia et applicata* **37,** 169–174.

Frank, J. H., Curtis, G. A. & Evans, H. J. (1976). On the bionomics of bromeliad-inhabiting mosquitoes. II. The relationship of bromeliad size to the number of immature *Wyeomyia vanduzeei* and *Wy. medioalbipes. Mosquito News* **37,** 180–192.

Frank, J. H., Curtis, G. A. & O'Meara, G. F. (1984). On the bionomics of bromeliad-inhabiting mosquitoes. X. *Toxorhynchites r. rutilus* as a predator of *Wyeomyia vanduzeei* (Diptera: Culicidae). *Journal of Medical Entomology* **21,** 149–158.

Freeman, P. (1956). A study of the Chironomidae (Diptera) of Africa south of the

Sahara, Part II. *Bulletin of the British Museum (Natural History), Entomology* **4**, 285–366.

Freeman, P. (1983). Sciarid flies, Diptera, Sciaridae. *Handbooks for the Identification of British Insects* **9**, 1–42.

Friedenreich, C. W. (1883). *Pentameria bromeliarum*, eine pentamere Halticide. *Stettiner Entomologische Zeitung*, **44**, 140–144.

Galindo, P. (1958). Bionomics of *Sabethes chloropterus* Humboldt, a vector of sylvan yellow fever in middle America. *American Journal of Tropical Medicine and Hygiene* **7**, 429–440.

Gallopin, G. C. (1972). Structural properties of food webs. In: *Systems Analysis and Simulation in Ecology* Vol. II. (ed. B. C. Patten), Academic Press, New York, pp. 241–282.

George, U. (1989). Venezuela's islands in time. *National Geographic* **175**, 526–561.

Gerberg, E. J. & Visser, W. M. (1978). Preliminary field trial for the biological control of *Aedes aegypti* by means of *Toxorhynchites brevipalpis*. *Mosquito News* **38**, 197–200.

Gilpin, M. & Hanski, I. (eds) (1991). *Metapopulation Dynamics: Empirical and Theoretical Investigations*. Academic Press, London.

Glasby, C. J., Kitching, R. L. & Ryan, P. A. (1990). Taxonomy of the arboreal polychaete *Lycastopsis catarractarum* Feuerborn (Namanereidinae: Nereididae), with a discussion of the feeding biology of the species. *Journal of Natural History* **24**, 341–350.

Glaw, F. & Vences, M. (1992). *Amphibians and Reptiles of Madagascar*. Vences and Glaw Verlags GbR, Cologne.

Gleick, J. (1987). *Chaos: making a new science*. Viking, New York.

Glenn-Lewin, D. C., Peet, R. K. & Veblen, T. T. (eds) (1992). *Plant Succession: Theory and Prediction*. Chapman and Hall, London.

Goodwin, J. T. & Murdoch, W. P. (1974). A study of some immature Neotropical Tabanidae (Diptera). *Annals of the Entomological Society of America* **67**, 85–133.

Goss, R. C., Whitlock, L. S. & Westrick, J. P. (1964). Isolation and ecological observations of *Panagrodontus* sp. (Nematoda: Cepalobidae) in pitcher plants (*Sarracenia sledgei*). *Proceedings of the Heminthological Society of Washington* **31**, 19–20.

Grandison, A. G. C. (1980). Aspects of breeding morphology in *Mertensophryne micranotis* (Anura: Bufonidae): secondary sexual characters, eggs and tadpoles. *Bulletin of the British Museum (Natural History) (Zoology)* **39**, 299–304.

Grandison, A. G. C. & Ashe, S. (1983). The distribution, behavioural ecology and breeding strategy of the pigmy toad, *Mertensophryne micranotis* (Lov.). *Bulletin of the British Museum (Natural History) (Zoology)* **45**, 85–93.

Grayson, A. J. & Jones, E. W. (1955). *Notes on the History of the Wytham Estate with Special Reference to the Woodlands*. Imperial Forestry Institute, Oxford.

Greenslade, J. (1983). Adversity selection and the habitat templet. *American Naturalist* **122**, 352–365.

Grjebine, A. (1979). Les moustiques des *Nepenthes* de Madagascar. Espèces nouvelles du genre *Uranotaenia* Lynch Arribalzaga (Diptera, Culicidae). *Annales de la Société Entomologique de France* (Nouvelle Série) **15**, 53–74.

Grime, J. P. (1979). *Plant Strategies and Vegetation Processes*. John Wiley & Sons, Chichester.

Grimstad, P. R., Garry, C. E. & DeFoliart, G. R. (1974). *Aedes hendersoni* and

Aedes triseriatus in Wisconsin: characterization of larvae, larval hybrids, and comparison of adult and hybrid mesoscutal patterns. *Annals of the Entomological Society of America* **67**, 795–804.

Grimstad, P. R., Craig, G. B., Ross, Q. E. & Yulil, T. M. (1977). *Aedes triseriatus* and LaCrosse virus: geographic variation in vector susceptibility and ability to transmit. *American Journal of Tropical Medicine and Hygiene* **26**, 990–996.

Günther, K. (1913). Die lebenden Bewöhner der Kannen der insektenfressenden Pflanze *Nepenthes distillatoria* auf Ceylon. *Zeitschrift für wissenchaftliche Insektenbiologie* **9**, 90–95, 122–130, 156–160, 198–207, 259–267.

Gurney, R. (1920). Notes on certain British freshwater Entomostraca. *Annals and Magazine of Natural History*. Series 9, **5**, 351–360.

Haddow, A. J. (1948). The mosquitoes of Bwamba County, Uganda. VI. Mosquito breeding in plant axils. *Bulletin of Entomological Research* **39**, 185–212.

Haenni, J-P. & Vaillant, F. (1994). Description of dendrolimnetobiotic larvae of Scatopsidae (Diptera) with a review of our knowledge of the preimaginal stages of the family. *Mitteilungen der Schweizerischen entomologischen Gesellschaft* **67**, 43–59.

Hamilton, A. G. (1904). Notes on the west Australian pitcher plant (*Cephalotus follicularis* LaBill.). *Proceedings of the Linnaean Society of New South Wales* **29**, 36–53.

Hammond, P. M. (1992). Species inventory. In: *Global Diversity: Status of the Earth's Living Resources* (ed. B. Groombridge), Chapman & Hall, London, pp. 17–39.

Hartley, J. C. (1961). A taxonomic account of the larvae of some British Syrphidae. *Proceedings of the Zoological Society of London*, **136**, 505–573.

Hartley, J. C. (1963). The cephalopharyngeal apparatus of syrphid larvae and its relationship to other Diptera. *Proceedings of the Zoological Society of London* **141**, 261–280.

Harvey, M. S. (1990). A review of the water-mite family Anisitsiellidae in Australia (Acarina). *Invertebrate Taxonomy* **3**, 629–646.

Hastings, A. & Harrison, S. (1994). Metapopulation dynamics and genetics. *Annual Review of Ecology and Systematics* **25**, 167–188.

Hauer, Von J. (1923). *Habrotrocha thienemanni* sp. n., ein in den Höhlungen der Buchen lebendes Rädertier. *Archiv für Hydrobiologie* **14**, 585–591.

Havens, K. E. (1993). Predator-prey relationships in natural community food webs. *Oikos* **68**, 117–124.

Havens, K. E. (1994). Experimental perturbation of a freshwater plankton community – a test of hypotheses regarding the effects of stress. *Oikos* **69**, 147–153.

Hawley, W. A. (1985a). A high fecundity aedine: factors affecting egg production of the Western treehole mosquito, *Aedes sierrensis* (Diptera: Culicidae). *Journal of Medical Entomology* **22**, 220–225.

Hawley, W. A. (1985b). Population dynamics of *Aedes sierrensis*. In: *Ecology of Mosquitoes: Proceedings of a Workshop* (eds. L. P. Lounibos, J. R. Rey & J. H. Frank), Florida Medical Entomology Laboratory, Vero Beach, pp. 167–184.

Hawley, W. A., Pumpuni, C. B., Brady, R. H. & Craig, G. B. (1989). Overwintering survival of *Aedes albopictus* (Diptera: Culicidae) eggs in Indiana. *Journal of Medical Entomology* **26**, 122–129.

Hayes, R. O. & Morland, H. B. (1957). Notes on *Aedes triseriatus* egg incubation and colonization. *Mosquito News* **17**, 33–36.

Hegner, R. W. (1926). The Protozoa of the pitcher plant *Sarracenia purpurea*. *Biological Bulletin* (Woods Hole) **50**, 271–276.

Heiss, E. M. (1938). A classification of the larvae and puparia of the Syrphidae of Illinois exclusive of aquatic forms. *Illinois Biological Monographs*, **16**, 1–142.

Hemmerlein, A. H. & Crans, W. J. (1968). A review of the biology of *Toxorhynchites rutilus septentrionalis* (Dyar & Knab), and some observations on the species in New Jersey. *Proceedings of the New Jersey Mosquito Extermination Association* **55**, 219–226.

Henderson, P. A. (1990). *Freshwater Ostracods*. Universal Book Services, Oegstgeest.

Hering, M. (1931). Eine in den Kannen von *Nepenthes* minierende *Phyllocnistis* (Lepidopt.) und ihre Parasit, eine neue *Coprodiplosis* (Diptera. Cecidom.). *Archiv für Hydrobiologie, Supplementumband* **8**, 50–70.

Hero, J-M. (1990). An illustrated key to tadpoles occurring in the central Amazon rainforest, Manaus, Amazonas Brasil. *Amazoniana* **11**, 201–262.

Heywood, V. H. (ed.) (1978). *Flowering Plants of the World*. Mayflower, New York.

Hirst, S. (1928). A new tyroglyphid mite, (*Zwickia nepenthesiana*, sp. n.) from the pitchers of *Nepenthes ampullaria*. *Journal of the Malayan Branch of the Royal Asiatic Society* **6**, 19–22.

Ho, B. C., Ewert, A. & Chew, L. M. (1989). Interspecific competition among *Aedes aegypti, Ae. albopictus* and *Ae. triseriatus* (Diptera: Culicidae): larval development in mixed cultures. *Journal of Medical Entomology* **26**, 615–623.

Hölldobler, B. & Wilson, E. O. (1990). *The Ants*. Belknap Press, Harvard, Cambridge.

Holling, C. S. (1964). The analysis of complex population processes. *Canadian Entomologist* **96**, 335–347.

Hoogmoed, M. S. & Cadle, J. E. (1991). Natural history and distribution of *Agalychnis craspedopus* (Funkhouser, 1957). (Amphibia: Anura: Hylidae). *Zoologische Mededelingen* **65**, 129–142.

Holzapfel, C. M. & Bradshaw, W. E. (1976). Rearing of *Toxorhynchites rutilus septentrionallis* (Diptera: Culicidae) from Florida and Pennsylvania with notes on their pre-diapause and pupal development. *Annals of the Entomological Society of America* **69**, 1062–1064.

Holzapfel, C. M. & Bradshaw, W. E. (1981). Geography of larval dormancy in the tree-hole mosquito, *Aedes triseriatus* (Say). *Canadian Journal of Zoology* **59**, 1014–1021.

Horsfall, W.R. (1955). *Mosquitoes: their Bionomics and Relation to Disease*. Constable, London.

Howard, L. O., Dyar, H. G. & Knab, F. (1913). *The Mosquitoes of North and Central America and the West Indies*. Carnegie Institution, Washington.

Huang, C.-M. & Cheng, X.-Y. (1993). Diptera: Syrphidae. In: *Bioresources Expedition to the Longqi Mountain Nature Reserve: animals of Longqi Mountain* (ed. Huang, C.-M.). China Forestry Publishing House, Beijing.

Hunter, P. E. & Hunter, C. A. (1964). A new *Anoetus* mite from pitcher plants. *Proceedings of the Entomological Society of Washington* **66**, 39–46.

Hutchinson, G. E. (1959). Homage to Santa Rosalia or why are there so many kinds of animals? *American Naturalist* **93**, 145–159.

Hynes, H. B. N. (1958). *A Key to the Adults and Nymphs of the British Stoneflies (Plecoptera) with Notes on their Ecology and Distribution*. Freshwater Biological Association, Ambleside.

Inger, R. F. (1954). Systematics and zoogeography of Philippine Amphibia. *Fieldiana (Zoology)* **34**, 389–424.

Inger, R. F. (1956). Some amphibians from the lowlands of North Borneo. *Fieldiana: Zoology* **34**, 389–424.

Inger, R. F. (1966). The systematics and zoogeography of the Amphibia of Borneo. *Fieldiana: Zoology* **52**, 1–402.

Inger, R. F. (1985). Tadpoles of the forested regions of Borneo. *Fieldiana: Zoology* New Series **26**, 1–89.

Inger, R. F. & Steubing, R. B. (1991). *Frogs of Sabah.* Sabah Parks, Kota Kinabalu.

Inger, R. F., Schaffer, H. B., Kosky, M. & Bakde, R. (1985). A report on a collection of amphibians and reptiles from Ponmudi, Kerala, South India. *Journal of the Bombay Natural History Society* **81**, 406–427.

Istock, C. A., Wasserman, S. S. & Zimmer, H. (1975). Ecology and evolution of the pitcher-plant mosquito: 1. Population dynamics and laboratory responses to food and population density. *Evolution* **29**, 296–312.

Istock, C. A., Vavra, K. J. & Zimmer, H. (1976a). Ecology and evolution of the pitcher-plant mosquito. 3. Resource tracking by a natural population. *Evolution* **30**, 548–557.

Istock, C. A., Zisfein, J. & Vavra, K. J. (1976b). Ecology and evolution of the pitcher-plant mosquito. 2. The substructure of fitness. *Evolution* **30**, 535–557.

Istock, C. A., Tanner, K. & Zimmer, H. (1983). Habitat selection by the pitcher plant mosquito, *Wyeomyia smithii:* behavioral and genetic aspects. In: *Ecology of Mosquitoes: Proceedings of a Workshop* (eds. L. P. Lounibos, J. R. Rey & J. H. Frank), Florida Medical entomology Laboratory, Vero Beach, pp. 191–204.

Jaffé, K., Michelangeli, F., Gonzalez, J. M., Miras, B. & Ruiz, M. C. (1992). Carnivory in pitcher plants of the genus *Heliamphora* (Sarraceniaceae). *New Phytologist* **122**, 733–744.

Jalil, M. (1974). Observations on the fecundity of *Aedes triseriatus* (Diptera: Culicidae). *Entomologia experimentia et applicata* **17**, 223–233.

Jeffries, M. J. & Lawton, J. H. (1985). Predator-prey ratios in communities of freshwater invertebrates: the role of enemy-free space. *Freshwater Biology* **15**, 105–112.

Jenkins, B. A. (1991). *A study of food webs of species that colonize artificial containers in subtropical rainforest in relation to local environmental heterogeneity.* PhD Thesis, University of New England, Armidale.

Jenkins, B. & Kitching, R. L. (1990). The ecology of water-filled treeholes in Australian rainforests: community re-assembly following disturbance. *Australian Journal of Ecology* **15**, 199–205.

Jenkins, B., Kitching, R. L. & Pimm, S. L. (1992). Productivity, disturbance and food web structure at a local spatial scale in experimental container habitats. *Oikos* **65**, 249–255.

Jenkins, D. W. & Carpenter, S. J. (1946). Ecology of the tree hole breeding mosquitoes of nearctic North America. *Ecological Monographs,* **16**, 31–47.

Jenner, C. E. & McCrary, A. B. (1964). Photoperiodic control of larval diapause in *Toxorhynchites rutilus. American Zoologist* **4**, 434.

Jensen, H. (1910). Nepenthes-tiere, II. Biologische Notizen. *Annales de Jardin botanique de Buitenxorg,* 193.

Johannsen, O. A. (1931). Ceratopogininae from the Malayan subregion of the Dutch East Indies. *Archiv für Hydrobiologie Supplementumband* **9**, 403–448.

Johannsen, O. A. (1932). Tanypodinae from the Malayan subregion of the Dutch East Indies. *Archiv für Hydrobiologie Supplementumband* **9**, 493–507.

Johannsen, O. A. (1935). Aquatic Diptera II. Orthorrhapha – Brachycera and Cyclorrhapha. *Cornell University Agricultural Experiment Station, Memoir* **177**, 1–62.

Jones, F. M. (1916). Two insects of the California pitcher-plant *Darlingtonia californica*. *Entomological News* **29**, 299–302.

Jones, F. M. (1918). *Dohrniphora venusta* Coq. in *Sarracenia flava*. *Entomological News* **29**, 299–302.

Judd, W. W. (1959). Studies of the Byron Bog in southwestern Ontario X. Inquilines and victims of the pitcher plant, *Sarracenia purpurea* L. *Canadian Entomologist* **91**, 171–180.

Juliano, S. A. & Reminger, L. (1992). The relationship between vulnerability to predation and the behavior of larval treehole mosquitoes: geographic and ontogenetic differences. *Oikos* **63**, 465–476.

Jungfer, K.-H. (1985). Beitrag zur Kenntnis von *Dendrobates speciosus* O. Schmidt, 1857 (Salientia: Dendrobatidae). *Salamandra* **21**, 263–280.

Kappus, K. D. & Venard, C. E. (1967). The effect of photoperiod and temperature on the induction of diapause in *Aedes triseriatus* (Say). *Journal of Insect Physiology* **13**, 1007–1019.

Karstens, E. & Pavlovskij, E. (1927). Analyse des eaux, prises dans les creux du frêne (*Fraxinus excelsior*) et habituées par les larves de l'*Anopheles plumbeus* Steph. *Comptes rendus de l'Academie des Sciences de l'URSS* **1927**, 293–295.

Kato, M. & Toriumi, M. (1951). Studies on the associative ecology of insects. IV. Synecological analysis of the larval association of mosquitoes in a bamboo thicket. *Science Reports, Tohuku University* Series IV (Biology) **19**, 152–160.

Kato, M., Hotta, M., Tamin, R. & Itino, T. (1993). Inter- and intra-specific variation in prey assemblages and inhabitant communities in Nepenthes pitchers in Sumatra. *Tropical Zoology* **6**, 11–25.

Keddy, P. A. (1989). *Competition*. Chapman & Hall, London.

Keilin, D. (1927). Fauna of a horse-chestnut tree (*Aesculus hippocastanum*). Dipterous larvae and their parasites. *Parasitology* **19**, 368–374.

Kenny, J. S. (1969). The Amphibia of Trinidad. *Studies on the Fauna of Curaçao and other Caribbean Islands. No. 108*, **29**, 1–79.

Khoo, K. C. (1984). The Australian species of *Cyamops* Melander (Diptera: Periscelidae). *Australian Journal of Zoology* **32**, 527–536.

Kingsolver, J. G. (1979). Thermal and hydric aspects of environmental heterogeneity in the pitcher plant mosquito. *Ecological Monographs* **49**, 357–376.

Kitching, R. L. (1969). *The fauna of tree holes in relation to environmental factors*. DPhil Thesis, University of Oxford.

Kitching, R. L. (1971). An ecological study of water filled tree-holes and their position in the woodland ecosystem. *Journal of Animal Ecology* **40**, 281-302.

Kitching, R. L. (1972a). Population studies of the immature stages of the tree-hole midge *Metriocnemus martinii* Thienemann (Diptera: Chironomidae). *Journal of Animal Ecology* **41**, 53–62.

Kitching, R. L. (1972b). The immature stages of *Dasyhelea dufouri* Laboulbène (Diptera: Ceratopogonidae) in water-filled tree-holes. *Journal of Entomology* (A), **47**, 109–114.

Kitching, R. L. (1977). Time, resources and population dynamics in insects. *Australian Journal of Ecology* **2**, 331–342.

Kitching, R. L. (1983). Community structure in water-filled treeholes in Europe and Australia – some comparisons and speculations. In: *Phytotelmata: Terrestrial*

Plants as Hosts of Aquatic Insect Communities (eds J. H. Frank & L. P. Lounibos), Plexus, Medford, 205–222.

Kitching, R. L. (1986a). A dendrolimnetic dragonfly from Sulawesi (Anisoptera: Libellulidae). *Odonatologica* **15**, 203–209.

Kitching, R. L. (1986b). Prey-predator interactions. In: *Community Ecology: Pattern and Process* (eds J. Kikkawa & D. J. Anderson), Blackwell Scientific Publications, Oxford, pp. 214–239.

Kitching, R. L. (1986c). Exotics in Australia – synopsis and strategies. In: *The Ecology of Exotic Animals and Plants* (ed. R. L. Kitching), Wiley Interscience, Brisbane, pp. 262–269.

Kitching, R. L. (1987a). Spatial and temporal variation in foodwebs from water-filled treeholes. *Oikos* **48**, 280–288.

Kitching, R. L. (1987b). A preliminary account of the metazoan food webs in phytotelmata in northern Sulawesi. *Malayan Nature Journal* **41**,1–12.

Kitching, R. L. (1990). Foodwebs from phytotelmata in Madang, Papua New Guinea. *Entomologist* **109**, 153–164.

Kitching, R. L. (1992). Modelling. In: *Research Methods for the Study of* Heliothis (eds. M. P. Zalucki & P. Twine), Springer Verlag, Berlin, pp. 177–189.

Kitching, R. L. & Allsopp, P. G. (1987). *Prionocyphon niger* sp. n. (Coleoptera: Scirtidae) from water-filled treeholes in Australia. *Journal of Australian Entomological Society* **26**, 73–79.

Kitching, R. L. & Beaver, R. (1990). Patchiness and community structure. In: *Living in a Patchy Environment* (eds N. C. Stenseth & I. R. Swingland), Oxford University Press, Oxford. pp.147–176.

Kitching, R. L. & Callaghan, C. (1982). The fauna of water-filled tree holes in box forest in south-east Queensland. *Australian Entomological Magazine* **8**, 61-70.

Kitching, R. L. & Orr, A. G. (1996). The foodweb from water-filled treeholes in Kuala Belalong, Brunei. *Raffles Bulletin of Zoology* **44**, 405–413.

Kitching, R. L. & Pimm, S. L. (1985). The length of food chains: phytotelmata in Australia and elsewhere. *Proceedings of the Ecological Society of Australia* **14**, 123–140.

Kitching, R. L. & Schofield, C. (1986). Every pitcher tells a story. *New Scientist* **109**, 48–50.

Kitching, R. L., Harvey, M. S. & Fashing, N. J. (In preparation) Mites in phytotelmata: a review with a description of the new species, *Anisitsielloides richeae* Harvey.

Kitching, R. L., Mitchell, H., Morse, G. & Thebaud, C. (1997). Determinants of species richness in assemblages of canopy arthropods in rainforests. In: *Canopy Arthropods* (eds N. E. Stork, R. Didhams & J. Adis) Chapman and Hall, London, pp. 131–150.

Klausnitzer, B. (1980). Ein neu Art der Gattung *Cyphon* Paykull von der Cocos-Insel (Coleoptera, Helodidae). *Reichenbachia*, **18**, 77–79.

Knab, F. (1913a). New moth flies (Psychodidae) bred from Bromeliaceae and other plants. *Proceedings of the United States National Museum*, **46**, 103–106.

Knab, F. (1913b). Some Neotropical Syrphidae. *Insectutor Inscitiae Menstruus*, **1**, 13–15.

Knab, F. & Malloch, J. R. (1912). A borborid from an epiphytic bromeliad (Diptera: fam. Borboridae). *Entomological News*, **23**, 413–415.

Kovac, D. (1994). Die Tierwelt des Bambus: ein Modell für komplexe tropische Lebensgemeinschaften. *Natur und Museum* **124**, 119–136.

Kovac, D. & Streit, B. (1996). The arthropod community of bamboo internodes in peninsular Malaysia: microzonation and trophic structure. In: *Tropical Rainforest Research – Current Issues* (eds. D. S. Edwards, W. E. Booth & S. C. Choy), Kluwer, Dordrecht, pp. 85–99.

Kovac, D. & Yong, H. S. (1992). *Abryna regispetri:* a bamboo long-horned beetle. *Nature Malaysiana* **17**, 92–98.

Kovac, D., Pont, A. C. & Skidmore, P. (1997). *Graphomya kovaci* n. sp. from bamboo phytotelmata in Indonesia: adult, immature stages, and biology. *Senckenbergiana biologica* **77**, 37–45.

Kryzhanovsky, O. L. (1976). Revised classification of the family Carabidae. *Entomological Review* **1**, 80–91.

Kurata, S. (1976). Nepenthes *of Mount Kinabalu*. Sabah National Parks, Kota Kinabalu.

Kurihara, Y. (1954). Synecological study on the relationship between the benthonic microorganism community and dipterous insect larvae in the bamboo container. *Science Reports, Tohuku University* Series IV (Biology) **20**, 130–138.

Kurihara, Y. (1959). Synecological analysis of the biotic community in microcosm. IV. Studies on the relations of Diptera larvae to pH in bamboo containers. *Science Reports, Tohuku University* Series IV (Biology) **25**, 165–171.

Kurihara, Y. (1983). The succession of aquatic dipterous larvae inhabiting bamboo phytotelmata. In: *Phytotelmata: Terrestrial Plants as Hosts of Aquatic Insect Communities* (eds J. H. Frank & L. P. Lounibos), Plexus, Medford, pp. 55–77.

Kurahashi, H. & Beaver, R.A. (1979). *Nepenthomyia malayana* gen. n., sp. n. a new calliphorid fly bred from pitchers of *Nepenthes ampullaria* in West Malaysia (Diptera, Calliphoridae). *Annales de la Société Entomologique de France* (Nouvelle Série), **15**, 25–30.

Lackey, J. B. (1940). The microscopic fauna and flora of tree-holes. *Ohio Journal of Science* **40**, 186–192.

Laessle, A. M. (1961). A micro-limnological study of Jamaican bromeliads. *Ecology* **42**, 499–517.

Laird, M. (1988). *The Natural History of Larval Mosquito Habitats*. Academic, London.

Lamb, R. J. & Smith, S. M. (1980). Comparison of egg size and related life-history characteristics for two predaceous tree-hole mosquitoes (*Toxorhynchites*). *Canadian Journal of Zoology* **58**, 2065–2070.

Lamberson, C., Pappas, C. D. & Pappas, L. G. (1992). Pupal taxonomy of the tree-hole *Culicoides* (Diptera: Ceratopogonidae) in eastern North America. *Annals of the Entomological Society of America* **85**, 111–120.

Lang, J. T. (1978). Relationship of fecundity to the nutritional quality of larval and adult diets of *Wyeomyia smithii*. *Mosquito News* **38**, 396–403.

Lang, J. T. & Ramos, A. C. (1981). Ecological studies of mosquitoes in banana leaf axils on central Luzon. *Mosquito News* **41**, 665–673.

Lannoo, M. J., Townsend, D. S. & Wassersug, R. J. (1987). Larval life in the leaves: arboreal tadpole types, with special attention to the morphology, ecology and behavior of the oophagous *Osteophilus brunneus* (Hylidae) larva. *Fieldiana, Zoology N. S.* **38**, 1–31.

Laurence, R. B. (1953). The larvae of *Ectaetia* (Dipt.: Scatopsidae). *Entomologist's Monthly Magazine* **89**, 204–205.

Laurent, R. F. (1964). Adaptive modifications in frogs of an isolated highland fauna in Central Africa. *Evolution* **18**, 458–467.

Lawton, J. H. & Warren, P. H. (1988). Static and dynamic explanations for patterns in food webs. *Trends in Ecology and Evolution* **3**, 242–245.

Lee, D. J. (1944). *An Atlas of the Mosquito Larvae of the Australasian Region. Tribes – Megarhinini and Culicini.* Austarlian Military Forces, Melbourne.

Lee, V. H. (1974). *Aedes (Stegomyia)* spp. utilizing *Euphorbia kamerunica* as a larval habitat in Nigeria. *Mosquito News* **44**, 229–231.

Leicester, G. F. (1903). A breeding place of certain forest mosquitoes in Malaysia. *Journal of Tropical Medicine and Hygiene* **6**, 291–293.

Lever, R. J. A. W. (1950). Mosquitoes from pitcher plants in the Cameron Highlands. *Malayan Nature Journal* **5**, 98–99.

Lever, R. J. A. W. (1956). Notes on some flies recorded from pitcher plants. *Malayan Nature Journal*, **10**, 109–110.

Levins, R. A. (1970). Extinction. *Lectures on Mathematics in the Life Sciences* **2**, 75–107.

Lieftinck, M. A. (1954). Handlist of Malaysian Odonata. *Treubia* **22** (Suppl.) 1–202.

Lieftinck, M. A. (1962). Odonata. *Insects of Micronesia* **5**, 1–95.

Liem, S. S. (1970). The morphology, systematics and evolution of the Old World tree frogs (Rhacophoridae and Hyperoliidae). *Fieldiana: Zoology* **57**, 1–145.

Lien, J. C. & Matsuki, K. (1979). On the larvae of two species of the genus *Lyriothemis* on Taiwan (Libellulidae; Odonata). *Nature and Insects* **14**, 57–60.

Little, T. J. & Hebert, P. D. N. (1996). Endemism and ecological islands: the ostracods from Jamaican bromeliads. *Freshwater Biology* **36**, 327–338.

Livdahl, T. P. (1982). Competition within and between hatching cohorts of a treehole mosquito. *Ecology* **63**, 1751–1760.

Livdahl, T. P. & Willey, M. S. (1991). Prospects for an invasion: competition between *Aedes albopictus* and native *Aedes triseriatus*. *Science* **253**, 189–191.

Lloyd, L. I., Graham, J. F. & Reynoldson, T. B. (1940). Materials for a study in animal competition. The fauna of sewage bacterial beds. *Annals of Applied Biology* **27**, 122–150.

Loden, M. S. (1974). Predation by chironomid larvae on oligochaetes. *Limnology and Oceanography*, **19**, 156–159.

Lounibos, L. P. (1978). Mosquito breeding and oviposition stimulant in fruit husks. *Ecological Entomology* **3**, 299–304.

Lounibos, L. P. (1979a). Mosquitoes occurring in the axils of *Pandanus rabaiensis* Rendle on the Kenya coast. *Cahiers O.R.S.T.O.M., Entomologie médicale et Parasitologie* **17**, 25–29.

Lounibos, L. P. (1979b). Temporal and spatial distribution, growth and predatory behaviour of *Toxorhynchites brevipalpis* (Diptera: Culicidae) on the Kenya coast. *Journal of Animal Ecology* **48**, 213–236.

Lounibos, L. P. (1980). The bionomics of three sympatric *Eratmopodites* (Diptera: Culicidae) at the Kenya coast. *Bulletin of Entomological Research* **70**, 309–320.

Lounibos, L. P. (1983). The mosquito community of treeholes on subtropical Florida. In: *Phytotelmata: Terrestrial Plants as Hosts of Aquatic Insect Communities* (eds. J. H. Frank & L. P. Lounibos), Plexus, Medford, 223–246.

Lounibos, L. P. & Bradshaw, W. E. (1974). A second diapause in *Wyeomyia smithii*: seasonal incidence and maintenance by photoperiod. *Canadian Journal of Zoology* **53**, 215–221.

Lounibos, L. P., Rey, J. R. & Frank, J. H. (1985). *Ecology of Mosquitoes: Proceedings of a Workshop*. Florida Medical Entomology Laboratory, Vero Beach, Florida.

Lounibos, L. P., Frank, J. H., Machado-Allison, C. E., Navarro, J. C. & Ocanto, P.

(1987a). Seasoniality, abundance and invertebrate associates of *Leptagrion siquierai* Santos in *Aechmea* bromeliads in Venezuelan rain forests (Zygoptera: Coenagrionidae). *Odonatologica* **16**, 193–199.

Lounibos, L. P., Frank, J. H., Machado-Allison, C. E., Ocanto, P. & Navarro, J. C. (1987b). Survival, development and predatory effects of mosquito larvae in Venezuelan phytotelmata. *Journal of Tropical Ecology* **3**, 221–242.

Louton, J., Gelhaus, J. & Bouchard, R. (1996). The aquatic macrofauna of water-filled bamboo (Poaceae: Bambusoideae: *Guardea*) internodes in a Peruvian tropical lowland forest. *Biotropica* **28**, 228–242.

Love, G. J. & Whelchel, J. G. (1955). Photoperiodism and the development of *Aedes triseriatus* (Diptera: Culicidae). *Ecology* **36**, 340–342.

Lu, P. L., Li, B. S., Xu, R. M. & Jiang, Y. Y. (1980). The composition of treehole breeding mosquitoes in the mountain forests of Diaolou, Hainan Island. (Diptera: Culicidae). *Bulletin of the Academy of Military Medical Science (Peking)* **1980**, 55–59.

Lutz, B. (1973). *Brazilian Species of* Hyla. University of Texas Press, Austin.

MacArthur, R. H. (1972). *Geographical Ecology: Patterns in the Distribution of Species*. Harper & Row, New York.

MacArthur, R. H. & Wilson, E. O. (1966). *The Theory of Island Biogeography*. Princeton University Press, Princeton.

Macdonald, W. W. & Traub, R. (1960). Malaysian parasites XXXVIII. On the systematics and ecology of *Armigeres* and *Leicesteria* (Diptera: Culicidae). *Studies of the Institute of Medical Research, Malaya* **29**, 79–110.

Macfie, J. W. S. & Ingram, A. (1923). Certain nurseries of insect life in West Africa. *Bulletin of Entomological Research* **13**, 291–294.

Machado, A. B. M. (1976). Fauna associada al agua das folhas da umbelíferas com observaçoes sobre a ninfa da *Roppaneura beckeri* Santos. *Ciência e Cultura* (São Paulo) **28**, 895–896.

Machado, A. B. M. & Martinez, A. (1982). Oviposition by egg-throwing in a zygopteran, *Mecistogaster jocaste* (Pseudostigmatidae). *Odonatologica* **11**, 15–22.

Machado-Allison, C. E., Rodriguez, D. J., Barrera, R. & Gomez, C. (1983). The insect community associated with inflorescences of *Heliconia caribaea* Lamarck in Venezuela. In: *Phytotelmata: Terrestrial Plants as Hosts of Aquatic Insect Communities* (eds. J. H. Frank & L. P. Lounibos), Plexus, Medford, pp. 247–270.

Maguire, B. (1970). Aquatic communities in bromeliad leaf axils and the influence of radiation. In: *A Tropical Rain Forest* (ed. H. T. Odum & R. F. Pigeon), Division of Technical Information, U.S. Atomic Energy Commission, Oak Ridge, Tennessee, pp. E.95–E.101.

Maguire, B., Belk, D. & Wells, G. (1968). Control of community structure by mosquito larvae. *Ecology* **49**, 207–210.

Malloch, J. R. (1914). Costa Rican Diptera collected by Philip P. Calvert, PhD., 1909–1910. Paper 1. A partial report on the Borboridae, Phoridae and Agromyzidae. *Transactions of the American Entomological Society* **40**, 8–36.

Marten, G. G. (1984). Impact of the copepod, *Mesocyclops leuckarti pilosa* and the green alga *Kirschneriella irregularis* upon larval *Aedes albopictus* (Diptera: Culicidae). *Bulletin of the Society for Vector Ecology* **9**, 1–5.

Martinez, N. D. (1992). Constant connectance in community food webs. *American Naturalist* **139**, 1208–1218.

Martinez, N. D. (1993). Effects of resolution on food web structure. *Oikos* **66**, 403–412.

Mather, T. N. (1981). Larvae of alderfly (Megaloptera: Sialidae) from pitcher plants. *Entomological News* **92**, 32.

Matthews, E. G. & Kitching, R. L. (1984). *Insect Ecology* (2nd Edition), University of Queensland Press, St Lucia.

Mattingly, P. F. (1969). *The Biology of Mosquito-borne Disease*, George Allen & Unwin, London.

Mattingly, F. (1981). Medical entomology studies XIV. The subgenera *Rachionotomyia, Tricholeptomyia* and *Tripteroides* (mabinii group) of genus *Tripteroides* in the Oriental region. *Contributions of the American Entomological Institute* **17**, 1–147.

Mattingly, P. F. & Brown, E. S. (1955). The mosquitoes (Diptera: Culicidae) of the Seychelles. *Bulletin of Entomological Research* **46**, 69–110.

Maier, C. J. (1982). Larval habitats and mate-seeking sites of flowerflies (Diptera: Syrphidae, Eristalinae). *Proceedings of the Entomological Society of Washington* **84**, 603–609.

May, R. M. (1972). Will a large complex system be stable? *Nature (London)* **238**, 413–414.

May, R. M. (1973). Time-delay versus stability in population models with two and three trophic levels. *Ecology* **54**, 315–325.

May, R. M. (1974). Biological populations with non-overlapping generations: stable points, stable cycles and chaos. *Science (New York)* **186**, 645–647.

May, R. M. (1975). *Stability and Complexity in Model Ecosystems*. Princeton University Press, Princeton.

May, R. M. (1983). The structure of food webs. *Nature (London)* **301**, 566–568.

May, R. M. (1990). How many species? *Philosophical Transactions of the Royal Society* Series B **330**, 293–304.

Mayer, K. (1938). Zur Kenntnis der Buchenhöhlenfauna. *Archiv für Hydrobiologie* **33**, 388–400.

McAlpine, D. K. (1978). Description and biology of a new genus of flies related to *Anthoclusia* and representing a new family (Diptera, Schizophora, Neurochaetidae). *Annals of the Natal Museum* **23**, 273–295.

McAlpine, D. K. (1988a). Studies in upside-down flies (Diptera: Neurochaetidae). Part I. Systematics and phylogeny. *Proceedings of the Linnaean Society of New South Wales* **110**, 31–58.

McAlpine, D. K. (1988b). Studies in upside-down flies (Diptera: Neurochaetidae). Part II. Biology, adaptations, and specific mating mechanisms. *Proceedings of the Linnaean Society of New South Wales* **110**, 59–82.

McAlpine, D.K. (1993). Review of the upside-down flies (Diptera: Neurochaetidae) of Madagascar and Africa, and evolution of neurochaetid host plant associations. *Records of the Australian Museum* **45**, 221–239.

McAlpine, J. F., Peterson, B. V., Shewell, G. E., Teskey, H. J., Vockeroth, J. R. & Wood, D. M. (Coords) (1981). *Diptera of the Nearctic Region*, Volume 1. Agriculture Canada, Hull, Quebec.

McDaniel, S. T. (1971). The genus *Sarracenia* (Sarraceniaceae). *Bulletin of the Tall Timbers Research Station* **9**, 1–36.

McDiarmid, R. W. & Foster, M. S. (1975). Unusual sites for two Neotropical tadpoles. *Journal of Herpetology* **9**, 264–265.

McDonald, W. J. F. & Whiteman, W. G. (1979). *Moreton Bay Vegetation Map Series: Murwillumbah*. Queensland Department of Primary Industries, Brisbane.

McIntyre, S. M., Kitching, R. L. & Jessup, L. W. (1994). Vegetation structure in rainforest plots at Cape Tribulation, North Queensland. *Proceedings of the Royal Society of Queensland* **104**, 25–41.

Means, R. (1972). Mosquito breeding in leaf axils of the teasel *Dipsacus laciniatus* (Linn.) in New York. *Mosquito News* **33**, 107–108.

Meijere, J. C. H de (1910). *Nepenthes* – Tiere. I. Systematik. *Annales du Jardin Botanique Buitenzorg*, Supplement, **3**, 917–940.

Mellanby, H. (1963). *Animal Life in Fresh Water*. 6th edition, Methuen, London.

Meillon, B. de & Wirth, W. W. (1979). A taxonomic review of the subgenus *Phytohelea* of *Forcipomyia* (Diptera: Ceratopogonidae). *Proceedings of the Entomological Society of Washington* **81**, 178–206.

Menzel, R. (1922). Beiträge zur Kenntnis der Mikrofauna von Nederländische-Ost-Indien. II. Über den tierischen Inhalt der Kannen von *Nepenthes melamphora* Reinw. mit besonderer Berücksichtigung der Nematoden. *Treubia* **3**, 116–126.

Merritt, R. W. & Craig, D. A. (1987). Larval mosquito (Diptera: Culicidae) feeding mechanisms: mucosubstance production for capture of fine particles. *Journal of Medical Entomology* **24**, 275–278.

Merritt, R. W. & Cummins, K. W. (1984). *An Introduction to the Aquatic Insects of North America*, 2nd edition. Kendall/Hunt, Duboque, Iowa.

Merritt, R. W., Dadd, R. H. & Walker, E. D. (1992). Feeding behavior, natural food, and nutritional relationships of larval mosquitoes. *Annual Review of Entomology* **37**, 349–376.

Miller, A. C. (1971). Observations on the Chironomidae (Diptera) inhabiting the leaf axils of two species of Bromeliaceae on St. John, U.S. Virgin Islands. *Canadian Entomologist* **103**, 391–396.

Mithen, S. J. & Lawton, J. H. (1986). Food-web models that generate constant predator-prey ratios. *Oecologia (Berlin)* **69**, 542–550.

Miyagi, I. & Toma, T. (1980). Studies on the mosquitoes in Yaeyama Islands. 5. Notes on the mosquitoes collected in forest areas of Iriomotejima. *Japanese Journal of Sanitary Zoology* **31**, 81–91.

Moeur, J. E. & Istock, C. A. (1980). Ecology and evolution of the pitcher-plant mosquito. 4. Larval influence over adult reproductive performance and longevity. *Journal of Animal Ecology* **49**, 775–792.

Mogi, M. (1984). Distribution and overcrowding effects in mosquito larvae (Diptera: Culicidae) inhabiting taro axils in the Ryukyus, Japan. *Journal of Medical Entomology* **21**, 63–68.

Mogi, M. & Chan, K. L. (1996). Predatory habits of dipteran larvae inhabiting *Nepenthes* pitchers. *Raffles Bulletin of Zoology* **44**, 233–245.

Mogi, M. & Chan, K. L. (1997). Variation in communities of dipterans in *Nepenthes* pitchers in Singapore: predators increase prey community diversity. *Annals of the Entomological Society of America* **90**, 177–183.

Mogi, M. & Sembel, D. T. (1996). Predator-prey system structure in patchy and ephemeral phytotelmata: aquatic communities in small aroid axils. *Researches in Population Ecology* **38**, 95–103.

Mogi, M. & Suzuki, H. (1983). The biotic community in the water-filled internode of bamboos in Nagasaki, Japan, with special reference to mosquito ecology. *Japanese Journal of Ecology* **33**, 271–279.

Mogi, M. & Yamamura, N. (1988). Population regulation of a mosquito *Armigeres theobaldi* with a description of the animal fauna of zingiberaceous inflorescences. *Researches in Population Ecology* **30**, 251–265.

Mogi, M. & Yong, H. S. (1992). Aquatic arthropod communities in *Nepenthes* pitchers: the role of niche differentiation, aggregation, predation and competition in community organization. *Oecologia* **90**, 172–184.

Mogi, M., Horio, M., Miyagi, I. & Cabrera, B. D. (1985). Succession, distribution, overcrowding and predation in the aquatic community in aroid axils, with special reference to mosquitoes. In: *Ecology of Mosquitoes* (eds L. P. Lounibos, J. R. Rey & J. H. Frank), Florida Medical Entomology Laboratory, Vero Beach, pp. 95–119.

Moore, M. (1988). Differential use of food resources by the instars of *Chaoborus punctipennis*. *Freshwater Biology* **19**, 249–268.

Moore, M. & Gilbert, J. J. (1987). Age-specific *Chaoborus* predation on rotifer prey. *Freshwater Biology* **17**, 223–236.

Müller, F. (1878). Sobre as casas construidas pelas larvas de insectos trichopteros da Provincia de Santa Catharina. *Archivos do Museu Nacional, Rio de Janeiro,* **3**, 99–134, 209–214.

Müller, F. (1879). Wasserthiere in den Wipfeln des Waldes. *Kosmos* **4**, 390–392.

Muspratt, J. (1951). The bionomics of an African *Megarhinus* (Dipt.: Culicidae) and its possible use in biological control. *Bulletin of Entomological Research* **42**, 55–370.

Myers, C. W., Daly, J. W. & Martinez, V. (1984). An arboreal poison frog (*Dendrobates*) from western Panama. *American Museum Novitates* **2783**, 1–20.

Nakata, G., Matsuo, K. & Ito, S. (1953). Ecological studies on mosquitoes about Kyoto City. I. On the successions of mosquito larvae breeding in minute inland waters found in graveyards and bamboo groves. *Eiseidobutsu* **4**, 62–72.

Naeem, S. (1990). Patterns of the distribution and abundance of competing species when resources are heterogeneous. *Ecology* **71**, 1422–1429.

Neboiss, A. (1991). Trichoptera (Caddis-flies, caddises). In: *Insects of Australia* 2nd Edition (ed. I. D. Naumann), Melbourne University Press, Melbourne, pp.787–816.

Nesbitt, H. H. J. (1954). A new mite, *Zwickia gibsoni* n. sp., Fam. Anoetidae, from the pitchers of *Sarracenia purpurea* L. *Canadian Entomologist* **86**, 193–197.

Nesbitt, H. H. J. (1979). A new anoetid (Acari) of the genus *Creutzeria* from the Seychelles. *Canadian Entomologist* **111**, 1201–1205.

Nesbitt, H. H. J. (1985). A new mite from bromeliad leaf-axils from Costa Rica (Acari: Acaridae). *International Journal of Acarology* **11**, 209–214.

Nesbitt, H. H. J. (1990). A second new mite from the leaf-axils of a bromeliad from Costa Rica (Rhizoglyphinae: Acaridae). *Acaralogia* **31**, 39–42.

Newman, C. M. & Cohen, J. E. (1986). A stochastic theory of community food webs IV. Theory of food chain length in large webs. *Proceedings of the Royal Society of London* B **228**, 355–377.

Noble, G. K. (1929). The adaptive modifications of the arboreal tadpoles of *Hoplophryne* and the torrent species of *Staurois*. *Bulletin of the American Museum of Natural History* **83**, 291–336.

O'Connor, B. M. (1994). Life-history modifications in astigmatid mites. In *Mites: Ecological and Evolutionary Analyses of Life-History Patterns* (ed. M. A. Houck), Chapman and Hall, New York, pp. 136–159.

Ogden, J. & Powell, J. A. (1979). A quantitative description of the forest vegetation on an altitudinal gradient in the Mount Field National Park, Tasmania, and a discussion of its history and dynamics. *Australian Journal of Ecology* **4**, 293–325.

Oldroyd, H. (1957). *The Horse-flies (Diptera: Tabanidae) of the Ethiopian Region. Vol. II. Subfamilies Chrysopinae, Scepsidinae and Pangoniinae and a revised classification*. British Museum (Natural History), London.

Oldroyd, H. (1964). *The Natural History of Flies.* Weidenfield & Nicholson, London.

Oliver, D. R. (1971). Life history of the Chironomidae. *Annual Review of Entomology* **16**, 211–230.

Orr, A. G. (1994). Life histories and ecology of Odonata breeding in phytotelmata in Bornean rainforest. *Odonatologica* **23**, 365–377.

Orton, G. L. (1944). *Studies on the systematic and phylogenetic significance of certain larval characters in the Amphibia salientia.* PhD Thesis, University of Michigan, Ann Arbor.

Oudemans, A. C. (1932). Opus 550. *Tidjschrift fur Entomologie* **75**, 207–210.

Paine, R. T. (1966). Food web complexity and species diversity. *American Naturalist* **100**, 65–75.

Paine, R. T. (1988). Food webs: road maps of interactions or grist for theoretical development? *Ecology* **69**, 1648–1654.

Paine, R. W. (1934). The introduction of *Megarhinus* mosquitoes into Fiji. *Bulletin of Entomological Research* **25**, 1–32.

Paradise, C. J. & Dunson, W. A. (1997). Insect species interactions and resource effects in treeholes: are helodid beetles bottom-up facilitators of midge populations? *Oecologia (Berlin)* **109**, 303–312.

Park, O., Auerbach, S. & Corley, G. (1950). The tree-hole habitat with emphasis on the pselaphid beetle fauna. *Bulletin of the Chicago Academy of Science* **9**, 19–56.

Parker, H. W. (1934). *A Monograph of the Frogs of the Family Microhylidae.* British Museum (Natural History), London.

Paterson, C. G. (1971). Overwintering ecology of the aquatic fauna associated with the pitcher plant, *Sarracenia purpurea* L. Canadian Journal of Zoology **49**, 1455–1459.

Paterson, C. G & Cameron, C. J. (1982). Seasonal dynamics and ecological strategies of the pitcher plant chironomid, *Metriocnemus knabi* Coq. (Diptera: Chironomidae), in southeast New Brunswick. *Canadian Journal of Zoology* **60**, 3075–3083.

Paulian, R. (1961). La zoogéographie de Madagascar et des îles voisines. *Faune de Madagascar*, **13**, 1–481.

Paulson, S. L. (1984). Oviposition and egg diapause in field populations of *Aedes triseriatus* and *Aedes hedersoni*. *Proceedings of the Indiana Vector Control Association* **8**, 1–8.

Peixota, O. L. (1981). Notas sobre a girino de *Crossodactylodes pintoi* Cochran (Amphibia, Anura, Leptodactylidae). *Revista Brasileira Biologia* **41**, 339–341.

Peixota, O. L. (1983). Two new species of *Crossodactylodes* of Santa Tereza, state of Espirito Santo, Brazil Cochran (Amphibia, Anura, Leptodactylidae). *Revista Brasileira Biologia* **42**, 619–626.

Pennak, R. W. (1953). *Freshwater Invertebrates of the United States.* Ronald Press, New York.

Perrett, J-L. (1962). La biologie d'*Acanthixalus spinosus* (Amphibia Salientia). *Recherches Études Camerounaises* **1**, 90–101.

Perrett, J.-L. (1966). Les amphibiens du Cameroun. *Zoologisches Jahrbuch der Systematich* **93**, 289–464.

Perry, I. & Stubbs, A. E. (1978). Some microhabitats. Dead wood and sap runs. In: *A Dipterist's Handbook* (eds A. Stubbs and P. Chandler), Amateur Entomological Society Special Publication **15**, London, pp. 65–73.

Peters, W. L. & Campbell, L. G. (1991). Ephemeroptera (Mayflies). In: *Insects of*

Australia 2nd Edition (ed. I. D. Naumann), Melbourne University Press, Melbourne, pp. 279–293.

Petersen, A. (1953). *Larvae of Insects. An Introduction to Nearctic Species. Part II: Coleoptera, Diptera, Neuroptera, Siphnaptera, Mecoptera, Trichoptera.* University of Ohio, Columbus.

Petersen J. J., Chapman, H. C. & Willis, O. R. (1969). Predation by *Anopheles barberi* Coquillet on first instar mosquito larvae. *Mosquito News* **29**, 134–135.

Peterson, A. (1951). *Larvae of Insects*, Volume 2. Privately published, Columbus, Ohio.

Phillips, A. & Lamb, A. (1996). *Pitcher Plants of Borneo.* Natural History Publications, Kota Kinabalu.

Pianka, E. R. (1975). Niche relations of desert lizards. In: *Ecology and Evolution of Communities*, (eds M. L. Cody & J. M. Diamond) Harvard University Press, Cambridge, Mass., pp. 292–314.

Picado, C. (1912). Les maires aeriennes de la forêt vièrges américaine: les bromeliacées. *Biologica* **11**, 110–115.

Picado, C. (1913). Les broméliacées épiphytes considerées comme milieu biologique. *Bulletin scientifique de la France et de la Belgique* **47**, 215–360.

Pimm, S. L. (1979). The structure of food webs. *Theoretical Population Biology* **16**, 144–158.

Pimm, S. L. (1980). Properties of food webs. *Ecology* **61**, 219–225.

Pimm, S. L. (1982). *Food Webs.* Chapman and Hall, London.

Pimm, S. L. (1992). *The Balance of Nature*, Chicago University Press, Chicago.

Pimm, S. L. & Kitching, R. L. (1987). The determinants of food chain lengths. *Oikos.* **50**, 302–307.

Pimm, S. L. & Lawton, J. H. (1977). Number of trophic levels in ecological communities. *Nature (London)* **268**, 329–331.

Pimm, S. L. & Lawton, J. H. (1978). On feeding at more than one trophic level. *Nature (London)* **275**, 542–544.

Pimm, S. L. & Lawton, J. H. (1980). Are food webs divided into compartments? *Journal of Animal Ecology* **49**, 879–898.

Pimm, S. L., Lawton, J. H. & Cohen, J. E. (1991). Food web patterns and their consequences. *Nature (London)* **350**, 669–674.

Pittendrigh, C. S. (1948). The bromeliad-*Anopheles*-malaria complex in Trinidad I. The bromeliad flora. *Evolution* **2**, 58–89.

Pittendrigh, C. S. (1950a). The ecological divergence of *Anopheles bellator* and *A. homunculus. Evolution* **4**, 43–63.

Pittendrigh, C. S. (1950b). The ectopic specialization of *Anopheles homunculus* and its relation to competition with *A. bellator. Evolution* **4**, 64–78.

Pittman, J. L., Turner, T. S., Frederick, L., Petersen, R. L., Poston, M. E., Mackenzie, M. & Duffield, R. M. (1996). Occurrence of alderfly larvae (Megaloptera) in a West Virginia population of the purple pitcher plant, *Sarracenia purpurea* L. (Sarraceniaceae). *Entomological News* **107**, 137–140.

Polis, G. A. (1994). Food webs, trophic cascades and community structure. *Australian Journal of Ecology* **19**, 121–136.

Poinar, G. O. (1996). A fossil stalk-winged damselfly, *Diceratobasis worki* spec. nov., from Dominican amber, with possible ovipositional behavior in tank bromeliads (Zygoptera: Coenagrionidae). *Odonatologia* **25**, 381–385.

Polhemus, J. T. & Pohemus, D. A. (1991). A review of the veliid fauna of bromeliads with a key and description of a new species (Heteroptera: Veliidae). *Journal of the New York Entomological Society* **99**, 204–216.

Poynton, J. C. & Broadley, D. G. (1988). Amphibia Zambesiaca 4. Bufonidae. *Annals of the Natal Museum* **29**, 447–490.

Price, P. W. (1975). *Insect Ecology*. Wiley Interscience, New York.

Price, R. D. (1958). Notes on the biology and laboratory colonization of *Wyeomyia smithii* (Coquillett) (Diptera: Culicidae). *Canadian Entomologist* **90**, 473–478.

Pugialli Domingues, R. A., Lima Pugialli, H. R. & Dietz, J. M. (1989). Densidade e diversidade de fauna fitotelmata em bromélias de quatro tipos de florestas degradadas. *Revista Brasileira de Biologia* **49**, 125–129.

Putman, R. J. (1994). *Community Ecology*. Chapman & Hall, London.

Quate, L. W. & Vockeroth, J. R. (1981). Psychodidae. In: *Manual of Nearctic Diptera*, Agriculture Canada, Ottawa, pp. 293–300.

Ratasirarson, J. & Silander, J. A. (1996). Structure and dynamics in *Nepenthes madagascariensis* pitcher plant micro-communities. *Biotropica* **28**, 218–227.

Razahelisoa, M. (1974). Contribution à l'étude des batrachiens de Madagascar. Écologie et développement larvaire de *Gephyromantis methueni* Angel, batracien à biotope végétal sur les Pandanus. *Bulletin Académie Malgaches* **51**, 113–128.

Reddingius, J. & den Boer, P. J. (1970). Simulation experiments illustrating stabilization of animal numbers by spreading of risk. *Oecologia (Berlin)* **5**, 240–284.

Reichle, D. E. (1970). *Analysis of Temperate Forest Ecosystems*. Springer-Verlag, Berlin, New York.

Reid, J. W. & Janetsky, W. (1996). Colonization of Jamaican bromeliads by *Tropocyclops jamaicensis* n. sp. (Crustacea: Copepoda: Cyclopoidea). *Invertebrate Biology* **115**, 305–320.

Reidel, M. P. (1921). Dipteren-ausbeute aus Paraguay. Nematocera polyneura (Family Tipulidae). *Archiv für Naturgeschichte, Abteilung A.* **3**, 123–127.

Rentz, D. C. F. (1991). Orthoptera (Grasshoppers, locusts, katydids, crickets). In: *Insects of Australia* 2nd Edition (ed. I. D. Naumann), Melbourne University Press, Melbourne, pp. 369–393.

Rettenmeyer, C. W. & Akre, R. D. (1968). Ectosymbiosis between phorid flies and army ants. *Annals of the Entomological Society of America* **61**, 1317–1326.

Reynoldson, T. B. (1964). Evidence for intra-specific competition in field populations of triclads. *Journal of Animal Ecology* **33**, (Supplement), 187–201.

Reynoldson, T. B. & Davies, R. W. (1970). Food niche and coexistence in lake-dwelling triclads. *Journal of Animal Ecology* **39**, 599–617.

Robbins, R. G., Saunders, J. C. & Pullen, R. (1976). Part VII. Vegetation and ecology. In: *Lands of the Ramu-Madang Area, Papua New Guinea* (ed. R. G. Robbins), CSIRO Land Research Series No. 37, Melbourne, pp. 96–109.

Robinson, D. C. (1977). Herpetofauna bromelicola Costarricense y renacuajas de *Hyla picadoi* Dunn. *Historia Natural de Costa Rica* **1**, 31–42.

Röhnert, U. (1950). Wassererfüllte Baumhöhlen und ihre Besiedlung. Ein Beitrag zur Fauna dendrolimnetica. *Archiv für Hydrobiologie* **44**, 475–516.

Root, R. B. (1967). The niche exploitation pattern of the blue-gray gnatcatcher. *Ecological Monographs* **37**, 317–350.

Rozkosny, R. (1982). *A Biosystematic Study of the European Stratiomyidae (Diptera)*. Junk, The Hague.

Rubio, Y., Rodriguez, D., Machado-Allison, C. E. & Leon, J. A. (1980). Algunos aspectos del comportomiento de *Toxorhynchites* (Diptera: Culicidae). *Acta Cientifica Venezolano* **31**, 345–351.

Russo, R. (1983). The functional response of *Toxorhynchites rutilus rutilus*

(Diptera: Culicidae), a predator on container-breeding mosquitoes. *Journal of Medical Entomology* **20**, 584–590.

Russo, R. (1986). Comparison of predatory behavior in five species of *Toxorhynchites* (Diptera: Culicidae). *Annals of the Entomological Society of America* **79**, 715–722.

Rymal, D. E. & Folkerts, G. W. (1982). Insects associated with pitcher-plants (*Sarracenia* Sarraceniaceae), and their relationship to pitcher-plant conservation. *Journal of the Alabama Academy of Science* **53**, 131–151.

Sack, P. (1931). Syrphidae (Diptera) der Deutschen limnologischen Sunda-Expedition. "Tropische Binnengewässer". *Archiv für Hyrdrobiologie Supplementumband* **8**, 585–592.

Santos, N. D. dos (1962). Fauna do Estado da Guanabara. L. Descrição de *Leptagrion perlongum* Calvert, 1909, fêmea, e notas sôbre outras espécies do gênero. *Boletim do Museo Nacional, Nova Serie, Rio de Janeiro, Zoologica* **233**, 1–8.

Santos, N. D. dos (1966a). Contribuição au conheciemento de região de Pocos de Caldas, MG. Brasil. *Roppaneura beckeri* gen. nov., sp. nov. (Odonata, Protoneuridae). *Boletim do Museo Nacional, Nova Serie, Rio de Janeiro, Zoologica* **256**, 1–4.

Santos, N. D. dos (1966b). Contribuição au conheciemento da fauna do Estado da Guanabara. 56. – Notas sobre coenagriideos (Odonata) que se criam em bromélias. *Atas de Sociedade de Biologia do Rio de Janeiro* **10**, 83–85.

Santos, N. D. dos (1968a). Descrição de *Leptagrion dardanoi* sp. n. (Odonata, Coenagrionidae). *Atas de Sociedade de Biologia do Rio de Janeiro* **12**, 63–65.

Santos, N. D. dos (1968b). Descrição de *Leptagrion siqueirai* sp. n. (Odonata, Coenagrionidae). *Atas de Sociedade de Biologia do Rio de Janeiro* **12**, 137–139.

Santos, N. D. dos (1978). Descrição de *Leptagrion vriesianum* sp. n. (Odonata, Coenagrionidae). *Boletim do Museo Nacional, Nova Serie, Rio de Janeiro, Zoologica* **292**, 1–6.

Santos, N. D. dos (1979). Descrição de *Leptagrion bocainense* Santos, 1978, coenagrionídea bromelícola (Odonata: Coenagrionidae). *Anais da Sociedade Entomologica do Brasil* **8**, 167–173.

Santos, N. D. dos (1981). Odonata. In: *Biota Aquatica da America do Sul Tropical* (ed. S. H. Hurlbert), San Diego University, California.

Sarasin, P. & Sarasin. F. (1905). *Reisen in Celebes ausgefahrt in den Jahren 1893–96 und 1902–3*. Band 1, Kriedel, Weisbaden.

Sattler, C. & Sattler, W. (1965). Decapode Krebse als Bewohner von Bromelien – "Zisternen". *Natur und Museum* **95**, 411–415.

Saul, S. H., Sinsko, M. J., Grimstad, P. R. & Craig, G. B. (1977). Identification of sibling species, *Aedes triseriatus* and *Ae. hendersoni* by electrophoresis. *Journal of Medical Entomology* **13**, 705–708.

Saul, S. H., Sinsko, M. J., Grimstad, P. R. & Craig, G. B. (1978). Population genetics of the mosquito *Aedes triseriatus:* genetic-ecological correlation at an esterase locus. *American Naturalist* **112**, 333–339.

Saunders, J. C. (1976). Part VIII. Forest resources. In: *Lands of the Ramu-Madang Area, Papua New Guinea* (ed. R. G. Robbins), CSIRO Land Research Series No. 37, Melbourne, pp. 110–125.

Saunders, L. G. (1956). Revision of the genus *Forcipomyia* based on characters of all stages (Diptera: Ceratopogonidae). *Canadian Journal of Zoology* **34**, 657–705.

Saunders, L. G. (1959). Methods for studying *Forcipomyia* midges, with special reference to cacao-pollinating species (Diptera: Ceratopogonidae). *Canadian Journal of Zoology* **37**, 33–51.

Savage, R. M. (1952). Ecological, physiological and anatomical observations on some species of anuran tadpoles. *Proceedings of the Zoological Society of London* **122**, 467–514.

Schmidt, K. P. (1933). New reptiles and amphibians from Honduras. *Zoological Series of the Field Museum of Natural History* **20**, 15–22.

Schmitz, P. H. & Villeneuve, J. de J. (1932). Contribution à l'étude de la faune népenthicole. *Natuurhistorische maandblad* **21**, 116–117.

Schoener, T. W. (1974). Competition and the form of habitat shift. *Theoretical Population Biology* **4**, 265–307.

Schoener, T. W. (1989). Food webs from the small to the large. *Ecology* **70**, 1559–1589.

Schoenly, K., Beaver, R. A. & Heumer, T. A. (1991). On the trophic relationships of insects: a food web approach. *American Naturalist* **137**, 597–638.

Scholl, P. J. & DeFoliart, G. R. (1977). *Aedes triseriatus* and *Aedes hendersoni*: vertical and temporal distribution as measured by oviposition. *Environmental Entomology* **6**, 355–357.

Scholl, P. J. & DeFoliart, G. R. (1978). The influence of seasonal sex ratio on the number of annual generations of *Aedes triseriatus*. *Annals of the Entomological Society of America* **71**, 677–679.

Schuitemaker. J. P. & Stärke, A. (1933). Contribution à l'étude de la faune nepenthicole. Art. III. Un nouveau *Campanotus* de Borneo, habitant les tiges cruises de *Nepenthes* récolté par J. P. Scuitemaker et décrit par A. Stärke, den Dolder. *Overdruck uit het Natuurhistorisch Maandblad* **22**, 29–31.

Scoble, M. J. (1992). *The Lepidoptera: Form, Function and Diversity*. Oxford University Press, Oxford.

Scott, H. (1912). A contribution to the knowledge of the fuana of Bromeliaceae. *Annals and Magazine of Natural History* (8 ser.) **10**, 424–438.

Scott, H. (1914). The fauna of "reservoir plants". *Zoologist* **18**, 183–195.

Scourfield, D. J. (1915). A new copepod found in water from hollows on tree trunks. *Journal of the Quekett Microscopical Club* Series 2. **12**, 431–440.

Seifert, R. P. (1975). Clumps of *Heliconia* as ecological islands. *Ecology* **56**, 1416–1422.

Seifert, R. P. (1980). Mosquito fauna of *Heliconia aurea*. *Journal of Animal Ecology* **49**, 687–697.

Seifert, R. P. (1982). Neotropical *Heliconia* insect communities. *Quarterly Review of Biology* **57**, 1–28.

Seifert, R. P. & Barrera, R. R. (1981). Cohort studies on mosquito (Diptera: Culicidae) larvae living in water-filled bracts of *Heliconia aurea* (Zingiberales: Musaceae). *Ecological Entomology* **6**, 191–197.

Seifert, R. P. & Seifert, F. H. (1976a). Natural history of insects living in inflorescences of two species of *Heliconia*. *Journal of the New York Entomological Society* **34**, 233–242.

Seifert, R. P. & Seifert, F. H. (1976b). A community matrix analysis of *Heliconia* insect communities. *American Naturalist* **84**, 233–242.

Seifert, R. P. & Seifert, F. H. (1979a). Utilization of *Heliconia* (Musaceae) by the beetle *Xenarescus monoceros* (Olivier) (Chrysomelidae: Hispinae) in a Venezuelan forest. *Biotropica* **11**, 51–59.

Seifert, R. P. & Seifert, F. H. (1979b). A *Heliconia* insect community in a Venezuelan cloud forest. *Ecology* **60**, 51–59.

Service, M. W. (1965). The ecology of the treehole breeding mosquitoes in the northern Guinea savanna of Nigeria. *Journal of Applied Ecology*, **2**, 1–16.

Service, M. W. (1976). *Mosquito Ecology: Field Sampling Methods*. John Wiley, New York.

Shinonaga, S. & Beaver, R. A. (1979). *Pierretia urceola* : a new species of sarcophagid fly found living in *Nepenthes* pitchers in West Malaysia (Diptera: Sarcophagidae). *Annales, de la Société entomologiques de France* (N. S.) **15**, 37–40.

Shroyer, D. A. & Craig, G. B. (1980). Egg hatchability and diapause in *Aedes triseriatus* (Diptera: Culicidae): temperature and photoperiod induced latencies. *Annals of the Entomological Society of America* **73**, 39–43.

Shroyer, D. A. & Craig, G. B. (1981). Seasonal variation in sex ratio of *Aedes triseriatus* (Diptera: Culicidae) and its dependence on egg hatching behaviour. *Environmental Entomology* **10**, 147–152.

Shroyer, D. A. & Craig, G. B. (1983). Egg diapause in the mosquito *Aedes triseriatus* (Diptera: Culicidae): geographic variation in photoperiodic response and factors affecting diapause termination. *Journal of Medical Entomology* **20**, 601–607.

Silverstone, P. A. (1973). Observations on the behavior and ecology of a Columbian poison-arrow frog, the kokoé-pá, (*Dendrobates histrionicus* Berthold). *Herpetologica* **29**, 295–301.

Silverstone, P. A. (1975). A revision of the poison arrow frogs of the genus *Dendrobates* Wagler. *Natural History Museum of Los Angeles County, Science Bulletin* **21**, 1–55.

Silverstone, P. A. (1976). A revision of the poison arrow frogs of the genus *Phyllobates* Bibron in Sagra (family Dendrobatidae). *Natural History Museum of Los Angeles County, Science Bulletin* **27**, 1–53.

Simberloff, D. & Dayan, T. (1991). The guild concept and the structure of ecological communities. *Annual Review of Ecology and Systematics* **22**, 115–143.

Sims, S. R. (1982). Larval diapause in the eastern tree-hole mosquito, *Aedes triseriatus:* latitudinal variation in induction and intensity. *Annals of the Entomological Society of America* **75**, 195–200.

Sinsko, M. & Craig, G. B. (1979). Dynamics of an isolated population of *Aedes triseriatus* (Diptera: Culicidae). I. Population size. *Journal of Medical Entomology* **15**, 89–98.

Sinsko, M. J. & Grimstad, P. R. (1977). Habitat separation by differential vertical oviposition of two treehole Aedes in Indiana. *Environmental Entomology* **6**, 485–487.

Skidmore, P. (1985). *The Biology of the Muscidae of the World*. Junk, Dordrecht.

Smart, J. (1938). Note on the insect fauna of the bromeliad *Brocchinia micrantha* (Baker) Mez of British Guiana. *Entomologist's Monthly Magazine* **74**, 198–200.

Smith, R. M. (1987). Zingiberaceae. *Flora of Australia* **45**, 19–34.

Smith, J. B. (1902). Life-history of *Aedes smithii* Coq. *Journal of the New York Entomological Society* **10**, 10–15.

Smith, H. M. (1941). Snakes, frogs and bromelia. *Chicago Naturalist* **4**, 35–43.

Smith, S. M. & Brust, R. A.(1971). Photoperiodic control of the maintenance and termination of larval diapause in *Wyeomyia smithii* Coq. (Diptera, Culicidae) with notes on oogenesis in the adult female. *Canadian Journal of Zoology* **49**, 1065–1073.

Snow, W. E. (1949). *The Arthropoda of wet tree holes.* Ph.D. Thesis, University of Illinois, Urbana.

Snow, W. E. (1958). Stratification of arthropods in a wet stump cavity. *Ecology* **39**, 83–88.

Solomon, M. E. (1949). The natural control of animal populations. *Journal of Animal Ecology* **18**, 1–35.

Soria, S. de J., Wirth, W. W. & Besemer, H. A. (1978). Breeding places and sites of collection of adults of *Forcipomyia* spp. midges (Diptera: Ceratopogonidae) in cacao plantations in Bahia, Brazil: a progress report. *Revista Theobroma* **8**, 21–29.

Sota, T. & Mogi, M. (1992). Interspecific variation in desiccation survival time of *Aedes* (*Stegomyia*) mosquito eggs is correlated with habitat and egg size. *Oecologia* **90**, 353–358.

Sota, T. (1996). Effects of capacity on resource input and the aquatic metazoan community structure in phytotelmata. *Researches on Population Ecology* **38**, 65–73.

Sota, T. (1998). Microhabitat size distribution affects local difference in community structure: metazoan community in treeholes. *Researches on Population Ecology*, **40**, 249–255.

Sota, T. & Mogi, M. (1996). Species richness and altitudinal variation in the aquatic metazoan community in bamboo phytotelmata from North Sulawesi. *Researches in Population Ecology* **38**, 275–281.

Sota, T., Mogi, M. & Hyamizu, E. (1992). Seasonal distribution and habitat selection by *Aedes albopictus* and *Ae. riversi* (Diptera: Culicidae) in northern Kyushu, Japan. *Journal of Medical Entomology* **29**, 296–304.

Sota, T., Mogi, M. & Kato, K. (1998). Local and regional food web structure in *Nepenthes alata* pitchers. *Biotropica*, **30**, 82–91.

Sousa, W. P. (1984). The rôle of disturbance in natural communities. *Annual Review of Ecology and Systematics*, **15**, 353–391.

Souza-Lopez, H. de (1958). Sarcophagidae, Honolulu. *Insects of Micronesia* **13**, 15–49.

Southwood, T. R. E. (1977). Habitat, the templet for ecological strategies? *Journal of Animal Ecology* **46**, 337–365.

Speight, M. (1974). *Phaonia exoleta* Mg. (Diptera: Muscidae) new to Ireland. *Entomologist's Record and Journal of Variation* **86**, 246.

Spence, J. R. (1979). Riparian carabid guilds – a spontaneous question generator. In:*Carabid Beetles: their Evolution, Natural History and Classification* (eds. T. L. Erwin, G. E. Ball, D. R. Whitehead & A. L. Halpern). Junk, the Hague, pp. 525–537.

Starrett, P. (1960). Descriptions of tadpoles of Middle American frogs. *Miscellaneous Publications of the Museum of Zoology, University of Michigan* **110**, 1–37.

Steffan, W. A. (1975). Systematics and biological control potential of *Toxorhynchites* (Diptera: Culicidae). *Bulletin of Entomological Research* **42**, 355–370.

Steffan, W. A. & Evenhuis, N. L. (1981). Biology of *Toxorhynchites*. *Annual Review of Entomology* **26**, 159–181.

Steffan, W. A., Stoaks, R. D. & Evenhuis, N. L. (1980). Biological observations on *Toxorhynchites amboinensis* in the laboratory. *Journal of Medical Entomology* **17**, 515–518.

Stehr, F. W. (1987). *Immature Insects,* Volume 2. Kendall/Hunt, Duboque, Iowa.

Steyskal, G. C. (1987). Richardiidae. *Manual of Nearctic Diptera (Agriculture Canada)*, **2**, 833–837.

Strenzke, K. (1950). Die Pflanzengewässer von *Scirpus silvaticus* und ihre Tierwelt. *Archiv für Hydrobiologie* **44**, 123–170.

Stribling, J. B. & Young, D. K. (1990). Descriptions of the larva and pupa of *Flavohelodes thoracica* (Guérin-Méneville) with notes on a phytotelma association (Coleoptera: Scirtidae). *Proceedings of the Entomological Society of Washington* **92**, 765–770.

Strong, D. A. (1984). Exorcising the ghost of competition past: phytophagous insects. In: *Ecological Communities: Conceptual Issues and the Evidence* (eds. D. A. Strong, D. Simberloff, L. G. Abele & A. B. Thistle), Princeton University Press, Princeton, pp. 28–41.

Strong, D. A., Simberloff, D. Abele, L. G. & Thistle, A. B. (eds) (1984). *Ecological Communities: conceptual issues and the evidence.* Princeton University Press, Princeton.

Stuart, L. C. (1948). The amphibians and reptiles of Alta Verapaz, Guatemala. *Miscellaneous Publications of the Museum of Zoology, University of Michigan* **69**, 1–109.

Sugihara, G. (1984). Graph theory, homology and food webs. *Proceedings of Symposia in Applied Mathematics* **30**, 83–101.

Surtees, G. (1959). On the distribution and seasonal incidence of culicine mosquitoes in southern Nigeria. *Proceedings of the Royal Entomological Society of London (A)* **34**, 110–120.

Swezey, O. H. (1932). *Stenomicra*, apparently *angustata* Coq. *Proceedings of the Hawaiian Entomological Society* **8**, 88–92.

Swezey, O. H. (1936). The insect fauna of ieie (*Freycinetia arborea*) in Hawaii. *Proceedings of the Hawaiian Entomological Society* **9**, 191–196.

Szerlip, S. L. (1975). Insect associates (Diptera: Chironomidae, Sphaeroceridae) of *Darlingtonia californica* (Sarraceniaceae) in California. *Pan-Pacific Entomologist* **51**, 169–170.

Tate, P. (1935). The larva of *Phaonia mirabilis* Ringdahl, predatory upon mosquito larvae (Diptera: Anthomyidae). *Parasitology, Cambridge* **27**, 556–560.

Taylor, E. H. (1939). A new bromeliad frog. *Copeia* **1939**, 97–100.

Taylor, E. H. (1940). Two new anuran amphibians from Mexico. *Proceedings of the United States National Museum* **89**, 43–47.

Taylor, E. H. (1954). Frog-egg-eating tadpoles of *Anotheca coronata* (Stejneger) (Salientia, Hylidae). *University of Kansas Science Bulletin* **36**, 589–596.

Taylor, E. H. (1962). The amphibian fauna of Thailand. *University of Kansas Science Bulletin* **43**, 265–599.

Teesdale, C. (1957). The genus *Musa* Linn. and its role in the breeding of *Aedes (Stegomyia) simpsoni* (Theo.) on the Kenya coast. *Bulletin of Entomological Research* **48**, 251–260.

Teskey, H. J. (1976). Diptera larvae associated with trees in North America. *Memoirs of the Entomological Society of Canada*, **100**, 1–53.

Theischinger, G. (1991). Plecoptera (Stoneflies). In: *Insects of Australia* 2nd Edition (ed. I. D. Naumann), Melbourne University Press, Melbourne, pp. 311–319.

Thiele, H., U-. (1977). *Carabid beetles in their Environments*. Springer-Verlag, Berlin.

Thienemann, A. (1932). Die Tierwelt der Nepenthes-Kannen. *Archiv für Hydrobiologie* Supplementum **3**, 1–54.

Thienemann, A. (1934). Der tierwelt der tropischen Planzengewässer. *Archiv für Hydrobiologie* Supplementum **13**, 1–91.

Thienemann, A. (1954). Chironomus: Leben, Verbreitung und wirtschäftliche Bedeutung der Chironomiden. *Binnengewässer* **20**, 1–834.

Tillyard, R. J. (1917). *The Biology of Dragonflies (Odonata or Paraneuroptera)*. Cambridge University Press, Cambridge.

Torales, G. J., Hack, W. H. & Turn, B. (1972). Criaderos de culícidos en bromeliáceas del NW de Corrientes. *Acta Zoologica Lilloana*, **29**, 293–308.

Tracey, J. G. (1982). *The Vegetation of the Humid Tropical Region of North Queensland.* CSIRO Melbourne, Australia.

Tressler, W. L. (1941). Ostracoda from Puerto Rican bromeliads. *Journal of the Washington Academy of Sciences* **31**, 264–269.

Tressler, W. L. (1956). Ostracoda from bromeliads in Jamaica and Florida. *Journal of the Washington Academy of Sciences* **46**, 333–336.

Trimble, R. M. (1979). Laboratory observations on oviposition by the predaceous tree-hole mosquito, *Toxorhynchites rutilus septentrionalis* (Diptera: Culicidae). *Canadian Journal of Zoology* **57**, 1104–1108.

Trimble, R. M. & Smith, S. M. (1975). A bibliography of *Toxorhynchites rutilus* (Coquillet) (Diptera: Culicidae). *Mosquito Systematics* **7**, 115–126.

Trimble, R. M.& Smith, S. M. (1978). Geographic variation in development time and predation in the tree hole mosquito, *Toxorhynchites rutilus septentrionalis* (Diptera: Culicidae). *Canadian Journal of Zoology* **56**, 2156–2165.

Trimble, R. M. & Smith, S. M. (1979). Geographic variation in the effects of temperature and photoperiod on dormancy induction, development time, and predation in the treehole mosquito, *Toxorhynchites rutilus septentrionalis* (Diptera: Culicidae). *Canadian Journal of Zoology* **57**, 1612–1618.

Trpis, M. (1972). Breeding of *Aedes aegypti* and *A. simpsoni* under the escarpment of the Tanzanian plateau. *Bulletin of the World Health Organisation* **47**, 77–82.

Trpis, M. (1973). Interaction between *Toxorhynchites brevipalpis* and its prey *Aedes aegypti*. *Bulletin of the World Health Organisation* **49**, 359–365.

Truman, J. E. & Craig, G. B. (1968). Hybridization between *Aedes hendersoni* and *Aedes triseriatus*. *Annals of the Entomological Society of America* **61**, 1020–1025.

Vaillant, F. (1971). 9d. Psychodidae-Psychodinae In: *Die Fliegen der Palaearktichen Region*, Lieferung 287 (ed. E. Lindner), Schweizerbart'sche Buchhandlung, Stuttgart, pp. 1–48.

van Oye, P. (1921). Zur Biologie de Kanne von *Nepenthes melamphora* Reinw. *Biologisches Zentralblatt* **41**, 529–534.

van Oye, P. (1923). De mikrofauna en flora der bladtrechters van Bromeliaceae. *Natuurwetenschappelijk Tijdschrift* **5**, 179–182.

Vandermeer, J., Addicott, J., Andersen, A., Kitasako, J., Pearson, D., Schnell, C. & Wilbur, H. (1972). Observations on *Paramecium* occupying arboreal standing water in Costa Rica. *Ecology* **53**, 291–293.

Varga, L. (1928). Ein interessanter Biotop der Biocönose von Wasserorganismen. *Biologisches Zentralblatt* **48**, 143–162.

Vavra, J. (1969). *Lankesteria barretti* n.sp. (Eugregarinida, Diplocystidae), a parasite of the mosquito. *Aedes triseriatus* (Say) and a review of the genus *Lankesteria* Mingazzini. *Journal of Protozoology* **16**, 546–570.

Veloso, H. P., Fontana, J., Klein, R. M. & Siquiera-Jaccoud, R. J. de (1956). Os anofelinos do sub-gênero *Kerteszia* em relação à distribuição das bromeliáceas em comunidades forestais do município de Brusque, Estado de Santa Catarina. *Memórias do Instituto Oswaldo Crux* **54**, 1–86.

Vitale, G., Wirth, W. W. & Aitken, T. H. G. (1981). New species and records of *Culicoides* from arboreal habitats in Panama, with a synopsis of the *debilipalpus* group (Diptera: Culicidae). *Proceedings of the Entomological Society of Washington,* **83**, 140–159.

Vitzhum, H. Graf (1931). Terrestriche Acarinen (unter Auschluß der Oribatiden und Ixodiden) der Deutschen Limnologische Sunda-Expedition. *Archiv für Hydrobiologie, Supplementumband* **9**, 59–134.

Vongtangswad, S. & Trpis, M. (1980). Changing patterns of prey consumption in the predatory larvae of *Toxorhynchites brevipalpus* (Diptera: Culicidae). *Journal of Medical Entomology* **17**, 439–441.

Walker, E. D. & Merritt, R. W. (1988). The significance of leaf detritus to mosquito (Diptera: Culicidae) productivity in treeholes. *Environmental Entomology* **17**, 199–206.

Walker, E. D. & Merritt, R. W. (1991). Behavior of larval *Aedes triseriatus* (Diptera: Culicidae). *Journal of Medical Entomology* **28**, 581–589.

Walker, E. D., Lawson, D. L., Merritt, R. W., Morgan, W. T. & Klug, M. J. (1991). Nutrient dynamics, bacterial populations, and mosquito productivity in tree hole ecosystems and microcosms. *Ecology* **72**, 1529–1546.

Wallace, A. R. (1869). *The Malay Archipelago.* Macmillan, London.

Wassersug, R. J., Frogner, K. J. & Inger, R. F.(1981). Adaptations for life in tree holes by rhacophorid tadpoles in Thailand. *Journal of Herpetology* **15**, 41–52.

Watson, J. A. L. & Dyce, A. L. (1978). The larval habitat of *Podopteryx selysi* (Odonata: Megapodagrionidae). *Journal of the Australian Entomological Society* **17**, 361–362.

Watson, J. A. L. & O'Farrell, A. E. (1991). Odonata (Dragonflies and Damselflies). In: *Insects of Australia* 2nd Edition (ed. I. D. Naumann), Melbourne University Press, Melbourne, pp. 294–310.

Watt, K. E. F. (1968). *Ecology and Resource Management.* McGraw Hill, New York.

Watts, D. M., Morris, C. D., Wright, R. E., DeFoliart, G. R. & Hanson, R. P. (1972). Transmission of LaCrosse virus (California encephalitis group) by the mosquito *Aedes triseriatus. Journal of Medical Entomology* **9**, 125–127.

Watts, R. B. & Smith, S. M. (1978). Oogenesis in *Toxorhynchites rutilus* (Diptera: Culicidae). *Canadian Journal of Zoology* **56**, 136–139.

Webb, L. J. (1959). A physiognomic classification of Australian rain forests. *Journal of Ecology* **47**, 551–570.

Wesenberg-Lund, C. (1920–21). Contributions to the biology of the Danish Culicidae. *Kongelige Danske Videnskabernes Selskab. Skrifter. Naturvidenskabelig og Mathematisk Afdeling.* **7**, 210 pp.

Weygoldt, P. (1980). Complex brood care and reproductive behaviour in captive poison-arrow frogs. *Behavioural Ecology and Sociobiology* **7**, 329–332.

White, M. E. (1986). *The Greening of Gondwana.* Reed, Frenchs Forest, NSW.

Whitmore, T. C. (1975). *Tropical Rain Forests of the Far East.* Clarendon Press, Oxford.

Whittaker, R. H. (1975). *Communities and Ecosystems.* 2nd Edn., Macmillan, New York.

Whittaker, R. H. & Levin, S. A. (eds) (1975). *Benchmark Papers in Ecology. Vol. 3. Niche: Theory and Applications.* Halstead, New York.

Whitten, A. L., Mustafa, M. & Henderson, G. S. (1987). *The Ecology of Sulawesi.* Gadjah Madha University Press, Yogyakarta.

Williams, F. X. (1936). Biological studies in Hawaiian water-loving insects. Part II. Odonata or dragonflies. *Proceedings of the Hawaiian Entomological Society,* **9**, 273–349.

Williams, F. X. (1944). Biological studies on Hawaiian water-loving insects. III. Diptera or flies. D. Culicidae, Chironomidae and Ceratopogonidae. *Proceedings of the Hawaiian Entomological Society* **12**, 149–180.

Williams, R. W. (1964). Observations on habitats of *Culicoides* larvae in Trinidad, W.I. (Diptera: Ceratopogonidae). *Annals of the Entomological Society of America,* **57**, 462–466.

Wilson, D. S. (1980). *The Natural Selection of Populations and Communities.* Benjamin/Cummings, Menlo Park, California.

Wilson, L. D., McCranie, J. R. & Williams, K. L. (1985). Two new species of fringe-limbed hylid frogs from nuclear Middle America. *Herpetologia* **41**, 141–150.

Wilton, D. P. (1967). *Oviposition site selection by the tree-hole mosquito,* Aedes triseriatus *(Say).* PhD Thesis, Cornell University.

Winder, J. A. (1977). Field observations on Ceratopogonidae and other Diptera (Nematocera) associated with cocoa flowers in Brazil. *Bulletin of Entomological Research* **61**, 651–655.

Winder, J. A. & Silva, P. (1972). Cacao pollination: microdiptera of cacao plantations and some of their breeding places. *Bulletin of Entomological Research* **61**, 651–655.

Wirth, M. O., Wirth, W. W. & Blanton, F. S. (1968). Plant materials as breeding places of Panama *Culicoides. Proceedings of the Entomological Society of Washington,* **70**, 132.

Wirth, W. W. (1952). Three new Nearctic species of *Systenus* with a description of the immature stages from tree cavities. *Proceedings of the Entomological Society of Washington* **54**, 236–244.

Wirth, W. W. (1975). Biological notes and new synonymy in *Forcipomyia* (Diptera: Ceratopogonidae). *Florida Entomologist* **58**, 243–245.

Wirth, W. W. (1976). A new species and new records of *Dasyhelea* from the Tonga Islands and Samoa (Diptera: Ceratopogonidae). *Proceedings of the Hawaiian Entomological Society* **22**, 381–384.

Wirth, W. W. & Beaver, R. A. (1979). The *Dasyhelea* biting midges living in pitchers of *Nepenthes* in southeast Asia (Diptera, Ceratopogonidae). *Annales de la Société Entomologiques de France (N. S.)* **15**, 41–52.

Wirth, W. W. & Blanton, F. S. (1968). A revision of the neotropical biting midges of the Hylas group of Culicoides (Diptera, Ceratopogonidae). *Florida Entomologist,* **51**, 201–215.

Wirth, W. W. & Hubert, A. A. (1972). A new oriental species of *Culicoides* breeding in tree rot cavities (Diptera: Ceratopogonidae). *Journal of the Washington Academy of Sciences* **62**, 41–42.

Wirth, W. W. & Soria, S. de J. (1981). Two *Culicoides* biting midges reared from inflorescences of *Calathea* in Brazil and Columbia, and a key to species of the *discrepans* group (Diptera: Ceratopogoniodae). *Revista Theobroma* **11**, 107–117.

Wirth, W. W. & Ratanaworabhan, N. C. (1976). A new species of parasitic midge (*Forcipomyia (Pterobosca)*) from Aldabra, with descriptions of its presumed larva and pupa and systematic notes on the subgenera of *Forcipomyia* (Ceratopogonidae). *Systematic Entomology* **1**, 241–245.

Wodzicki, K. (1968). *An Ecological Survey of Rats and Other Vertebrates of the Tokelau Islands.* Government Printer, Wellington.

Woodley, N. E. (1982). Two new species of *Neurochaeta* McAlpine (Diptera: Neurochaetidae) with notes on cladistic relationships within the genus. *Memoirs of the Entomological Society of Washington* **10**, 211–218.

Woodward, D. L., Colwell, A. E. & Anderson, N. L. (1988). The aquatic insect communities of tree holes in northern California oak woodlands. *Bulletin of the Society for Vector Ecology* **13**, 221–234.

Wray, D. L. & Brimley, C. S. (1943). The insect inquilines and victims of pitcher plants in North Carolina. *Annals of the Entomological Society of America* **36**, 128–137.

Wright, J. E. & Venard, C. E. (1971). Diapause induction in larvae of *Aedes triseriatus*. *Annals of the Entomological Society of America* **64**, 11–14.

Yates, M. G. (1979). The biology of the tree-hole breeding mosquito *Aedes geniculatus* (Oliver) (Diptera: Culicidae) in southern England. *Bulletin of Entomological Research*, **69**, 611–628.

Yeates, D. (1992). Immature stages of the apterous fly *Badisis ambulans* McAlpine (Diptera: Micropezidae). *Journal of Natural History* **26**, 417–424.

Yeates, D. K., de Souza Lopez, H. & Monteith, G. B. (1989). A commensal sarcophagid (Diptera; Sarcophagidae) in *Nepenthes mirabilis* (Nepenthaceae) pitchers in Australia. *Australian Entomological Magazine* **16**, 33–39.

Young, A. M. (1980). Feeding and oviposition in the giant tropical damselfly, *Megaloprepus coerulatus* (Drury) in Costa Rica. *Biotropica* **12**, 237–239.

Young, A. M. (1979). Arboreal movement and tadpole-carrying behavior of *Dendrobates pumilio* Schmidt (Dendrobatidae) in northeastern Costa Rica. *Biotropica* **11**, 238–239.

Young, S. (1997). All life is here. *New Scientist* **153** (2073), 24–26.

Zahl, P. A. (1975). Hidden worlds in the heart of a plant. *National Geographic* **147**, 389–397.

Zaragoza, C. S. (1974). Coleópteros de algunas bromelias epifitas y doce nuevos registros de especies para la fauna Mexicana. *Anales del Instituto de Biologia, Universidad Nacional autonoma de Mexico Serie Zoologia*, **45**, 111–118.

Zavrel, J. (1933). Larven und Puppen der Tanypodinae von Sumatra und Java. *Archiv für Hydrobiologie* Supplementband, **11**, 604–624.

Zavrel, J. (1935). Ze zivota v nejmensich tunkach. *Priroda* **28**, 166–169.

Zavrel, J. (1941). Chironomidarium larvae et nymphae IV. (Genus *Metrocnemus* v. d. Wulp). *Acta Societatis Scientiarum naturalium Moravicae* **13**, 1–28.

Zimmerman, B. L. & Rodriguez, M. T. (1990). Frogs, snakes and lizards of the INPA-WWF reserves near Manaus, Brazil. In: *Four Neotropical Forests* (ed. A. H. Gentry), Yale University Press, New Haven, pp. 426–454.

Zimmermann, E. (1982). Durch Nachzucht erhalten: Blattsteigerfröscher *Phyllobates vitatus* und *P. lugubris*. *Aquarien Magazin* **16**, 109–112.

Zimmermann, E. & Zimmermann, H. (1985). Brutflegestrategien bei Pfeilgiftfröschen (Dendrobatidae). *Verhandlungen der Deutschen Zoologischen Gesellschaft* **8**, 526–531.

Zimmermann, H. (1985). Die Aufzucht der Goldbaumsteigers *Dendrobates auratus*. *Aquarien Magazin* **8**, 526–531.

Zimmermann, H. & Zimmermann, E. (1980). Durch Nachzucht erhalten: der Baumsteiger *Dendrobates leucomelas*. *Aquarien Magazin* **14**, 211–217.

Zimmermann, H. & Zimmermann, E. (1984). Durch Nachzucht erhalten: Baumsteigerfröscher *Dendrobates quinquevittatus* und *D. reticulaltus*. *Aquarien Magazin* **18**, 35–41.

Index

This index gives access to all proper names and key concepts in the text. Where multiple information is available within Tables then the whole Table is indexed rather than its separate contents. A few near ubiquitous references (such as 'tree hole' and 'Australia') are omitted and should be accessed via the Contents page.

419